30106019406460

AF795804

[redacted]

Stamp indicates date for return

This item may be renewed if not required by another cu

To renew, use the library catalogue or tel: 0113 343 566

Fines will be charged for late returns

Edward Boyle

14 FE

WATER RESOURCES MANAGEMENT

Water resources management

DAVID STEPHENSON
University of the Witwatersrand

A.A. BALKEMA PUBLISHERS / LISSE / ABINGDON / EXTON (PA) / TOKYO

Printed in the Netherlands by Krips The Print Force, Meppel, The Netherlands

Copyright © 2003 Swets & Zeitlinger B.V., Lisse, The Netherlands

All rights reserved. No part of this publication or the information contained herein may be reproduced, stored in a retrieval system, or transmitted in any form or by any means, electronic, mechanical, by photocopying, recording or otherwise, without written prior permission from the publishers.

Although all care is taken to ensure the integrity and quality of this publication and the information herein, no responsibility is assumed by the publishers nor the author for any damage to property or persons as a result of operation or use of this publication and/or the information contained herein.

ISBN 90 5809 573 8

PREFACE

Mankind requires water for life, industry and agriculture. We are at the same time threatened by floods, droughts and water pollution. It is inevitable that we build dams, pipelines and water works to provide us with the water we need and to protect ourselves. That is, to provide safe, secure and adequate water for our reasonable needs. Water can also provide us with pollution-free power, with navigation ways, recreation and home for fish and aquatic plants.

In the past, hydraulic structures have been the most obvious, if not most important components in a water supply system. Many structures have been built which have had effects on the environment and water resources that were not fully comprehended at the time of design. We have become increasingly aware of shortcomings in designs as we hear of lost habitats and natural resources. Previously voiceless communities have indicated their concerns. Planning of water resources development is now expected to take a more holistic view and the term Integrated Water Resources Management emerged. New projects will have to be viewed from more angles and by more stakeholders and disciplines. We will pay more attention to environmental, social and legal implications of new projects.

Sustainability is not a new subject and has been recognized in many cultures. The Kanji symbol for eternal (naga-i) stems from that for water (Japanese; mizu);

水 永

Mizu Naga-i

Over the 20th century emphasis in the water industry transgressed from design and construction to management. Water is managed in time and space by means of dams and conduits. The rate of construction of dams and pipelines is slowing and we are likely to see a period of consolidation wherein we manage what we have in a more efficient way rather than forge ahead with new construction. Aging infrastructure will also require more attention, to reduce losses, to extend its life and to provide better service. The management of infrastructure and the environment must go hand in hand with managing the water resource.

Different societies will have different perspectives on water resources development. Arid countries will be prepared to spend more per cubic metre of water than water-rich countries. But developing countries will not be able to afford the higher standards being set by developed nations with regard to preservation and quality of water. We cannot set inflexible rules, we should take cognizance of the different stages of development of different regions and assist where possible in alleviating international shortfalls.

This book attempts to set out the problems and integrated solutions in using our water resources, be they from rivers, aquifers, wastewater, the atmosphere or the sea. It looks at different aspects requiring management, ranging from floods to droughts. The interaction between catchment development and runoff is analyzed as well as the effect of water resources development on the catchment. The book also considers water quality management, the environment, asset management and economics of development. Water uses and ways of saving water are investigated. Tools for water resources management, including computer modelling, hydrology and hydraulics are explained at a technical level.

The chapters in this book stem from graduate courses in water management presented by the author at the University of the Witwatersrand, University of Stuttgart and Tokyo Institute of Technology. The course has also been presented over the Internet. The text for the book was typed, edited and formatted by April Thompson.

Because of the author's active participation in research and consulting, many of the case studies involve personal experiences. The book can therefore serve as the basis for teaching engineers and hydrologists as well as being of use to practicing engineers, scientists and planners in developing and managing water resources. It may well be asked whether such a book as this serves a purpose. There are good books available on hydrology, hydraulics, water resources planning and economics of planning. But many of these books are written from the analysis point of view for the academic, or design point of view for the practitioner. For both, special skills are used. In this book it is hoped that sufficient skills are given to enable a graduate of an engineering or science course to learn the techniques for managing water and the associated structures without having to complete an advanced course in hydrology or water engineering.

CONTENTS

CHAPTER 1

Water, a Multi-Dimensional Resource

1.1 INTRODUCTION

In the past, water was regarded as an inexhaustible resource such as air. However, in more recent years we have begun to value water as a limited resource. This may be because of increasing demands but more probably due to the fact that we have realized there are demands on water other than for human consumption. Water, especially clean water, is an integral necessity of the environment, but when people abstract the water from rivers or aquifers the previous regime changes. And over centuries nature has reached a quasi equilibrium with water. For example, rivers become lined with vegetation that can withstand the floods and droughts experienced along the river. There may have been changes in the long term but these are not visible to man. Similarly river bed and bank configuration has changed over long periods. But we are now reaching an acceleration in those changes where the runoff has changed, due to man's using the water and developing the catchment.

Water forms part of the hydrological cycle and is continuously moving through this cycle. Whether it be as runoff from the surface, flow in rivers, groundwater flow in aquifers or evaporation or transpiration, water is constantly being recycled, and during this recycling process, there are changes in the quantity and quality of the water which are caused by natural and human intervention. It is because we have begun to realize we are changing nature by abstracting water that we have begun to look at water resources in a much broader picture.

By using water, we manage it in time and space. We store water and we divert it. This interferes with the natural cycle and we should heed other requirements in our management strategy. By incorporating demands other than physical abstraction, we are integrating a number of disciplines and requirements. We put pollutants in the waters and thereby affect the environment and the potability of the water.

Over the last century our efforts in supplying water have swung from philosophic to scientific analysis, then engineering design, construction and now management of facilities and resources. And we can no longer ignore boundaries, be they political, topographic, disciplinary or economic. That is, we have to look at other's shortages as well as our own. We can manage the water resources or the supply, or both, to meet demands. The following subjects need to be considered:

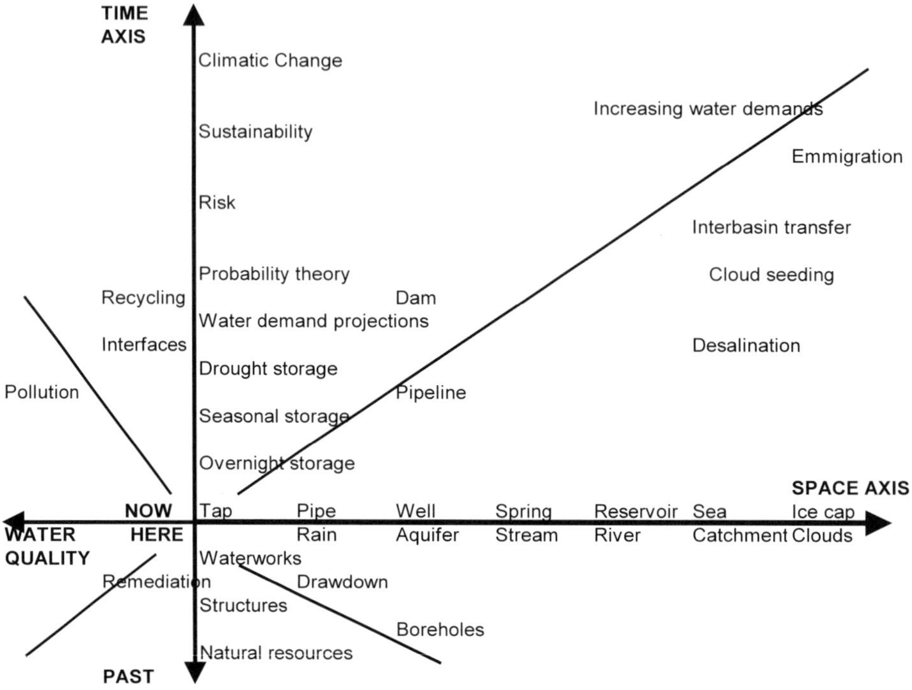

Figure 1.1 Management of water in time and space

Water Resources

- Assessment of water resources
- Interaction of various locations of water, i.e. atmospheric, sea, surface and ground
- Drought management
- Flood management
- Impacts of catchment and river development
- Impact on the environment
- Water quality

Water Supply:

- Demand management and loss control
- Hydraulics
- Costs and tariffs
- Economics and development
- Risk and vulnerability
- Asset management
- Administration and public-private partnerships

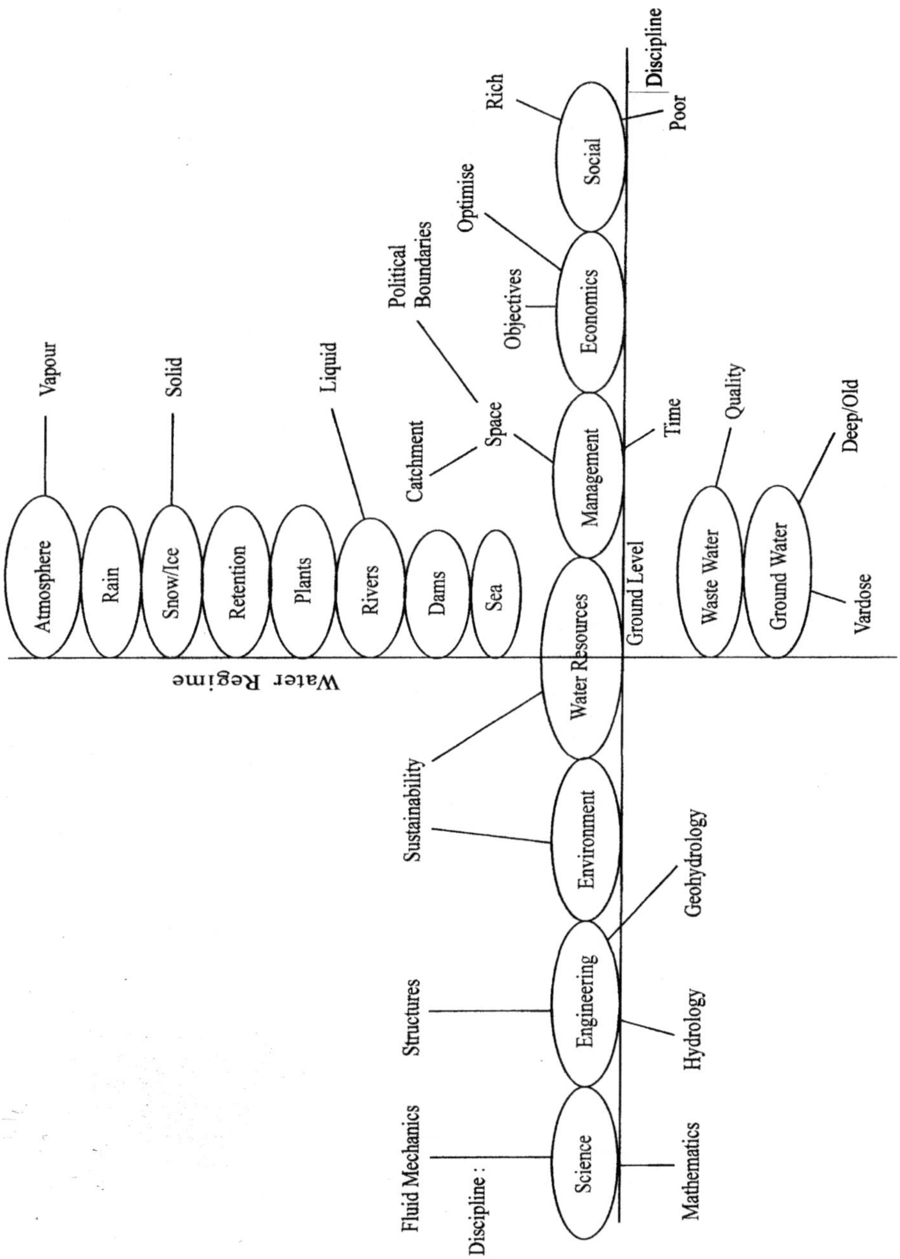

Figure 1.2 Multi-disciplinarity of water resources planning

1.2 WATER RESOURCES PLANNING

The traditional planning approach for water resources projects was for the supplier to identify demands, select the most convenient source, and design the least cost system to meet demands. Due largely to external pressures, the planning process was expanded to facilitate input by users and external groups. Multiple criteria emerged, including conservation and social issues. A long-term strategy was adopted. The process therefore became known as *Integrated Resources Planning.*

The process became cumbersome and planning became bogged down by major or minor inputs. More recently the process has been streamlined and the term *Least Cost Planning* originated. A comparison of the methods is made in Table 1.1.

LCP differs from traditional planning by considering long term and lifecycle costs. It costs all factors including demand management and risk. It is a transparent and comprehensive procedure. Integrated resources planning is a broader approach but does not necessarily lead to an engineering solution. It uses economics in decision making in the broader sense by attempting to balance multiple objectives. *Integrated Water Resources Management* is more focused and may consider a catchment, or a specific problem such as flood control. Nevertheless it is aimed at the end user rather than the utility.

Table 1.1 Comparison of Planning Process

	Type of Process		
Dimension	Traditional Planning	LCP	IRP
Resources options	Supply options (demand is taken as a given)	Demand and supply options (demand can be manipulated)	Demand and supply options (demand can be manipulated)
Resources diversity	Utility owned and centralised	Diversity of resources , demand-side management	Diversity of resources , demand-side management
Resources ownership	All infrastructure or resources owned by utility	Most infrastructure or resources owned by utility	Much owned by other utilities, other producers
Resource selections criteria	Minimise rates and maintain system reliability	Diverse criteria, including risk, technology, environmental, economics	Diverse criteria, including risk, technology, environmental, economics
Focus of economic cost analysis	Ratepayers	Multiple groups (participants, ratepayers, individuals, etc.)	Multiple groups (participants, ratepayers, individuals, etc.)
Conduct of planning	Internal to utility, mainly operational , financial planning	Internal to the water industry, planning by professionals	Several departments as well as non-utility experts, staff, public
Planning horizon	Five to ten years	Minimum of ten years	Minimum of twenty-five years
Role of public groups	Intervenors	Advisors	Participants
Judgement	Vague	More defined	Well defined
Preferences	Vague	Well defined	Well defined
Objectives	Single	Multiple	Multiple
Reliability	Constraint	Decision variable	Decision variable
Environmental quality	Constraint	Constraint	Objective
Risk	Should be avoided	Should be managed	Should be managed

Evaluation is over the lifecycle and should include non-targeted objectives, i.e. environment, legal and socio-economic issues. Integration implies the involvement of all stakeholders and disciplines. It also embraces all phases of water in the balance, i.e. precipitation, surface and groundwater as well as water quality and efficiency of use. Long term sustainability and short term requirements are balanced, so a multi objective solution is required.

1.3 MULTI-DIMENSIONAL MANAGEMENT

Whereas in the past, we would pump water from the nearest source and, after treating it, distribute it to consumers as part of the engineer's brief, we now realize this requires both management of the water in time and space.

Storage reservoirs are for management of water in time. That is, they store water in wet periods in order to provide us with water during the dry periods. This is indicated on the time scale vertically in Figure 1.1. On a smaller time scale, we provide balances in reservoir storage for meeting peak demands in consumption. On the larger time scale, we store water for meeting demands during the dry season and on a still larger scale we store water from year to year for overcoming droughts. At this scale we have to consider probability, as the time series of river flow is a variable that we do not understand completely and therefore tend to treat it as a stochastic variable. On a larger time scale we should consider the sustainability of the river flow and on an ever greater time scale, we need to consider cycles in the weather and climatic change due to interference with the atmosphere.

Geographically, we distribute water from watercourses to consumers. On a small scale, we supply water from reservoirs to taps, and on a bigger scale from river basins across catchment boundaries and international boundaries. Similarly the natural water cycle can be inspected on a small scale such as we do when we look at the transport of water or pollution in aquifers, or on a larger scale across catchments in streams and rivers. On an even larger scale, we look at international rivers and the oceans, and on a global scale we need to consider the ice caps and water vapour in the atmosphere.

We could consider water quality on similarly increasing scales. Engineers did not give much attention to the larger scales in the past in order to perform proper balances. Such scales need to be considered for long term sustainability. This may mean multi-disciplinary thinking. In Figure 1.2 are indicated a further range of phases of the water cycle with the disciplines involved, ranging from mathematicians doing modeling to engineers doing planning, natural scientists considering the environment in which the water is, economists considering the costs and benefits, and lawyers and politicians considering water rights.

As an example of the consideration of the sustainability of water in time, Figure 1.3 shows the changing demands over years accompanied by the changing availability. Whereas water demand increases with population and standard of living, resources may even be decreasing owing to abuse of the environment or climatic change. However bearing in mind that our usage does not necessarily lose the water, there is return flow and, depending on the quality the return flow, it may be added to the naturally available source to give a greater total available. Theoretically where the usage equals the availability is the limit of our consumption. However, we now realize that there are changing targets and that our planning numbers are merely guides. Demand management can reduce usage, and economic pressures on usage can be decreased.

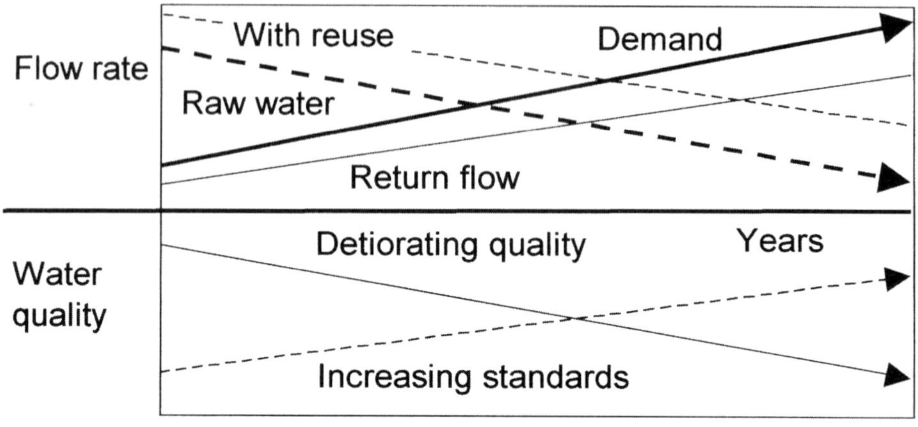

Figure 1.3 Sustainability of water resources

Similarly with more resourcefulness the availability of resources can be increased and provided long-term planning is conducted, we may use the resource in the best way possible. On the bottom axis of Figure 1.3 is water quality which deteriorates over time, i.e. with pollution. But scientists are continuously developing new standards which may mean that our water quality may limit the availability unless we move the goal posts.

Thus, from the water engineers' initial domain we have expanded our disciplinary involvement. The engineer needs to plan dams and conduits, i.e. design hydraulic structures, and the authorities operate these facilities to provide water. The construction phase would be the major expenditure. Nowadays, not only would the resources have to be managed but also the catchment and indeed the entire system including the use of the water needs consideration (Fig. 1.4).

We are also becoming more aware of socio-economic interaction. That is, water requirements depend on demography and economics (see Fig 1.5) as well as awareness of the users with regard to other water requirements in nature. Jansen et al, in Thomson Klein et al (2001) discuss the problem of conflict between long term societal goals and short term commercial objectives. The gap has to be bridged within the 21st century for sustainability to be reasonably assured. Interdisciplinary barriers have to be broken and we need a new mindset to solve the dilemma. Fortunately scientists are becoming more open minded in the faster and faster moving technological era. New research methodology is needed. Backcasting from needs to products and interdisciplinary networks are envisaged. Changes in methods and expectations can be anticipated in the coming century. We should not project the past into the future. A key word for the 21st century is Sustainability.

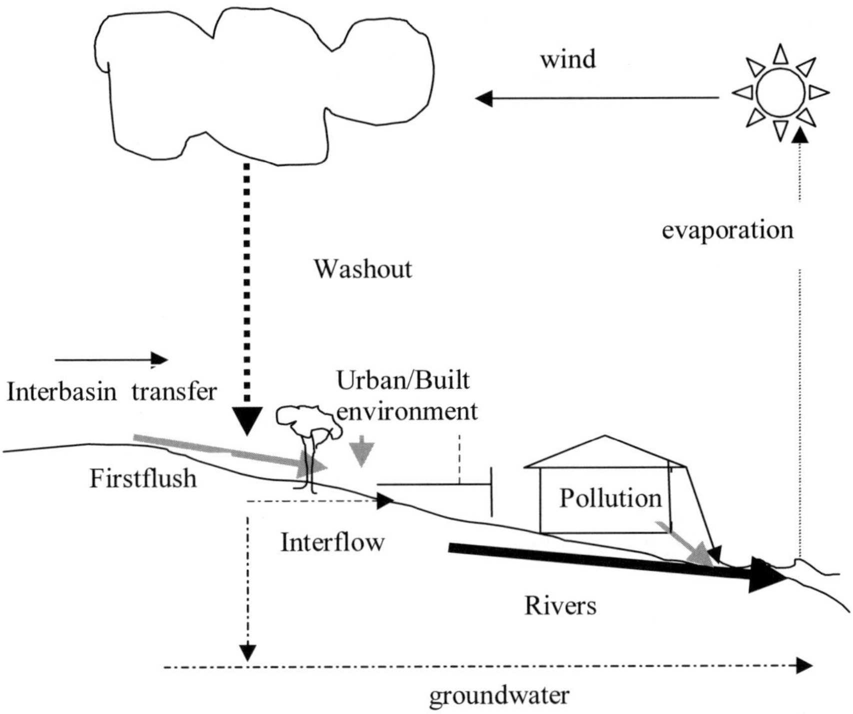

Figure 1.4 Hydrological cycle with factors affecting water quality

Long term monitoring of the new research is needed but time is pressing. Management and team building must therefore commence to solve the problems. Life cycle analysis must be performed not from the economic point of view but on a grander scale, from the environmental side.

Our world tends to evolve about ourselves, i.e. us as people, and we tend to look at resources such as land and water as available for our use. We have developed technology to use these resources and we regard conservation as secondary to our preservation (see Fig. 1.6).

We need to consider not only the uses to which we put water but also other requirements of nature. The subject of hydropower, agriculture, domestic and urban water uses, recreation and purification are man-involving activities. On the other hand, ecological requirements, both for fauna and flora, are related to the ecosystem as it has developed. Thus wetlands may thrive with water abundance and scarcer vegetation may exist in the arid environment. Similarly, animals, fish and bird life depends on the availability and quality of the water.

There is a natural purification process including aeration and solar radiation which sustains water quality and life, but if we tax these processes to the limit by over-polluting the system may break down. Similarly, the transpiration processes of plants and the dump areas of salts (the seas) can change if the quality is changed from what it was in the past. We therefore need to bear in mind that the ecosystem which contains the water may change and we should in fact consider the long-term changes and not necessarily try to resist them but reach a balance between abstraction for man and the requirements of nature. We should bear in mind that the hydrological cycle is a closed cycle and changing one component in nature may change others. Thus over-exploitation of water may degrade vegetation which in turn will increase runoff and decrease natural storage of water.

1.4 THE AGE OF MAN

It is only over the last century that we have begun to reach the limit of our resources (see Fig. 1.7). Figure 1.8 shows world population numbers and factors which have affected population growth rate are regionally orientated (Fig. 1.9). However, resource consumption is not uniform over the world's population. Indeed the power of people is measured in terms of wealth.

It is only the richest 20% that have surplus wealth for affecting world development (Fig. 1.10). It is these people who have power to control less wealthy people, and they often try to impose their desires on developing counties. This led to a change in emphasis at the 2002 World Summit on Sustainable Development in that ideas to help alleviate water and economic problems emanated (UNESCO, 2002).

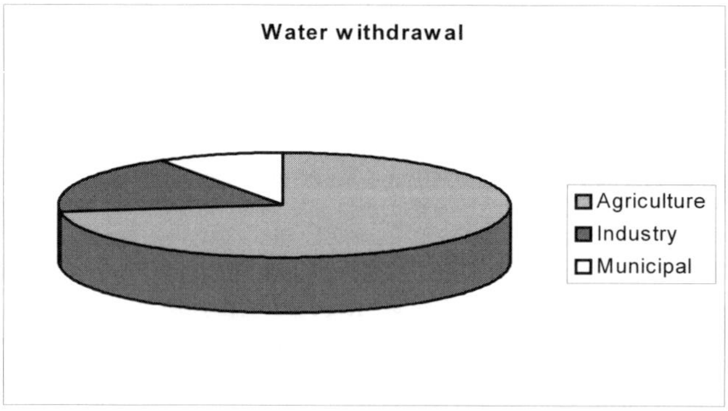

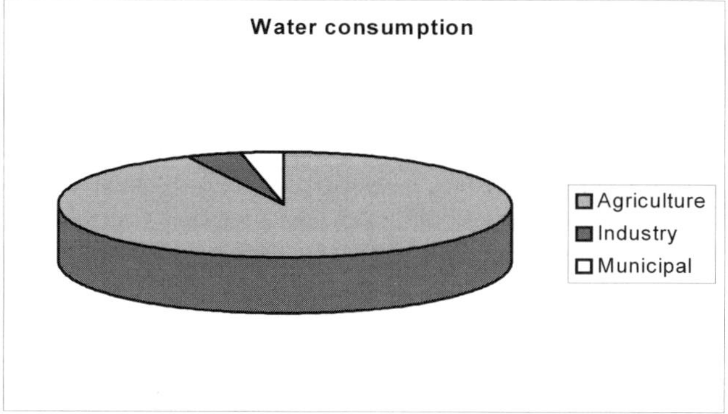

Figure 1.5 Water withdrawal and consumption by sector

However, we cannot even manage our world economy properly and attempts to redistribute wealth over the last century have not met with success and we see an ever widening gap between the rich and the poor. Thus the poor are unable to manage their resources properly and there is degradation which will affect mankind in the long-term. Thus we need to manage world wealth as well as its resources if we are to achieve a sustainable world. At present, there is large scale deterioration from all points-of-view in Africa, and to a decreasing extent in Asia, so that the world's resources, in particular water, are diminishing owing to the inevitable destruction of the habitat.

Mankind in many of these countries is also becoming more desperate and their ability to sustain themselves is decreasing. Both water resources and agricultural land are reaching their limits of availability within the next few decades in many of these countries. Many countries in the poorer continents import food and other resources merely to maintain the population. The creation of wealth, which in turn will lead to better management of other resources, is not in sight yet in most of Africa. This, coupled with overstressing the environment due to changing climate, again influenced by man, is causing loss of land and other resources leading to desertification and mining of resources including groundwater.

The competitive economy we have created has led to technological advances at the expense of nature. We mine natural resources on a huge scale, limiting man's life as we know it to only hundreds of years. We discharge large volumes of waste back to the earth often not realizing it affects water quality. And the competitive nature has made us not stop to think and even brought out dishonesty to our fellow mankind and to our descendants. Illegal discharges are a huge pollution problem, as not only are they often dangerous, they also go uncontrolled into the ground and water sources.

Perhaps fortunately the heavy industrial era is closing. Fossil fuels and even many fast replenishing natural resources are depleting to the extent new sources of energy and food have to be found rapidly for man to even survive in the numbers existing. Is it this ominous future which is resulting in population declines in developed countries?

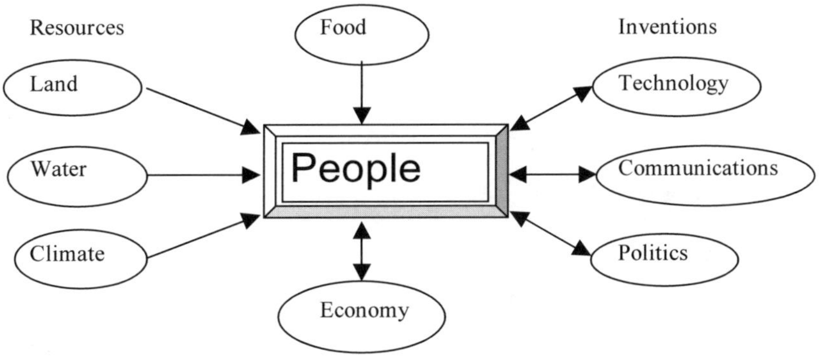

Figure 1.6 People's perspective of the world

Technological advance is another factor changing man's perspective. Since the industrial evolution not much more than a century ago, we have had numerous crises. International wars and terrorism are possible. The energy shortage frightened us for only a short period but even greater shortages in fossil fuels will recur within the next century. The relative cost of renewable energy will change modes of transport and work. But advances in electronics and computing are having just as great an impact on our lives. Many people will no longer need to commute to offices but will work where most convenient.

Urbanisation has occurred for convenience and because people seek employment and wealth. It is assumed by immigrants that community services exist in and around cities, i.e. electricity, water, waste collection and transport. But that mindset may change and people may move away from cities when it is not necessary to go in to an office to do a job. This may happen with more informal businesses and internet access. Water and waste disposal may be more economical in rural areas and these may become more significant relative to communications costs in the future. A healthier lifestyle may emerge.

1.5 STRESS

Water is emerging as the most stressed natural resource and some see the 21st Century as one of water stress. Water is more important than oil and fuel for life. Even at the turn of the twentieth century, half the world's 6 billion population did not have adequate water. By 2100 the world population is expected to have stabilized at 10 billion, but if the standard of living of the poor can be improved water consumption by industry and humans will more than double.

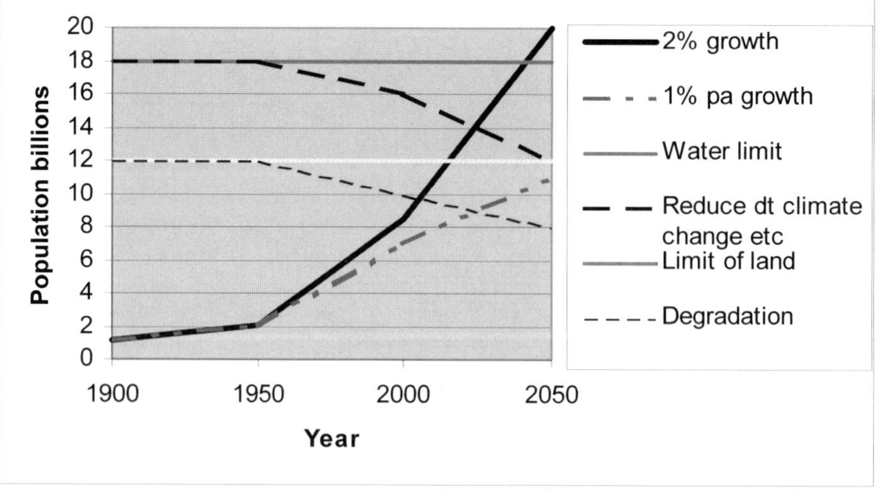

Figure 1.7 Limits to growth

Irrigation is the big unknown. At present it accounts for 70% of water use and most of it is consumed, i.e. evaporated. Future water sources will become more expensive and more polluted. Water will become an expensive commodity and the subject of regional and international disputes. This will be compounded by climate change, more extreme hydrology, worse droughts and worse water quality.

It is not water resources development causing land stress although they often go hand in hand. Both resources, land and water, are stressed by the short-sightedness of mankind. To alleviate the stress on resources would require international co-operation. People would have to be educated in correct ways, or relocated or provided with facilities. These are not likely in any significant level until all the world experiences the stresses. This book attempts to assist in developing and managing water resources to minimize the impacts.

Apart from a gradually reducing water availability and an increasing water demand, particularly in developing areas, the loss of arable land due to climatic change, over-exploitation of resources and disregard is threatening an ever increasing land area, particularly in poorer countries. Figure 1.11 shows areas which are being threatened by soil erosion. In Africa, vast tracts of land south of the Sahara and north of the Kalahari have decreased their carrying capacity by over 50% in the last 20 years. This is in addition to the large proportion of the continent which is in any case unsuitable for rain-fed agricultural production.

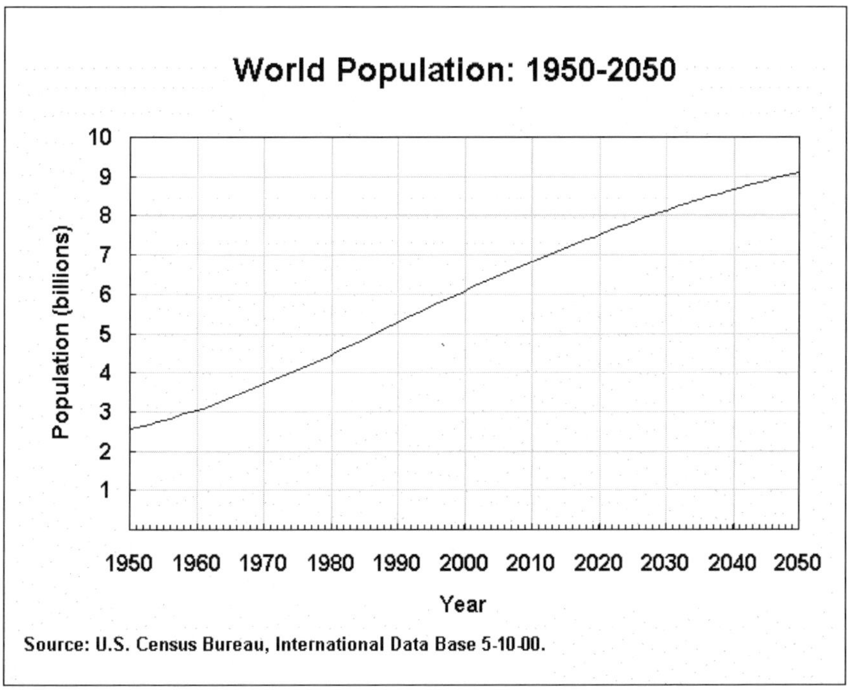

Figure 1.8 World population

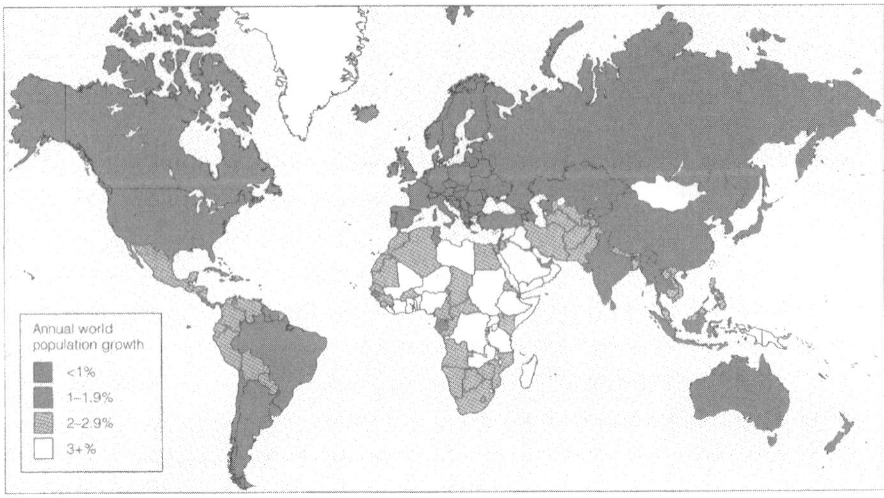

Figure 1.9 Average annual rate of population increase (natural) in 1996. (Data from Population Reference Bureau). (Miller, 1998)

The desertification and erosion threats are largely due to overstressing the environment by the increasing demands of mankind. These areas were particularly sensitive and climatic change, i.e. wider fluctuations in meteorology such as more intense droughts and floods, have leeched the soils out and receded the water tables. In Australia, salinization of soils has compounded the problem.

In the interior of the African continent, the equatorial forests have not yet been damaged to the same extent as those in South America. This could be due to poor accessibility and uncertain governance. The soils of the Amazon forests, which are being cleared at the rate of hundreds of hectares a day, will gradually recede in the same cycle. This is exacerbated by the changes in micro-climate which will come about because of reduced evapotranspiration and more intense soil erosion owing to the removal of the soil cover.

The desertification threat is very akin to the erosion threat. That is, once soil erosion starts the soil is more rapidly made barren.

Even with greater education of farmers and subsistence farmers, the erosion rate will not easily be reversed. This is because many of the populations are living from hand to mouth and with political and military threats, conservation and sustainability are not in their sights.

It is generally only possible to control farming methods in the case of large-scale commercial farmers. This is more achievable if the water resources are controlled, i.e. irrigation projects. Then ploughing methods, control of reaping and burning and crop planting can be taught or enforced. Similarly, wind erosion can be reduced by ensuring the planting of trees or other crops which protect the soil. Water erosion can be minimized by channeling runoff into lined or protected gullies and streams. Water can be harvested by providing infiltration areas or catchbasins. Then loss by evaporation will change the water balance. It may also be noted that eradication of alien vegetation is seen by some as a way of restoring natural balance.

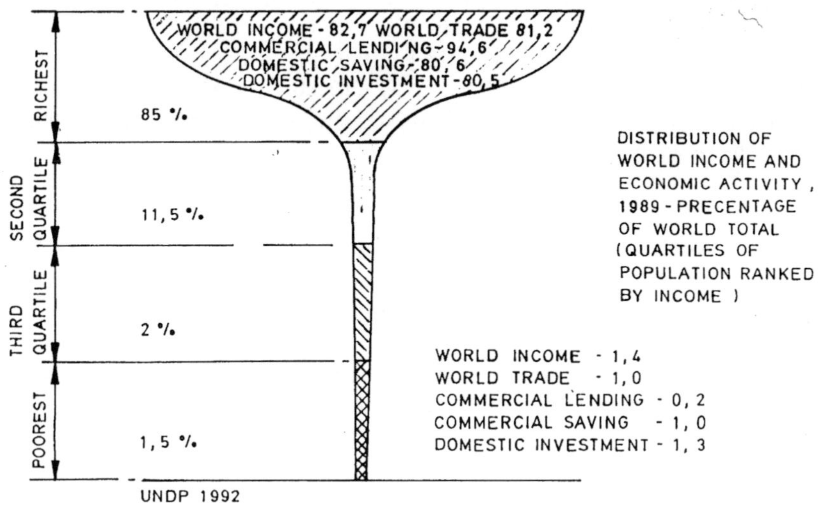

Figure 1.10 Global incomes and income disparities

Invasive water consuming plants and trees are being eradicated in South Africa and legislation to prevent their planting has been promulgated. On the other hand, the felling of Eucalyptus trees in Australia has been blamed for rising water table and salinization of soils.

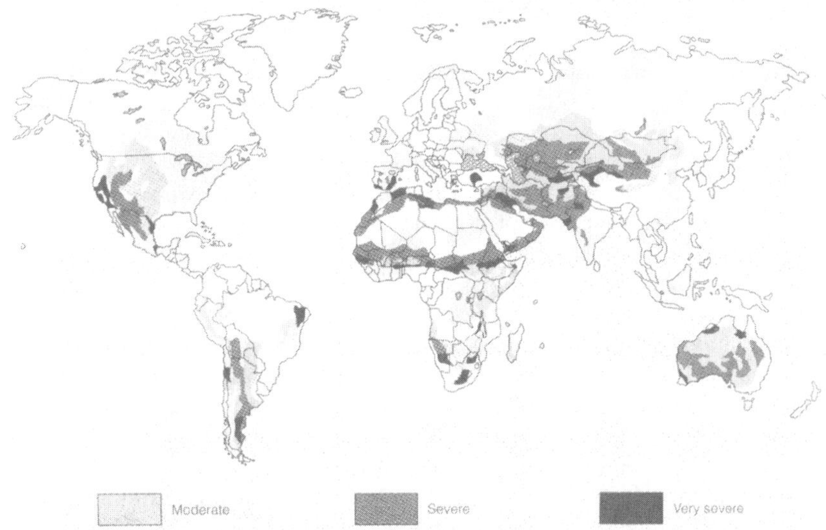

Figure 1.11 Global desertification of arid and semi-arid lands. (from UN Environmental Programme; Miller, 1998)

It is not just the decreasing average rate of rainfall which causes plants to wither and die and eventually the soils to wash away. The variability in climate may also exert a strong influence on the sustainability of the earth. It takes decades for plants to adapt to extreme climates and the desertification is probably extending faster than the adaptation of vegetation to long periods of drought, high or low temperatures, and intense storms when they occur.

1.6 INTERNATIONAL POLICY

The World Bank (2002) set out its policy on Water Resources Management based on the "Dublin Principles":

The *Ecological Principle* states that independent management of water by different sectors is not appropriate, and the river basin should become the unit of analysis. Land and water need to be managed together and much greater attention needs to be paid to the environment.

The *Institutional Principle* indicates that water resources management is best accomplished when all stakeholders participate. This means state, private sector and society, including women. Actions should be taken at the lowest appropriate level to ensure sustainability.

The *Instrument Principal* indicates that water is a scarce resource and greater use needs to be made of incentives and economic principles in improving allocation and enhancing quality.

The World Bank originally considered development of dams and other hydraulic infrastructure as synonymous with water resources management. It gradually refocused on the need to develop infrastructure and institutions. Now ineptitude and corruption in institutions has indicated good governance is important. Above all it is realized that development is a slow process and political setbacks can occur particularly where the country is developing.

The World Bank realizes that water resources have a significant role in national economies of developing countries. Water is one of the few sustainable free resources which can be developed or controlled. Irrigation and hydro power are relatively low-tech applications easily established and resulting in large economic benefits. Whereas irrigation can provide additional employment on the other hand hydro power is not labour intensive. But both rely on international free trade for viability. Agricultural products in developed countries are sometimes subsidized for political reasons, and the cost of alternative fuels in the way of oil and coal can also be manipulated by developed countries.

Devastating floods cost Moçambique 23% of its GDP. Drought in Zimbabwe in the early 1990s cost 11% of its GDP. 75% of income in some African counties is agriculture related and this prime source has a spin off effect through the economy. The level of schooling in India is directly related to water availability for agriculture. On the other hand water must be correctly valued. In Mexico proper valuation of water in 1992 reduced groundwater overexploitation, and re-evaluation of which crops to irrigate, as well as increased employment on high value crops. Water resources development can also unite countries. The Indus scheme and Lesotho Highlands schemes are such examples.

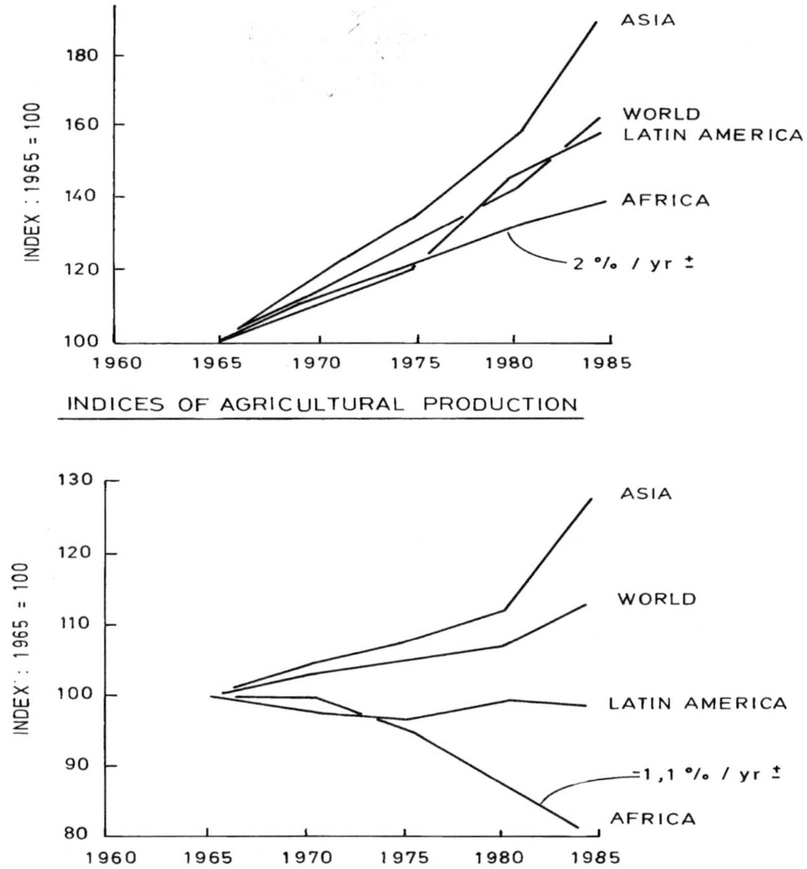

Figure 1.12 World Indices of Agricultural production: (a) Gross, (b) Per capita

The cooperation of the communities in developing countries is seen as important for success of water resource projects. Social interventions and training are seen as part of a successful project now. Irrigation projects on the Ganges showed that watershed management resulted in less erosion, improved water tables and in the end more employment. There will also be fewer obstacles to projects if the community see benefits, So communication is vitally important.

But there is a long way to go before the world is self sufficient, let alone sustainability or reparations. Whereas food production is increasing in many parts of the world, it is not necessarily keeping up with population increase. Figure 1.12 shows that per capita production in Africa is dropping. This is due to deteriorating soils and climate, poor access to water, political unrest and poor management.

1.7 CLIMATE CHANGE

The effects of climate change may be exaggerated in surface water runoff. The effects appear to be long-term rather than immediate. An intergovernmental panel on climate change (Watson, 2001) tried to answer 6 questions about climate change and its effect:

1. What constitutes dangerous anthropogenic interference in the climate systems? The answers appeared subjective and inconclusive.
2. Assess and where possible attribute observed change in climate and ecosystem since the pre-industrial era. They found a 31% increase in atmospheric CO_2 with corresponding decrease in terrestrial level, a 150% increase in N_2O, and an unquantified increase in HFCs and similar. O_3 was found to have increased in the troposphere and decreased in the stratosphere. A temperature increase of 0.6^o had been observed, in the Northern Hemisphere mainly. Precipitation had increased 5 to 10 % in the North but decreased in North and West Africa. Figure 1.13 depicts the observed and projected emissions for different scenarios.
3. Assess the impacts of future emissions of greenhouse gases and sulfate aerosols on climate, including variability and extreme events in ecology and socio-economic systems, with no interventions. By the year 2100, CO_2 concentration is estimated to be between 540 and 970 ppm compared with 280 for the pre-industrial era and 370 in 2000. Average temperatures are expected to go up 1 to 6°C in the same period. Precipitation will increase and changes will be in the range 5 to 20%. Sea level will increase up to 1m. Figure 1.14 shows the trends anticipated. Considerable hardship and stress would result, leading to much greater proportions than these figures.
4. What will be the effects on frequency and magnitude of variations? The variations are sensitive to model assumptions but it is agreed that the effects on biosystems will be large.
5. What inertia effects will be present in climate, ecology and socio-economics? Even if emissions are stabilized at present levels CO_2 will increase in the atmosphere for centuries, but other greenhouse gases will stabilize. Some effects will be irreversible e.g. melting of icecaps. Figure 1.15 illustrates the long-term effects.
6. Assess the implications of stabilization of emissions. To stabilize CO_2 at 450ppm will require emissions to drop below 1990 levels within a decade whereas temperatures and sea levels will continue to increase for centuries.
7. What are the technological, economic and time implications of reducing greenhouse gas emissions? Costs cannot be estimated until the technology has been discovered. Up to 5Gt of C of emissions could be achieved by 2020 at a cost of up to USD100 per t C.
8. The interaction of effects of emissions and other human induced changes needs identification. The latter include air pollution, desertification and ozone depletion. Human demands for food and energy are difficult to curb and they have a big effect on the environment.
9. Identify the most robust findings. These include certainty about temperature increase, greenhouse gas emissions, long term effect, rise in sea level and more intense hydrological extremes. Most adverse effects are expected to be felt by developing countries.

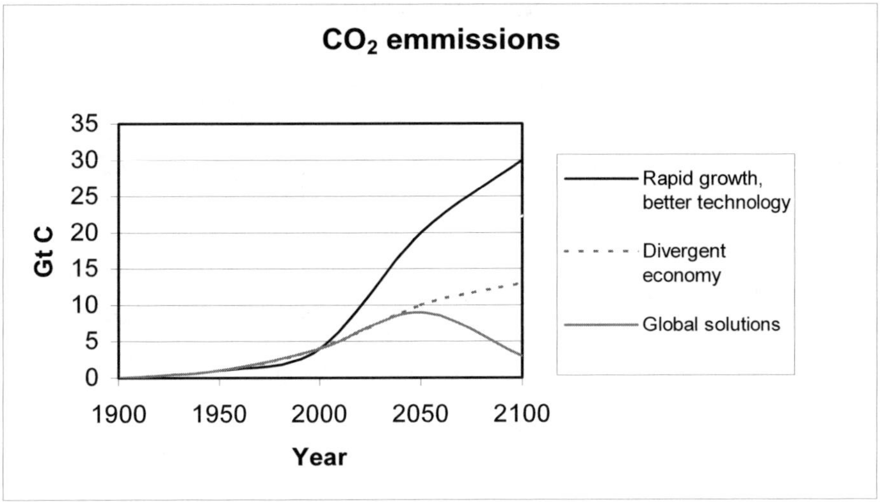

Figure 1.13 Greenhouse gas emissions

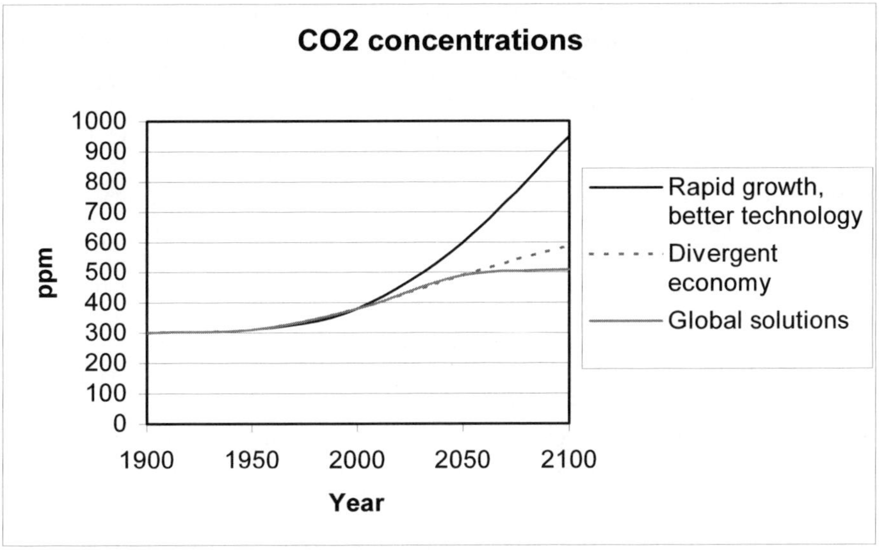

Figure 1.14 Greenhouse gas concentrations

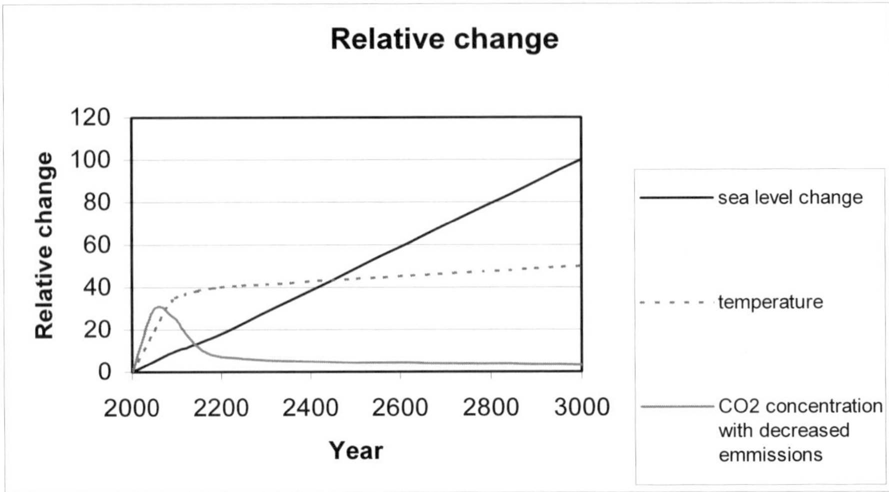

Figure 1.15 Relative changes due to greenhouse gas emissions

1.8 THE OCEANS

Although 70% of the world is covered by sea, this may become a future source for survival. The main problem of seawater is its high salinity, i.e. typically 35,000 mg/ℓ. This level of salinity makes it undrinkable, no use for agriculture and highly corrosive. The sea level is a base level as far as energy potential is concerned and energy can only be tapped from waves, tidal fluctuations and movement. Yet there are many uses made of the seas and more use could be made of them when surface resources become more stressed and we develop more applicable technology for the sea. The following lists some applications of the oceans:

- *Transport*. Ships are an economical form of bulk transport, although somewhat slow. Shipping densities are low and few rules exist, although international laws prevail. Counties with long coastlines benefit from access but are vulnerable too.
- *Communications*. Cables and pipelines cross the oceans.
- *Fishing*. Large stocks of fish and crustaceans exist in the sea. But care is needed to ensure sustainability, particularly in coastal regions.
- *Aquaculture*. Seaweed, oysters and pearls are some of the resources farmed in the oceans. There is an enormous potential for sourcing food from the oceans, owing to the large areas, high mineral and plankton loads in the water and continuous circulation.
- *Water*. At present desalination makes the sea an unattractive source of water to many. But icebergs, evaporation and the huge quantity are always there if we need it.
- *Recreation*. Including boating, skiing, travel, diving and fishing.
- *Minerals, oil and gas*. The costs are high but the resources are there. The often great depths to the bed of the oceans makes access difficult.

- *Biotechnology*. Pharmaceuticals, food and chemicals as well as processes and fields.
- *Energy*. Wave, tidal, circulation, thermal, solar, wind and salinity potential exist but all are not yet economically attractive.
- *Storage*. Nuclear waste, large inert containers.
- *Dumping*. From ships and coast. This should be controlled but offshore levels are too dilute to monitor. It is in the coastal regions that raw sewage and chemical wastes have already caused damage to marine life.
- *Environmental stabilization*. Temperature influences, evaporation and winds originating over seas control climate on land to a large degree. Thermal effects such as El Nino are likely to be of greater concern in the future.

Research into marine resources is probably too low. Compared with space research the oceans offer bigger and more profitable solutions. There is little man can do to alter the oceans and the resources are vast. The management of those resources warrants attention. Ownership issues and prospecting need addressing. International cooperation is needed for this although the UN Division of Ocean Affairs is addressing the subject.

REFERENCES

Department of Water Affairs and Forestry (2002). *Eradication of Alien Vegetation*. Pretoria

Middleton, N. (1991). *Desertification*. In the series Contemporary issues in Geography. Oxford University Press, 48pp.

Miller, G.T., Jnr. (1998). *Living in the Environment*, 10th Ed. Wadsworth Publishing Co., 761pp.

Stephenson, D. and Petersen M.S. (1991). *Water Resources Development in Developing Countries*, Elsevier, Amsterdam , 289pp

Thompson Klein. J. et al. (2001). *Transdisciplinarity: Joint Problem solving among Science, Technology and Society*. Birkhauser Verlag, Basel.

UNESCO (2002). *Proceedings of the World Summit on Sustainable Development*, Johannesburg.

Watson, R.T. (2001). *Climate Change 2001: Synthesis report*. Cambridge Univ. Press.

World Bank (2002). *Water Resources Sector Strategy: Strategic Directions for World Bank Engagement*. Washington.

CHAPTER 2

Water Resources Assessment

2.1 INTRODUCTION

In order to plan the development of a source of water or to plan for meeting a demand, accurate data is required concerning the volumes of water in the system and the flow rates, including the replenishment rate. The variability of flow, particularly river flow, is important from the point of view of risk of drought or flood and sizing of storage reservoirs. Rivers therefore need to be gauged over a period of time in order to understand the flow variations, and the longer the record, the better. More particularly, accurate and sufficient flow measurements are needed. The locality of the flow measurements is also important from the point of view of accuracy, but more particularly of relevance to the proposed development of the river. The same applies to other sources, e.g. groundwater. Extensive data is needed to assess the extent and yield of aquifers.

From the catchment management point of view, flows, both surface and subsurface, transport pollutants, dilute them and disperse them. To study and model these processes needs an extensive gauging network. Note that the term *catchment* is used here to denote the natural area in which precipitation collects and flows to one common point. The term *river basin* is also used for this, particularly in American textbooks. In this text the term *watershed* refers to the ridge which is a diving line between water which flows to the catchment or away from it. Again in American texts the term watershed may refer to the catchment.

Most national water authorities maintain some sort of network and database. Measurements most often made are those of water stage or depth, and these readings may be taken manually or by mechanical or electronic means. The raw data should be stored in archives, either as paper records or electronic records or preferably both. Hydrologists process this data and convert the water levels into flow rates. The processed data is published or released more frequently than the raw gauging data.

The user of the data needs to be aware of the various steps in obtaining this data and the limitations associated with obtaining the data and processing it, as there is a risk associated with incorrect data. In some cases, the observations are made at infrequent intervals missing important data. In other instances flood flows may not be measured because of limitations in the height of the gauge. Or base flows may be in the bed of the river and not measurable. The gauge observer may miss readings or insert incorrect readings into the record books. Therefore, quality assessment of the gauge and the observations as well as the processing is required. Calibration of gauging stations to indicate the correct flow rate versus water level requires field or computational work.

2.2 NETWORK DESIGN

There is a fine balance between expenditure on data which may be of use and over-expenditure on a gauging network. However, when one weighs up the value of incorrect data, gauging networks generally under-expend. That is the cost of lack of data or inaccurate data is far greater than the cost of the investment in getting better data to reduce incorrect expenditure on water resources development works. Both under-design and over-design of waterworks can be costly. Over-design, i.e. a larger capacity than can be justified with the correct data, wastes money. Under-design means that additional works may have to be built to cope with the demand.

Not only data accuracy is required, but also a lengthy record. Impatient politicians or hydrological engineers may plan dams based on scanty data and be unaware of extreme droughts or floods which could occur which were excluded from the limited records available. The incorrect location of gauging stations or the scarcity of gauging stations also means that the data has to be extrapolated to dam sites or river crossings and this also increases the uncertainty.

Many projects are designed on extrapolated time series. This is because the records may be short and the statistical parameters obtained from the brief record. The mean and deviation are used to build a mathematical model of the flows. The parameters used on stochastic models range from simply the mean and standard deviation such as in Markov chain models (Brittan, 1961) to more sophisticated models including serial correlation, skewness and trend. The unknown random component is generated with random numbers, but as we learn more about nature we may be able to predict rainfall and model river flow more accurately.

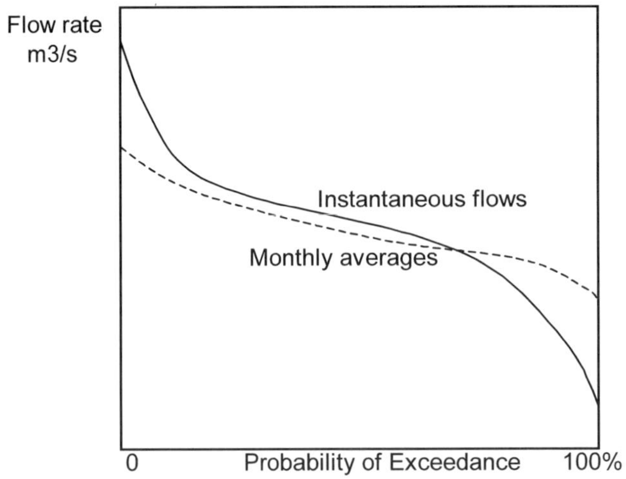

Figure 2.1 Comparison of instantaneous and monthly flow distributions

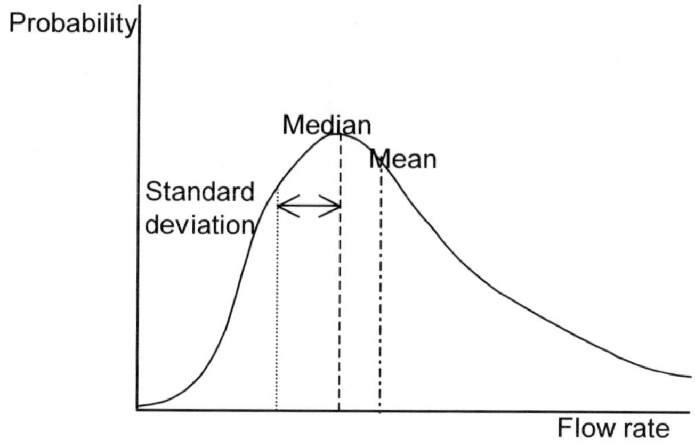

Figure 2.2 Probability distribution of flows

The distribution of flows depends on the time scale (see Figs. 2.1 and 2.2).

It is considered more reliable to extend flow records using deterministic rainfall-runoff modeling than to use statistical projections. Rainfall records are generally more extensive and easier to extrapolate from one position to another than streamflow records. They can be used in runoff models to reproduce the lacking records. The models may also be used to patch data during limited records. The runoff models, if based on hydraulic principles, have to be operated with a fairly short time interval even though monthly flow totals may be all that are required. Some regression-type or empirical models can be relatively simply applied but the true hydraulics of short-term storms and partial runoff of floods and soil penetration need a smaller time scale and an understanding of the physics. The RAFLER model (Stephenson, 2002) is able to zoom in on storm periods and therefore calculate hydraulically the runoff rates even when rainfall durations are considerably less than a month.

Many streamflow networks tend to be developed for specific purposes and, frequently, appear to have little relevance to national concerns with water resources. For example, the networks for flood forecasting normally include gauges primarily at forecasting points (usually locations with high potential flood losses) and may not be appropriate for irrigation, municipal water supply, industrial uses or navigation. Kovacs (1986) has demonstrated that runoff variability tends to increase with decreasing basin size. Also, the sampling interval is critical in determining data accuracy. Figure 2.3 relates drainage basin area to sampling interval. Nevertheless, for areas larger than 2,000 km^2, sampling should probably be at least twice each day since the error can become fairly large in estimating daily runoff.

Moss and Marshall (1979) discussed a useful categorization in network design which employs a user-oriented taxonomy as follows:

1. Project operations,
2. Project design,
3. Resource planning,
4. Water policy development and evaluation, and
5. Research.

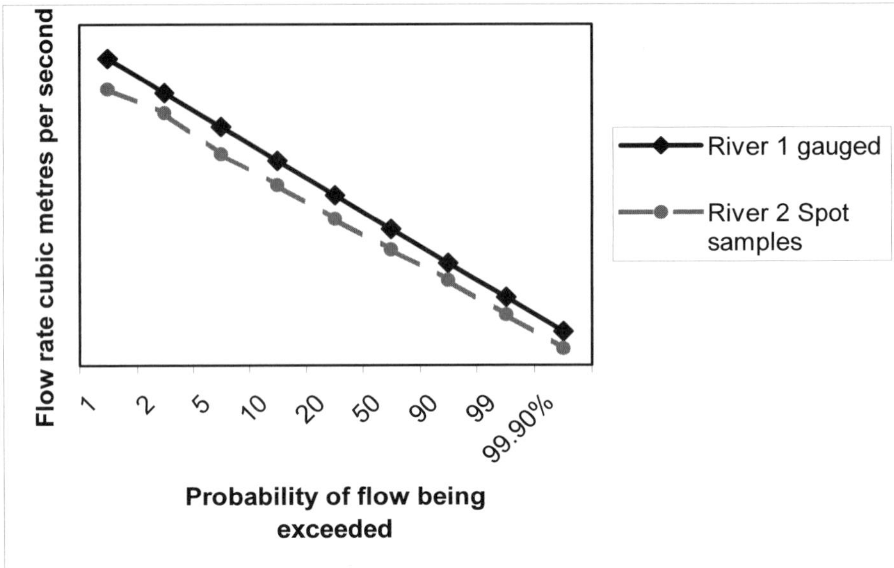

Figure 2.3 Extrapolation of river flows from one gauge to another (Logarithmic flow scale)

Except for research, the above list is ordered by the general intensity of data demands; project operation generally requires more information than project design or resource planning. One of the primary purposes at many stream gauging stations today is related to water quality monitoring, e.g. in Kansas, USA, seventy-nine of one hundred and nineteen stations (of course, other related uses are included at the same gauge – water availability, flood hazards, etc.).

The WMO (1972) has recommended that minimum gauging densities include:

1. For temperate and tropical climates, plains regions 1 gauge per 3,000-5,000 km^2
2. For mountainous basins in temperate and tropical zones 1 gauge per 1,000 km^2
3. For desert regions and polar zones 1 gauge per 1,000 km^2

These densities are, of course, very tentative in that all networks are also subject to economic considerations and specific data requirements.

Water gauging is not only carried out to measure flow rate or volume. Many other parameters are needed. The spacing and sampling interval will always depend on the purpose of the data. Design of gauging networks vary considerable on what data is to collected and when, such as:

1. Water quality
2. Sediment measurements
3. Groundwater table depth
4. Precipitation
5. Humidity
6. Temperature
7. Solar radiation
8. Atmospheric wind speed and direction
9. Soil moisture

10. Evaporation and evapotranspiration
11. Water temperature, and
12. Ice on rivers and lakes

Networks for a few gauge types are discussed below. Most of these networks are of a special type and network densities vary considerably – depending on data requirements, e.g. groundwater networks are frequently very dense in areas of extensive pumping. Even so the nature of aquifers cannot be estimated exactly and statistical methods are frequently used owing to uncertainties.

Rainfall data is generally more readily available than streamflow data. The data can be transposed more readily than stream data and generally can be patched if there are missing records, from adjacent gauges, with minimal extrapolation. Raingauges are cheaper and less complicated to install than streamgauges. Raingauges are used for many purposes, i.e. agriculture, meteorology, tourism and hydrology. It is therefore understandable that the time of readings may not be what the hydrologist always wants.

But the depth of rain is always there. The most common type of raingauge is the cone which is used by householders and has an accuracy consistent with the volume of precipitation. They are generally read once a day but monthly totals are often all that is available. Tipping buckets or siphon tubes are more accurate and can be automated. The data can be logged on an EPROM (Erasable programmable read only memory), or transmitted to a computer for direct processing.

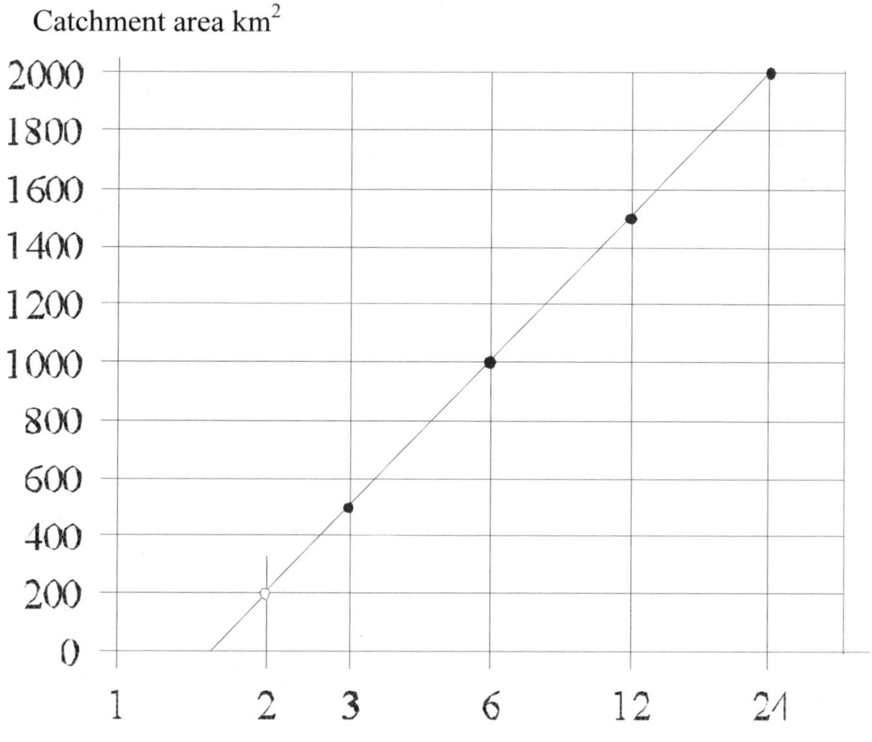

Figure 2.4 Relation between a reasonable sampling interval and drainage basin area (Kovacs, 1986)

The optimum spacing of raingauges depends on the study area. The larger the catchment the longer the time interval needed to model the runoff and the more an average rain is adequate. Generally about 5 or 6 gauges will ensure an accuracy of 5% (Johansen, 1971). The precipitation type, i.e. thunderstorm versus air transported, will affect the accuracy too but unless a subdivision of the catchment is required 10 raingauges is an upper limit required.

Weather stations can be classified into groups:

- Weather stations. These will be for accurate meteorological gauging and high standards of installation are adhered to. Masts for wind speed measurement will be 10m high. Topography should be such that there is no effect on wind and radiation. Radiation, temperature, humidity and atmospheric pressure are recorded. Snow measurement is measured separately from rain.
- Hydrometeorology stations. Installed for hydrological purposes and may include autographic precipitation gauge, evaporation pan, temperature and humidity. Raingauges. Raging from cones to autographic tipping buckets.
- Atmospheric, from balloons for climate studies.
- Remote sensing, e.g. radar, satellite imagery. These are subject to research as raindrop size can influence radar reflections. Thermal emissions and vegetation covers can be measured thus, both of which contribute to evaporation (Schmugge, 1985).

Evaporation is an important component in the water balance of any catchment. It is generally the second biggest input/output after precipitation. Hydrologists tend to use evaporation pans, whereas climatologists calculate evaporation from radiation, humidity and wind speed (e.g. Penman, 1948). Evaporation pan readings should be corrected by a factor of approximately 0.7 in transposing readings to a reservoir or lake, to account for different radiation level. A correction is also made for precipitation.

The method used to collect data can be manual in the case of remote poor regions, or on clockwork driven charts which require subsequent processing, or on battery driven charts or Eproms or transmitted to central computers by satellite link or radio transmitter. Obviously if the data is required for real time control, immediate direct transmission is necessary.

Water quality is a complex parameter to measure because there are many variables. It depends largely on the purpose of monitoring. The following need selection:

- Parameters to be detected.
- Accuracy required and base level in the case of automatic starting instruments.
- Time scale and interval
- Spacing of gauges
- Specification of output, e.g. total, peak or range, together with statistical ranges.

Laboratory tests for selected parameters are expensive and slow so the number of tests and parameters needs to be selected with care.

Sediment is a special quality parameter affecting reservoir capacity, fertility and water quality. It cannot be measured easily because it is suspended and transported at different densities at different depths below the surface of a river. It is generally measured indirectly by sonar surveys of reservoirs that contain sediment.

2.3 STREAMFLOW GAUGING

The objective in developing a stream gauge network is to improve the data available for water resources evaluation and flood estimation. A practical limit should be set on the number of proposed gauges, bearing in mind the difficulty of obtaining and processing many gauges and the cost of construction of weirs or even the equipment for measuring and recording water levels or flows. Gauge stations should be spaced to obtain a reasonable indication of flows from various types of catchments, and some gauges are established specifically for selected rivers. Gauge stations must be sited bearing in mind access and the suitability of the river for either construction of a weir or for rating the cross section (which should be reasonably stable). A preference for weirs as opposed to rated natural sections is made where trained hydrologists for rating natural rivers are lacking and because weirs generally give more accurate data, especially at low flows. Access is important for the purpose of retrieving data as well as construction. The decision as to whether or not to construct a weir should be based on the importance of the site, but also on whether a weir can be practically satisfactory for rating of the natural river cross section. Owing to the flat gradient of rivers near the coast, tidal effects and backwater rule out accurate gauging. The most important areas with respect to water utilization are the interior region and the coastal escarpment region where there are possible hydroelectric sites. In general, the sections should also be straight and uniform upstream, and a pool should be created behind the weir to ensure low velocities over the weir. There should be no bends or obstacles downstream to cause backwater or non-uniform flow profiles.

Where measured data are to be digitized and fed into computers, the software should be available and, in fact, the effort and cost can be significant. Data processing will make users aware of problems in data collection and made the data more meaningful to the processor when he sees summaries of flows. Some gauging sites can be influenced by downstream conditions especially at high flows, and this and other effects can affect theoretical rating curves. For this reason, it is advisable to calibrate some sections using both water profile computations and current meters.

A national gauging network may be established to assess water resources, to research into runoff and to study the environment. For these purposes sites will be selected with spacing of stations commensurate with the budget and norms in the industry. On the other hand gauges may be established for specific purpose, for example at a proposed dam site, for a diversion or to monitor discharges from the quantity or quality point of view.

The easiest but not the most accurate way of estimating flow rate is to measure depth of water in a channel, possibly even in a natural river channel, and use a rating curve to convert the flow depth to flow rate. However, natural channels may not have stable beds, i.e. may be liable to erosion or deposition over time. The method of converting flow depths to flow rates is also important. In a long straight channel reach, it may be possible to assume uniform flow and use a simple resistance equation to calculate flow rate. Obviously cross-sectional surveys will have to be made and the roughness will have to be assessed.

In many situations, the river bed changes in gradient from point to point and there are obstructions and deviations in the alignment of the river and therefore a water profile may have to be determined for alternative flow rates using backwater calculations. Then a stage- discharge curve should be prepared using the gradient of the energy line rather than bed of the river channel which is used for the simple method indicated above. Therefore the slope-area method requires observations of water level over a reach of river. This may be done theoretically using water profile calculations. The methods of measuring flow in rivers are numerous and depend on the budget available for establishing a gauging station, the purpose for which the readings are needed, the technical competence of the person establishing the gauging station and the facilities available for collecting and processing the data.

Current gauging using a propeller type meter is a direct method of measuring flow rate instead of flow depth. The cross section is divided into sections and flow velocity measured at each section. The current gauge can be hand held by wading across the river or held over the edge of a boat. If the water is deep or dangerous this can be done from bridges, but in other cases a cable is required with a bosun chair or mechanically operated gauge, to traverse the section, especially during flood flows. Salt dilution methods are also possible but expensive as are radioactive tracer methods. Rotating current gauges can be purchased in various sizes from a few millimetre diameter to over 200mm diameter. The larger units have heavy torpedo shaped ballast to align them facing the flow despite hanging from a cable.

Tracers can be used to obtain the flow rate by measuring the tracer concentration downstream of an injection point and relating this to the amount of dye or tracer injected upstream. The latter method is usually only useful for low flows because large volumes of tracer would be required for higher flows. In addition it is necessary to allow a considerable length of river to ensure good mixing before samples are taken.

Unsteady flow can also influence the flow rate versus stage relationship. Where large rivers are involved then a considerable length of river is needed for the measurements. The hydrograph could rise and fall within the time that measurements are made and the rate of rise and fall would also influence the mechanics, i.e. an equilibrium may not have been achieved.

It is more accurate to construct a measuring structure such as a weir or a flume in the river. These can be calibrated accurately theoretically and are stable. The shape of the weir or flume will depend on the shape of the river channel, the foundations and the importance of the readings over a range of flows. Many weirs are only designed to cater for low flows or small floods and larger floods overtop the banks of the weir which may require further calibration.

2.3.1 Weir design

Weirs offer a reliable and accurate way of measuring stream flow (e.g. Ackers et al., 1978; Herschey, 1985; WMO, 1980, 1981).The stage-discharge relationship is relatively independent of the downstream water level and the structure is stable, accurately dimensioned and easy to construct. Water level measurement is simple and accurate, and the accuracy can be controlled by selecting notch width and shape. Rectangular notches are generally preferred, in tiers, but triangular notches can be used for low flows or large variations in flow range. The alternative to a sharp crested weir is a flume or throat wherein water level is dropped by constricting the width of flow.

Older weirs are often sharp crested rectangular compound weirs. There is a problem of siltation upstream of weirs in some cases. However, due to high silt loads in many rivers this is difficult to avoid unless the river reaches are selected carefully or alternative designs are used. The Crump weir is now favoured for de-silting, but it requires a good foundation as it is very broad. Most river beds are alluvial and heavy grouting or excavation would be required; also the concrete volume required is considerably greater than for sharp crested weirs. There is also the problem of peak floods which will overtop any reasonable weir, and the actual design of the weir is not all that relevant during high flows because of inundation (e.g. Charlton, 1978).

In general, weirs are designed to monitor accurately flows up to approximately the 1-year flood without overtopping the flanks. Above this, the cross section of the valley will also have to be considered, and rating curves may be required for estimating higher flows. Concrete weir sections are generally grouted and founded on bedrock as seepage under the structure and around the flanks in alluvial material can be a real problem. It is suggested that the bottom crest be a minimum height above bed level to avoid inundation or backwater effects during the 1-year flood flow through the section. Similar criteria should apply to each crest. The maximum rise of each step should be 400 mm except in a steep sided valley; then dividing walls should be used to ensure accuracy of the calibration curves. The crest and sides of each level should be mounted with 100 x 100 x 10 mm angle irons. Weirs should be located in reaches relatively straight upstream to ensure symmetrical flow. Relatively steep sections are preferred (British Standards, 1986) because:

1. They assist in suspending silt and scouring it out upstream of the weir
2. They minimize the problems of backwatering and even tidal effects which inundate the weir.

Supercritical flow should, however, be avoided upstream of the weir. An upstream pool for stilling flow is also required, although these frequently silt up and require recalibration of the weir. The distance from the weir to the gauge is important and the water level should be unaffected by the drawdown curve. Bottom and side contractions can occur in the nappe as it flows over the crest and correction factors are necessary to the calibration equation.

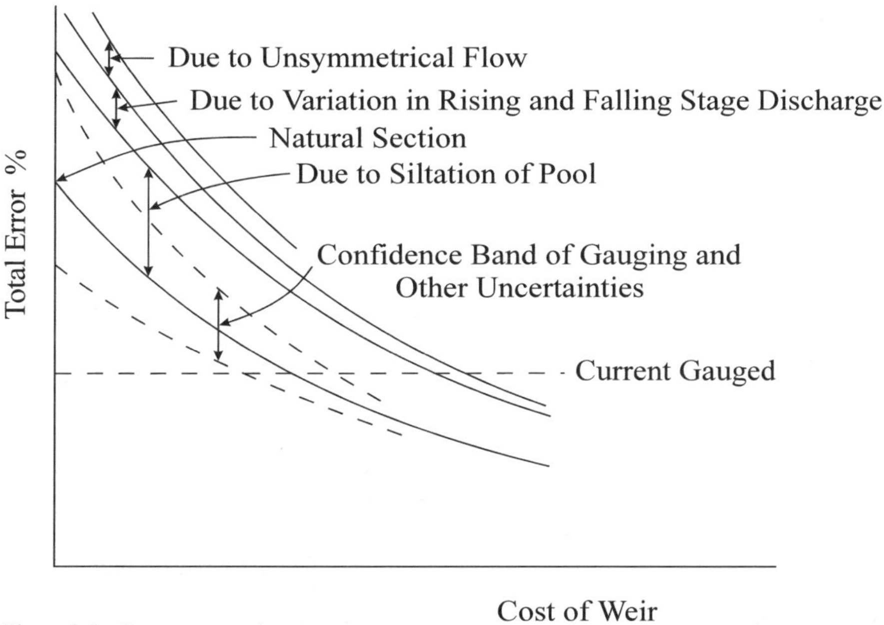

Figure 2.5 Cost versus weir accuracy

A correction factor is also necessary for inundation due to downstream flooding, and if the banks can overflow during flood an estimate of the overbank flow is necessary.

Supercritical flow should, however, be avoided upstream of the weir. An upstream pool for stilling flow is also required, although these frequently silt up and require recalibration of the weir. The distance from the weir to the gauge is important and the water level should be unaffected by the drawdown curve. Bottom and side contractions can occur in the nappe as it flows over the crest and correction factors are necessary to the calibration equation. A correction factor is also necessary for inundation due to downstream flooding, and if the banks can overflow during flood an estimate of the overbank flow is necessary.

Weirs are generally constructed of concrete so a sound impermeable bed and stable banks have to be found. The width of the valley affects the cost of the weir. On the crest may be fixed steel iron plates to form a sharp crest to ensure accurate rating.

The older type of float operated water level gauges are reliable and robust but at the same time piezoelectric or similar electronic water level recorders are much cheaper. These are more economical than float systems also in data collection and processing, and more foolproof. Experience and robustness are, however, limited in electronic data collection, and it is suggested that one stage of such water level recorders be installed initially in order to prove their efficiency and discover the various problems. It is therefore suggested that provision be made for float chambers; i.e. 150mm diameter pipes should be constructed from just upstream of the lowest weir crest into a float chamber on the bank of the river. The chamber shaft should be constructed of masonry or other rigid material to avoid being washed away in floods and to a level which ensures that the housing of the recorder is not washed away with floods of less than, say, the 100 year flood (this will require a risk analysis).

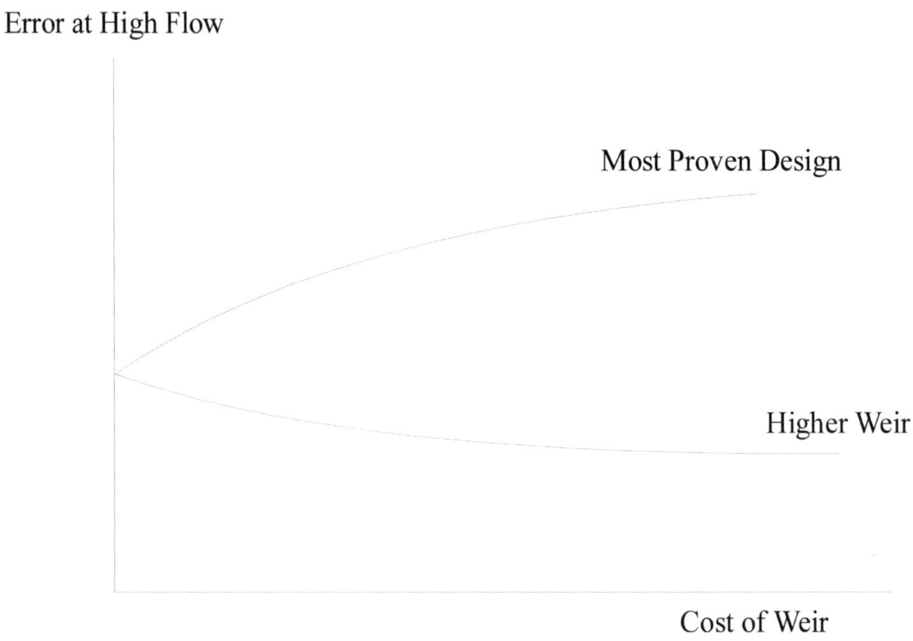

Figure 2.6 Effect of weir construction on gauging accuracy

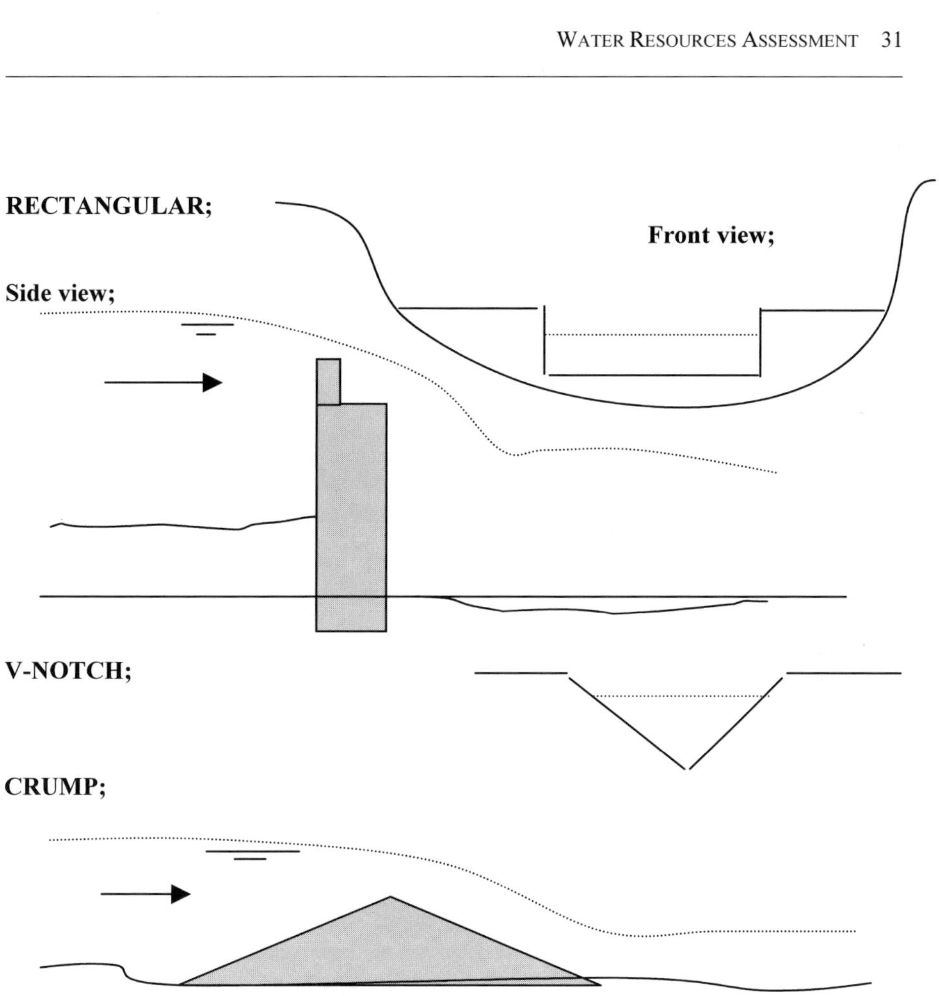

Figure 2.7 Examples of weir shapes

Housings for mechanical recorders need to be above flood level, weather proof and secure against vandalism. Provision should be made for floats and chart recorders together with the shaft down to low water level. The housings for electronic data collection systems should be weatherproof secure enclosed boxes away from the stream.

The rating of weirs can be carried out theoretically or in the field, and preferably using a number of ways. The rating curve or table, or equation for automatic processing, can be prepared with the assistance of computer programmes. Theoretical computations will allow for effects of side contractions, upstream pool depth, siltation, and estimate backwater effect for submergence of the weir. Each crest level can, thus, be selected to avoid submergence which could affect flow interpretation significantly. Above the top of the weir, the transition to full flood flow will be likewise estimated. In order to minimize error in reading there is a certain minimum width that can be used for the notches:

Weir notch sizing

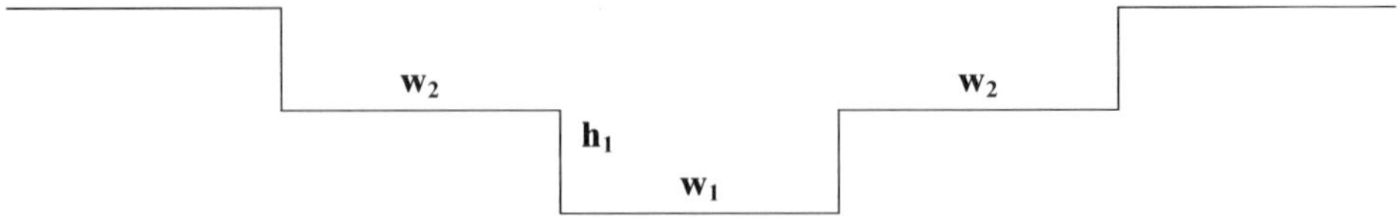

Figure 2.8 Front elevation of rectangular notch weir

Based on a minimum flow of Q_1 for notch 1 (say $0.1m^3/s$) and minimum accuracy desired for flow of dQ (m^3/s).

$$Q = kwy^{3/2} \approx 2wy^{3/2}, \quad \therefore y = \left(\frac{Q}{2w}\right)^{2/3} \tag{2.1}$$

$$DQ = 2w.\frac{3}{2}.y^{1/2}dy \tag{2.2}$$

$$= 3wy^{1/2}\,dy = 3wdy\left(\frac{Q}{2w}\right)^{1/3} \tag{2.3}$$

Solving for w: $$w = \left[\frac{dQ}{3dyQ^{1/3}/2^{1/3}}\right]^{3/2} \tag{2.4}$$

e.g. accuracy of measurement 10mm and minimum flow of $0.1m^3/s$

$$w = \left(\frac{0.1}{3x0.01x1^{1/3}/2^{1/3}}\right)^{3/2} = 8.6\text{m (or less)} \tag{2.5}$$

For notch 2, decide on height h_1 from slope of bank or construction ease, and calculate Q_2.

Continue to at least the 1-2 year flood level.

Check the average volume lost by overtopping using a flow duration extrapolation.

2.3.2 Gauges

The simplest form of a streamgauge is the staff gauge (see Fig. 2.9) These are usually of steel, enameled and bolted to a fixed iron post. They must be read manually. This is subject to the availability of a gauge observer at least once or twice a day, and more frequently during flood events. Education of the observer should go hand in hand with the establishment of a gauging site. The staff gauges can also be used to check automatic recorders. And there positioning should be checked to correct for sediment deposits or erosion which may affect the calibration.

Automated gauges are of several types – float type gauges requiring stilling wells and those utilizing some other type of device which measures water pressure (e.g. nitrogen bubbler gauges and other pressure measuring devices) or water level, e.g. sonar waves. In recent years, many of the automated gauges have been equipped with communication equipment (viz. telephones or radios utilizing relays where necessary or satellites). All of the above only measure water stage and rely on a stage-discharge relationship to convert recorded readings to streamflow.

Theoretical relationships between water level and flow rate in a river can be derived for weirs or river sections. The assumption of uniform flow will simplify the calculations, but the effects of obstructions such as bends, bridges and changes in alignment should be determined using a water surface profile computational program. And in the case of rapid flow changes a slope area approach may be necessary. That is the river bed slope should not be used for the energy gradient. The estimation of channel roughness is also not simple. It may vary with water stage. It is therefore good to observe water lines along the river bank and gauge the flow, to obtain calibration curves. Many assumptions must be made for these calculations so it is a good policy to actually measure flow rate in the field at some different water levels. This is the most accurate way at low flows.

2.3.3 Current gauging

An accurate method of measuring flow is using a current meter. A current meter is a propeller which has the revolutions counted electronically as it rotates. A rating equation gives the water velocity in m/s, and flow in cubic metres per second is obtained by multiplying by the area. The channel cross section should be divided into vertical strips using a tape measure or knotted line. For deep channels the current gauge can be lowered from a bridge or boat or pulley. In the latter case, a correction is needed for the angle of the suspension rope. For shallow water, span a rope across and walk along it. Take readings at measured distances and at depths of 0.2 and 0.8 x water depth from the bottom, or 0.6 x depth from the bottom for very shallow flow. The average velocity from three readings should be multiplied by depth and width to get flow.

Distance is measured across the river and the readings are taken every 1 meter or more. When taking readings, first measure the depth of the river Allow the propeller to run for 40 seconds or so and then take the reading on the counter. To get the level of the meter, read water depth and multiply water depth by 0.2 and 0.8 for the gauge depth. Take the mean of the two readings. For shallow water it may only be possible to take one reading, preferably at 0.6xdepth from the bottom. Take the readings for each meter or so across and integrate the results of velocity x depth x width. Use waders to keep dry, a life jacket for safety and a rope tied around observer's chest to shore in case of emergency.

Figure 2.9 Staff gauge and current meter

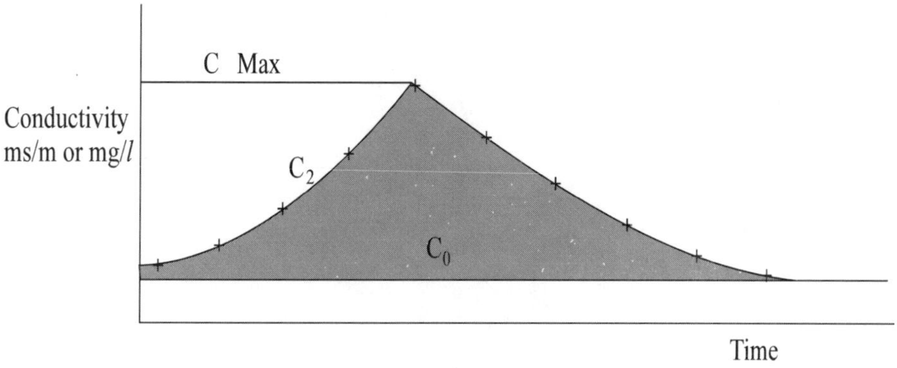

Figure 2.10 Plotting salt concentrations

2.3.4 Salt dilution

Salt dilution or tracers can be used to estimate flows for small rivers. Either a radioactive tracer or colour dye or soluble salt can be used. Either continuous dosing or gulp dilution can be used. The easiest method is to dissolve salt (sodium chloride) in water and then dump a measured mass of the solution into a river and measure concentrations downstream.

The flow rate of the river is calculated as follows:

$$Q = \frac{C_1 Vol}{\int_0^\infty (C_2 - C_0)dt} = \frac{C_1 Vol}{A} \quad \frac{mg/\ell.\ell}{mg/\ell.s} = \frac{\ell}{s} \tag{2.6}$$

Where:

Q = flow of river in m^3/s
C_1 = concentration or conductivity of initial concentrated solution
Vol = volume (m^3) of concentrated solution
C_o = background concentration
t = time, seconds

The quantity of salt to be used should be at least 20 kg per m^3/s of estimated river flow. A quantity of about 20 kg salt is dissolved in 80ℓ of water in a bucket. The concentration of water without salt and also with salt is measured, so that both can be used as data. Conductivity (in mS/m) can be used instead of salt concentration (TDS in mg/ℓ)). Plot concentrations of salt measured, or conductivity, and integrate (add) the area under the curve. Correct for background concentration.

This method needs a team of three people. One to take samples across the river downstream, one to pour the solution upstream, and one to be timekeeper for injecting and taking samples downstream. Sample at least 100 m downstream per 10 m of river width. Waterproof felt-type pens are needed to mark the plastic bottles (at least 50 number 200ml bottles are needed).

CALCULATIONS:

$$C_1V = Q\int_0^{\infty}.(C_2 - C_0)dt$$

$$Q = (C_1V) / (\int_0^{\infty}(C_2 - C_0)dt)$$

$C_1 = 415000$ mg/l
$V = 80$ litres $= 0.08$ m^3

$$Q = (33280) / (\int_0^{\infty}(C_2 - C_0)dt)$$

$C_0 = 320$ mg/l,

$\int$ Series 1 = 33 150 mg/l s

$\int$ Series 2 = 33 900

$\int$ Series 3 = 50 400

Mean = 39 150

$$Q = (33280) / (39150) = 0.85 \text{ m}^3/\text{s} \qquad (2.7)$$

2.4 HYDROLOGICAL MODELLING

River gauging records are rarely sufficiently long to enable reasonable assessments to be made of risk or extreme hydrological events to be estimated. It is therefore common to extend records, generate synthetic time series which can be used in computer models to study the operation of water resources systems. Some applications of models are described in Chapter 13. Computer models are used for:

- Patching records, for example filling in missing data using rainfall runoff correlation.
- Extending rating curves. For example, when measuring weirs are inundated, the flow can be estimated from an hydraulic model of the river reach.
- Generating flow time series using rainfall-runoff models. These models can be empirical or deterministic. Methods range from time-area to hydrodynamic and the catchment can be discretised using modules, finite differences or finite elements. It is useful to have some records to verify the model.
- Hydraulic based flow computation models require small time steps so totalization may be required for monthly flows, which are generally used for reservoir modeling.
- Synthetic flow series generation. The statistical properties of the river could be obtained from limited records or correlation with adjacent rivers. The mean, standard deviation, skewness, serial correlation, seasonal fluctuation and trend can be used in a stochastic model with a random number generator to generate flow series.
- Pre-programmed models such as unit hydrographs which can be used with limited rainfall data to generate runoff.
- Catchment water balance modeling with as many components as necessary included.
- Flow forecasting by getting data from Global climate models.

- Event modelling, e.g. the flood hydrograph for a known storm.
- Water quality modelling.
- Groundwater modeling, for drawdown tests to aquifer models.
- Groundwater pollution and remediation modeling.
- Reservoir operation modeling.

The complexity of a model should suit the purpose. There are large catchment models and drainage models available that require a learning curve to understand and calibrate them. On the other hand the data available may not warrant sophistication. Then a simple spreadsheet or correlation graph may suffice.

Although considerable research has been performed on simplistic methods such as time-are methods, unit hydrographs and Muskingum river routing, it is no longer necessary to use these empirical methods. More accurate results can be obtained with less effort using the appropriate software.

The linking of hydrological models to river models has also advanced considerably. River profiles can be obtained from digital terrain maps and used in hydraulic models to calculate water levels, flood risk, erosion potential and optimum bridge configuration. Then water lines and hazard zones can be plotted on maps with GIS software.

Table 2.1. Typical Data Input sheet for an Hydrological Event Model

RAFLS Event runoff model

18-Nov-02 | Title: Jukskei river - no reservoirs

Simulation duration	24	(Hrs)
Time interval	0.05	(Hrs)
Aquifer penetration	4	(m)
Gravity	9.81	(m/s^2)
Evaporation	2	(m/yr)
No. of modules	23	
Print at module...	23	

Calculate

	5y3h	50y3h		
HYETOGRAPHS				4
Hour	#1 (mm/h)	#2 (mm/h)	#3 (mm/h)	#4 (mm/h)
1	19	38	25	0
2	19	38	25	0.4
3	19	38	25	1.6
4				0
5				0
6				0
7				0
8				0
9				0
10				4.4
11				0
12				0
13				0
14				0
15				0
16				0
17				13
18				0.4

Module type	Mod. no	D.S. mod.	Length (m)	Width (m)	Gradient	Manning's N		FOR CATCHMENTS ONLY: Rill Ratio	Direct Runoff	Vert Perm.	Horiz. Perm.	Por.	Soil Suc.	
1 = Catchment							Hyetograph #	(frac.)	(%)	(mm/h)	(mm/h)	(frac.)	(m)	km2
2 = Conduit							No data							101.8405
3 = Channel							Side slope							Total Area
4 = Reservoir					Initial y (m)	Spillway ht (m)	(Horiz/Vert)							m^2
1	1	3	5000	2200	0.01	0.1	2	0.2	50	3	0.05	0.2	0.1	11000000
1	2	3	5000	1500	0.054	0.1	2	0.2	20	3	0.05	0.2	0.1	7500000
3	3	4	4500	5	0.02	0.05	2	0.2	30	3		0.2	0.1	22500
4	4	7	300	20	5	5	20	0.2	30	3		0.2	0.1	6000
1	5	7	2000	4500	0.05	0.1	2	0.2	30	3	0.05	0.2	0.1	9000000
1	6	7	3000	3500	0.03	0.1	2	0.2	30	3	0.05	0.2	0.1	10500000
3	7	8	3500	10	0.012	0.05	13	0.2	30	3		0.2	0.1	35000
4	8	15	300	20	3	3	15	0.2	20	3		0.2	0.1	6000
1	9	15	2300	3800	0.032	0.1	2	0.2	20	3	0.05	0.2	0.1	8740000
1	10	11	6000	1700	0.032	0.1	2	0.2	50	3	0.05	0.2	0.1	10200000
3	11	13	5000	9	0.015	0.05	10	0.2	30	3		0.2	0.1	45000
1	12	13	7000	1400	0.058	0.1	2	0.2	50	3	0.05	0.2	0.1	9800000
3	13	15	6000	9	0.015	0.05	13	0.2	30	3		0.2	0.1	54000
1	14	15	1500	2700	0.052	0.1	2	0.2	10	3	0.05	0.2	0.1	4050000
3	15	16	5000	9	0.008	0.05	8	0.2	30	3		0.2	0.1	45000
4	16	19	450	20	6	6	15	0.2	30	3		0.2	0.1	9000
1	17	19	3500	2700	0.043	0.1	2	0.2	30	3	0.05	0.2	0.1	9450000
1	18	19	2500	2400	0.063	0.1	2	0.2	60	3	0.05	0.2	0.1	6000000
3	19	21	4000	10	0.009	0.05	7.5	0.2	30	3		0.2	0.1	40000
1	20	21	4500	1800	0.055	0.1	2	0.2	10	3	0.05	0.2	0.1	8100000
3	21	23	5500	10	0.004	0.05	18	0.2	30	3		0.2	0.1	55000
1	22	23	5500	1300	0.075	0.1	2	0.2	10	3	0.05	0.2	0.1	7150000
3	23	24	3300	10	0.004	0.05	13							33000

LEEDS UNIVERSITY LIBRARY

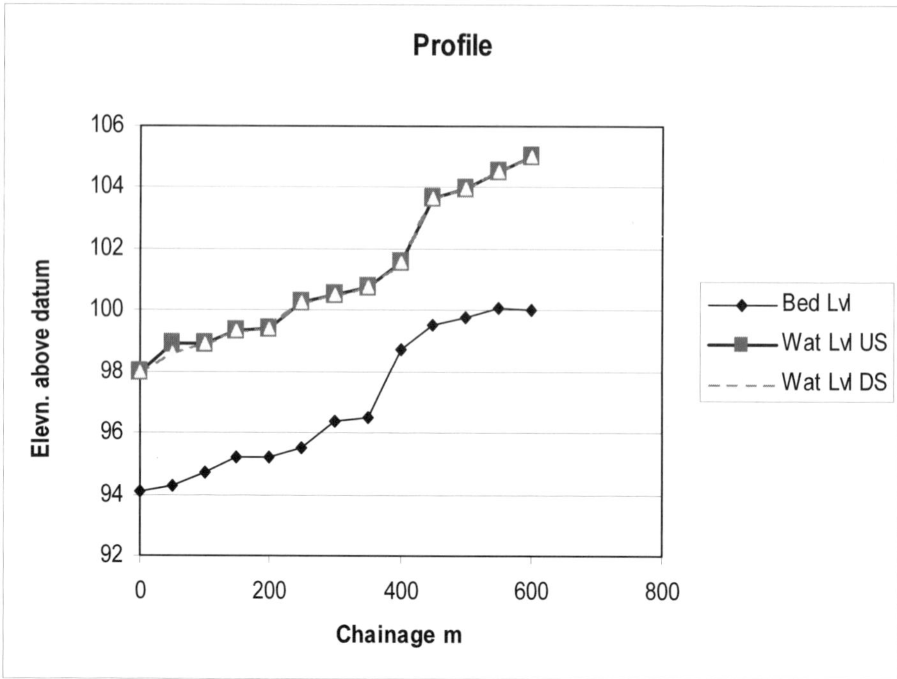

Figure 2.11 Output from hydraulic river model.

REFERENCES

Ackers, P., White, W.R., Perkins, J.A. and Harrison, A.J.M. (1978). *Weirs and Flumes for Flow Measurement.* John Wiley.

British Standard 3680: Part 2B (1986). Integration (gulp injection) method for the measurement of steady flow.

Brittan, M.R. (1961). Probability analysis to the development of synthetic hydrology for the Colorado River. In *Past and Probable Future Variations in Streamflow in the Upper Colorado River, IV*. Univ. Colorado.

Charlton, F.G. (1978). *Measuring Flows in Open Channels*, Report 75, CIRIA, London.

Herschy, R.W. (1985). *Streamflow Measurement*. Elsevier Applied Science, London.

Johansen, R.C. (1971). *Precipitation network requirements for streamflow estimation*. Dept. Civil Engg., Stanford Univ. Tech. Report 147.

Kovacs, G. (1986). Time and space scales in the design of hydrological networks. *Proc. Budapest Symp. IAHS*, 158, 283-294.

Moss, M. and Marshall, E. (1979). Some basic considerations in the design of hydrological data networks. *Water Resources Research*, 15 (6), 9.

Penman, H.L. (1948). Natural evaporation from open water, bare soil and grass. *Proc Royal Soc.* No.1032, 193.

Schmugge, T. (1985). Remote sensing of soil moisture, Chap. .5 in *Hydrological Forecasting*, Anderson, M.G. and Burt, T.P. (Eds.), John Wiley and Sons.

Stephenson, D. (2002). Modular kinematic runoff model (RAFLER). In Singh, V.P. (Ed) *Modelling of Small Watersheds*, Water Resource Publications, Boulder, Colorado.

World Meteorological Organization (1972). *Casebook on hydrological networks*, WMO Publ. 324, Geneva.

World Meteorological Organization (1981). No. 168, Guide to hydrological practices, Vol. 1, Data acquisition and processing, Vol. 2, Analysis forecasting and other applications.

World Meteorological Organization (1980). No. 519. Manual on stream gauging, Vol. 1, Fieldwork, Vol. 2, Computation of discharge.

CHAPTER 3

Drought Management

3.1 DEFINITION OF DROUGHT

A drought occurs when the flow of water is less than normal. This is distinct from seasonal or short-term fluctuations which are repetitive. There are different concepts of drought depending on the viewer:

- A meteorological drought occurs when precipitation is less than normal, generally assessed over the rain season.
- A hydrological drought is when river flow is less than normal. This may not occur simultaneously with meteorological drought, since soil moisture condition and upstream abstractions affect river flow.
- A groundwater drought may occur due to overuse of the aquifer and could be unrelated to recharge as the time scale is so big.
- An agricultural drought is said to exist if plants reach wilting point.
- An economic drought occurs if there is insufficient water to maintain production.

The level of deficit should be significant to warrant it being called a drought. Various indices exist for classifying drought severity in time and space. The intensity of deficit, or flow below average is one parameter used in the indices, and the other is the duration of the deficit. Together these produce a volume of water shortfall which could be expressed per unit area of catchment. Various indices have been suggested for indicating intensity of drought, for example the Percent of normal, or the decile. The Palmer drought severity index or PDSI (Palmer, 1965) has been adopted in the past. It is calculated from the departure in moisture from norm and is used in the US to trigger drought relief. More recently the Standardized Precipitation Index (SPI), developed by McKee et al (1993), has been adopted because it has an anticipatory component. It is calculated from the statistical properties of the precipitation (mean, standard deviation and skewness) and a value below zero signifies low flow, whilst a value below –1 signifies drought and below –2 extreme drought. The time scale can be introduced and SPI calculated for 3, 6, 12, 24 and 48 months.

Many developed countries have been accustomed to receiving all the water they require. Historically, the cost may have been relatively low due largely to bulk supplies. The quality has been good and the reliability acceptable. As the cost of tapping additional sources increases (Stephenson, 1995) due to greater distances and pumping lifts and increasing costs of purifying, we need to re-examine our standards. It may become more economical to suffer restrictions during drought because the average yield of surface reservoirs can thus be increased. The cost of suffering due to limited restriction can be less than the increased cost of new reliable sources (see e.g. Riley and Scherer, 1979).

Similarly in the future, the quality we expect may not always be warranted. Herold (1980) indicated that over 75% of urban water is returned to water courses. Wastewater is biologically and physically purified in municipal wastewater works before being discharged to streams although mineral concentrations increase during the cycle because they are more expensive to eradicate. Instead of tapping fresh surface water resources further away, it may be economical to recycle and tolerate lower quality for the bulk of our supply. In the extreme, potable water (which accounts for only 10% of the supply by water boards) could be supplied in a separate system or in containers.

This chapter looks at how to manage the water resources in the best way to mitigate the effects of drought, as an alternative to ensuring high reliability in surface water projects (Stephenson, 1996). The bulk of water in many countries is supplied from surface sources or rivers, which are notoriously erratic over seasons and cycles. Therefore the cost of storage is high. Alternatively water must be piped from distant, more abundant surface sources. The alternatives of groundwater have proved limited in some areas and recycled water is expensive to purify.

It was customary to ensure safety of supply during the worst recorded drought. Then the concept of recurrence interval, e.g. 50 years average between failures, became fashionable. More recently, Basson et al. (1994) proposed risk methods. By analysis the stochastic nature of river flow records, long-term simulations could be performed and the frequency of the necessity to reduce (ration) supply could be calculated. This varied depending on the target draft and level at which rationing commenced.

The optimum storage level at which rationing should commence and to what level to ration is addressed, bearing in mind the economics. That is, the cost of reliable supplies is balanced against the economic cost of rationing water. The sooner drought rationing is commenced, the less chance there is of running dry. This is addressed in the section on probability and again under operating rules. Then when the best operating rule has been selected, the usage has to be limited to the design figure. One way of rationing water is by means of water tariffs. A number of ways of controlling water consumption are discussed in Chapter 9.

3.2 RESERVOIR YIELD ANALYSIS

Reservoirs are generally sized to meet demands during critical droughts. These may be the worst droughts on record (then the draft could be referred to as the safe or reliable draft), or based on some selected recurrence interval. In the latter case, it is necessary to have an extended series of inflows, either from records or synthesized sequences extrapolated from rainfall correlation or statistically generated synthetic series.

Graphical methods, in particular, are useful for depicting reservoir storage state histories. The disadvantage of the graphical method is that an inflow series is necessary. However, synthetic records, probability matrix methods or analytical methods can be employed if records are limited, but the data can only be as valuable as the initial data set, in particular the flow record.

Analysis based on existing records was referred to by McMahon and Mein (1978) as Critical Period analysis, because the critical dry period in the records dictates the storage volume required.

Generally, methods of determining storage capacity at a particular river reach or site can be divided into four types:

1. Critical period techniques; based on historic flow records. Graphical or numerical analysis will reveal storage requirements as a function of draft, e.g. mass flow curves or simulation (Hufschmidt et al., 1966).
2. Probability matrix methods; based on statistical properties of the flow variations and independent sequence (See Brittan, 1961; Fiering, 1967; Gumbel, 1964).
3. Synthetic flow sequence, either from statistics or rainfall generation and simulation or analysis of the resulting flows, or use of statistical properties to derive deficits (Maass et al., 1962).
4. Analytical method using equations, theoretical or empirical (e.g. Alexander, 1962).

These methods are used for planning reservoir sizes, calculating yields of reservoirs, or deriving operating rules or risk.

Flow data can be recorded, estimated or synthesized. By estimation is meant a deterministic analysis of the available data in order to reproduce as accurately and as chronologically correct as possible, the flow sequences experienced. In synthetic flow patterns, on the other hand, only the statistical properties of the records are retained. The statistical properties are often sufficient to estimate storage requirements based on uniform drafts. On the other hand, frequently one wishes to consider variable draft operating rules, or the fluctuations in reservoir level. In such cases, a flow record can be synthesized. A random component is input to generate a synthetic record.

3.2.1 Definitions

A *reservoir* is a volume of water used to draw on in times of shortfall in the river flow.

Real reservoir *storage capacity* will be limited by the topography and dam wall height and has a maximum and minimum.

Some theories assume the reservoir is *infinite*, i.e. the reservoir can empty but never spill. This means that all the flood water is stored. Other theories assume a semi-infinite storage, i.e. one which can spill but never run dry (Moran, 1959).

Active storage is that above dead storage where dead storage is inaccessible due to the level of the offtake, or is allowed for silt accumulation.

Flood storage is the capacity provided, often above spillway level, for attenuation of floods and cannot be relied upon for increasing the yield.

Carry-over is the amount of water stored from one time period to the next.

The *time period* is the interval between storage computations, usually monthly for water supply reservoirs where there is seasonal variation in inflow, or annually for areas subject to large annual flow fluctuations.

Critical period is that in which the storage goes from full to empty without spilling.

Inflow is measured in cubic metres per month, or some other suitable units. Data may be either from past flow records, estimated from rainfall and related data, or synthesized from statistics derived from existing data. A number of ways exist for calculating storage once an inflow series exists (e.g. Hazen, 1914; Stall, 1962: Sudler, 1927).

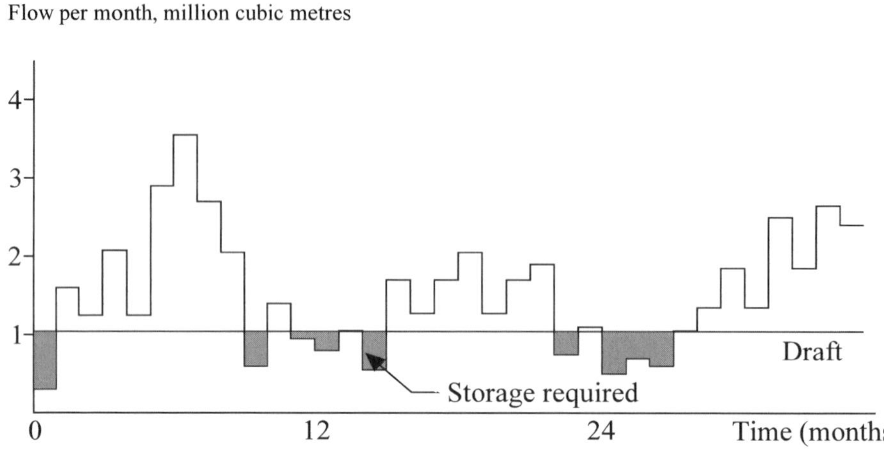

Figure 3.1 Flow rate versus time curve

3.2.2 Mass flow methods

A plot of river inflow rate versus time with a draft line superimposed will reveal when storage is required and when there is surplus water (see Fig. 3.1).

It is difficult to decide from Figure 3.1, however, what storage capacity is required, since carried-over shortfalls must be added. I.e. the total storage required is not just the area of one shaded portion, it is the sum of successive shaded portions, less inflow, provided net inflow does not exceed the reservoir capacity. For this reason it is easier to estimate the total maximum storage requirement from a massed flow curve such as Figure 3.2. This type of analysis is attributed to Rippl (1883).

The method for calculating storage is as follows: Plot cumulative inflow vertically against time from the start of the record on the horizontal axis. The slope of the curve represents the rate of inflow .A constant draft would also be represented by a positive sloping line.

If storage is just depleted at the end of a dry period, then the draft line will touch the inflow (point A). If the draft line is projected backwards, the difference between the ordinates of the draft line and the inflow line represents storage at any point of time. The maximum such difference (B) represents the storage required to meet the chosen draft for that drought period.

Extending the draft line further back indicates stored volume is increasing with time here, i.e. the inflow exceeds outflow. In fact, the reservoir would start filling at point D and may be full by some point E. Between B and E there would be spill.

The technique may be extended to allow for evaporation or any other loss. Net evaporation loss is calculated for each time period (usually monthly) by multiplying net evaporation rate by reservoir water surface area existing at the time. This is added to draft to make total loss each month, and the storage required now becomes CG. The slope of the line AB represents a constant draft over time. By varying the slope, the draft varies, e.g. a flat slope near A would represent restrictions in water use as the reservoir empties. The method is not very amenable to selecting the best operating rule, however. The same approach may be used in tabular or computer solutions.

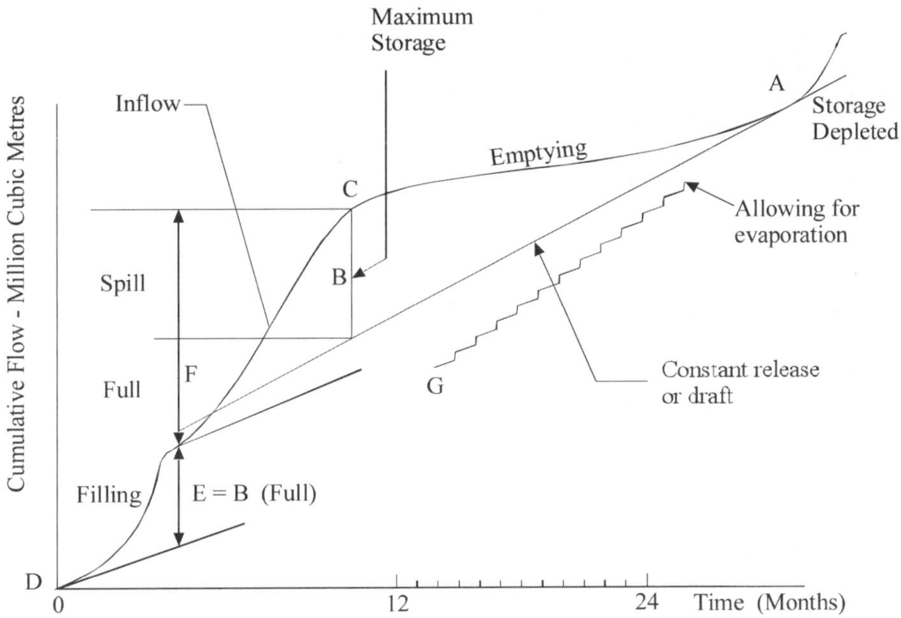

Figure 3.2 Massed flow curve for storage analysis

3.2.3 Simulation of reservoir operation

In order to investigate the adequacy of a water resource system design, operation of the system can be reproduced, or simulated, numerically. The analysis can be done by hand, studying only a short critical length of stream-flow record, or by digital or analogue computer capable of handling large amounts of data with relative ease.

A computer program may be devised for performing reservoir simulations by repetitive application of the hydrological equation.

$$S_{n+1} = S_n + I_n - U_n - E_n - [F_n] \tag{3.1}$$

where: S_n is storage at the beginning of a month, n,
I_n is inflow for month n,
U_n is release for month n,
E_n is evaporation, which is a function of S_n,
F_n = Flood overflow. [ ] implies this term is omitted if not positive.

It is usual to work with monthly flow records for draft studies, and with daily or even hourly readings for flood studies. Inflows I for a series of months may be obtained from records and fed into the computer. Draft U is specified as a constant value or as a mathematical function of storage state.

The simulation procedure is commenced by specifying an initial storage, then working through the available records month by month. Water demand level is held constant for a particular time horizon and historic flows routed through the system. The process may be repeated for demand levels associated with different time horizons.

It is specified in the program that if storage state reaches the capacity of the reservoir, spill will occur. Storage state at each month can be printed or plotted by the computer. If at any stage the storage is emptied, the design is deemed inadequate and storage capacity should be increased.

An optimum system design can be derived by analysing a number of combinations of reservoir capacity and draft. Net benefits are computed for each case, and that trial which indicates optimum benefit is selected. Variable draft operating rules can readily be accommodated in the program. A complete picture of storage states and drafts results from the analysis.

If a system comprising a number of reservoirs is to be simulated, then a large number of possible designs may exist, and many computations may be needed if adequate coverage is to be achieved. It is in such instances that the advantage of computers becomes evident.

When simulation methods are used, upper and lower bounds to storage can readily be accommodated. For example, draft can be stopped when storage is empty. This is not the case if the reservoir is assumed to have semi-infinite or even infinite storage. In the case of infinite storage, all flood water is retained and the calculations will overestimate yield. If semi-infinite storage is assumed, then the probability or frequency of failure is overestimated.

3.2.4 Storage-draft-frequency-analysis

The storage required to secure a desired yield from a river can be determined by iterative solutions of the hydrological equation. These may be performed either graphically or numerically. The graphical method is often referred to as the mass flow technique, usually attributed to Rippl (1883), and the latter as the water budget technique (see Klemes, 1979).

If the record of river flow at a proposed reservoir site is long enough to be reasonably representative of the full range of flow conditions likely to be experienced, the storage-yield relationship resulting from a straightforward mass flow analysis of the records might be considered acceptable as a basis for reservoir design, but would be subject to the proviso that drought sequences of the future would not be more severe than any in the records on which the analysis was based. In this method, the frequency with which a given storage would fail to meet a given yield would not be revealed. Methods have therefore been developed whereby the recurrent interval of deficient flow during the critical period associated with a desired yield from storage can be determined, and in this way the failure frequency aspect of storage design can be introduced.

Analysis of the mass flow curve for alternative drafts will yield a storage-draft relationship for the particular time series analyzed. Generally, short droughts will dictate storage requirement and longer droughts will dictate the large draft storage requirements. That is, the shorter droughts may be more intense than the average low inflow over a longer period. If there was a sufficiently long record or time series available, different droughts could be studied for any selected draft. There would be one drought resulting in maximum storage requirement and another indicating the second highest storage requirement and so on.

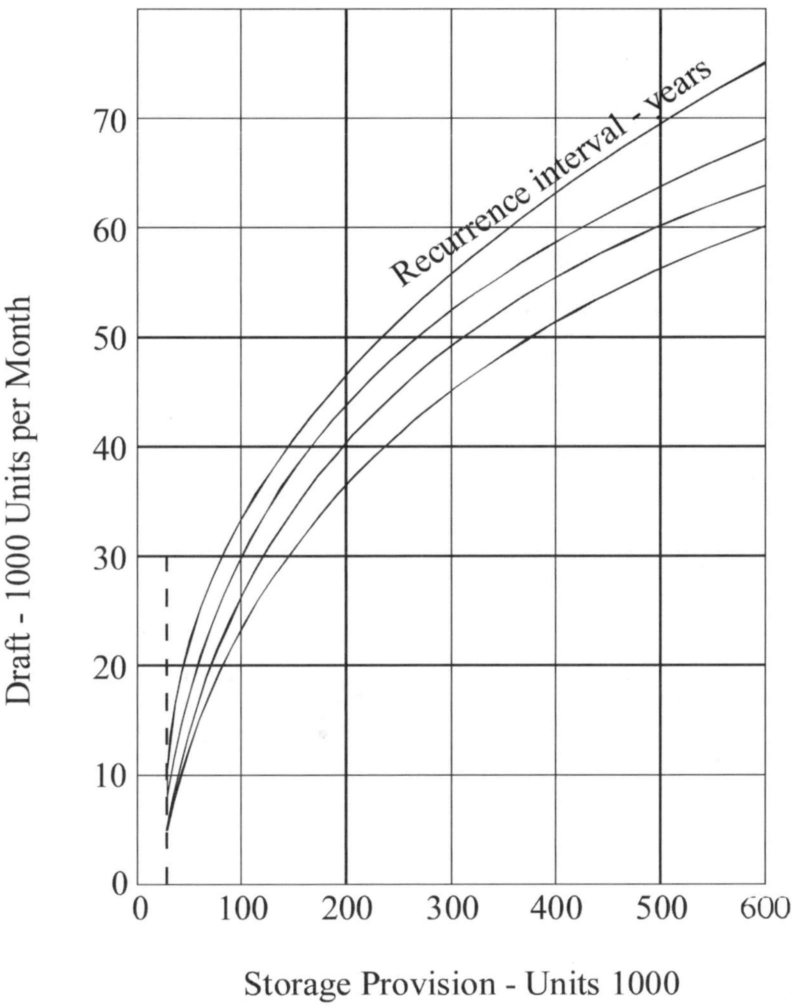

Figure 3.3 Storage - draft - frequency curves

The droughts, or critical time series, could be ranked and attached to recurrent intervals or frequencies. In that way, a table of storage versus draft and frequency could be established and followed as in Figure 3.3. The time series often has to be extended synthetically (using stochastic models, or deterministic models if long rainfall records were available).

The draft could also be plotted against probability (Fig. 3.4) (Basson et al., 1994). The horizontal lines are the so-called target draft. The higher the draft, the more frequently it is necessary to curtail draft. Once rationing is necessary, the draft can be decreased according to the water available or based on a derived operating rule.

The compilation of a number of simulations produced a generalized characteristic graph (Fig. 3.5).

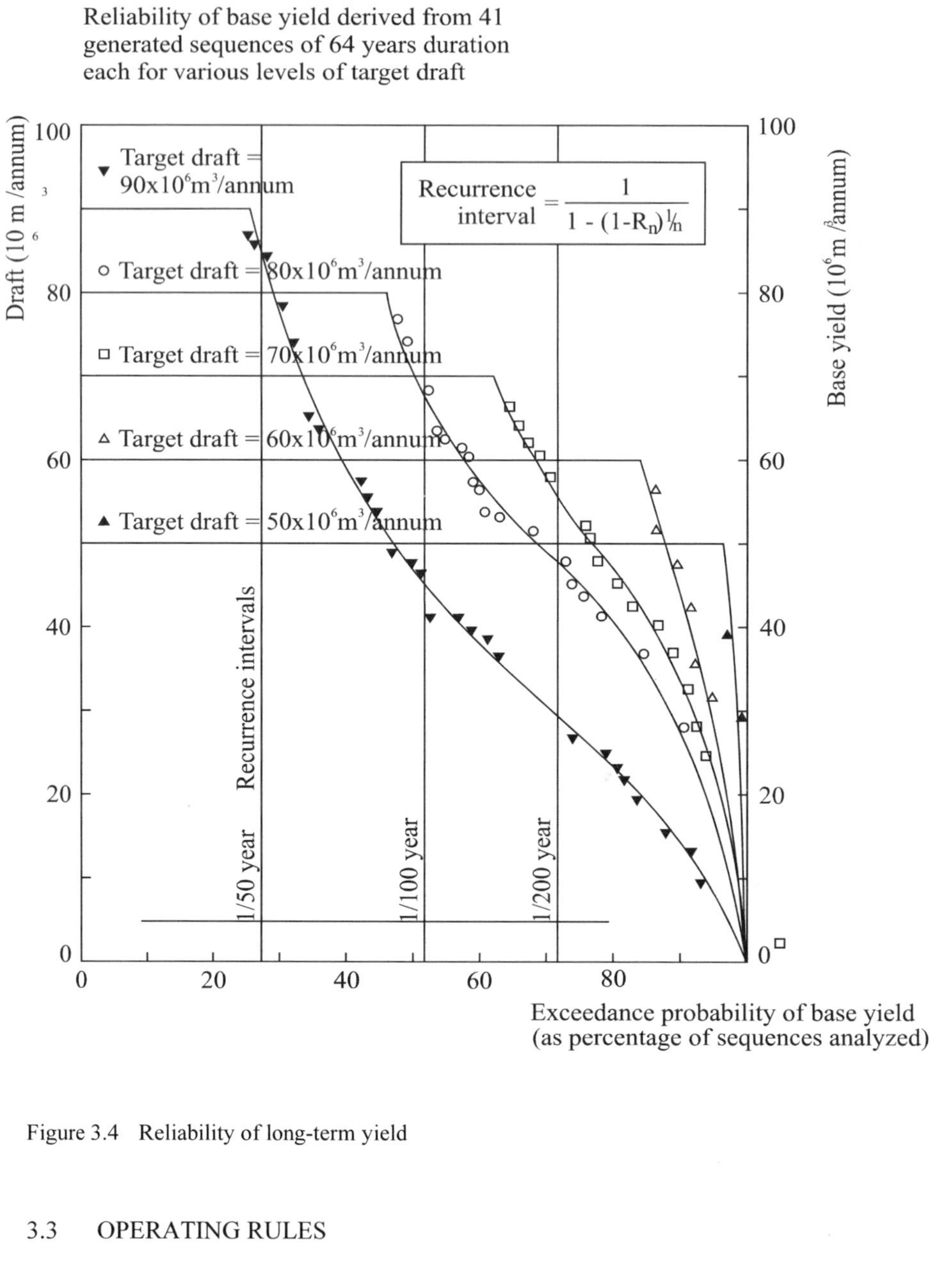

Figure 3.4 Reliability of long-term yield

3.3 OPERATING RULES

The concept of variable draft from reservoirs can increase the total yield, and postpone additional water schemes, thereby saving costs. There are many bases on which water may be rationed, including hydrological, economical, political, equitability or tradeoffs. Whether it is the total supply of water or the consequences of rationing which is to be optimized depends on whether there are conflicting demands.

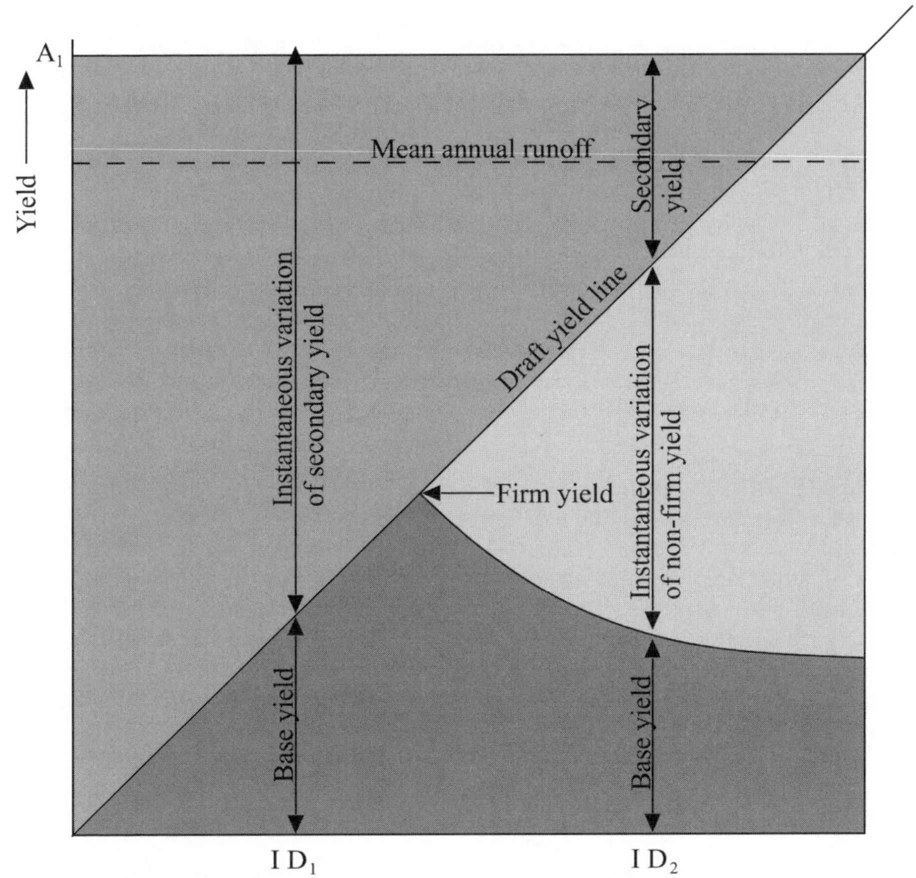

Figure 3.5 Graphical characterization of reservoir characteristics (Basson et al., 1994).

Alternative operating procedures or objectives have been proposed for optimizing the yield of reservoirs, e.g.

- Maximum total yield (Chen, 1972)
- Minimum economic loss (Stephenson, 1996)
- Continuous hedging (Shih and ReVelle, 1994) (see Figure 3.6)
- Proportional risk (Basson et al., 1994)
- Sharing (Dudley, 1990)
- Capacity allocation (Lund and Reed, 1995)
- Variable draft (Midgley and Pitman, 1967)

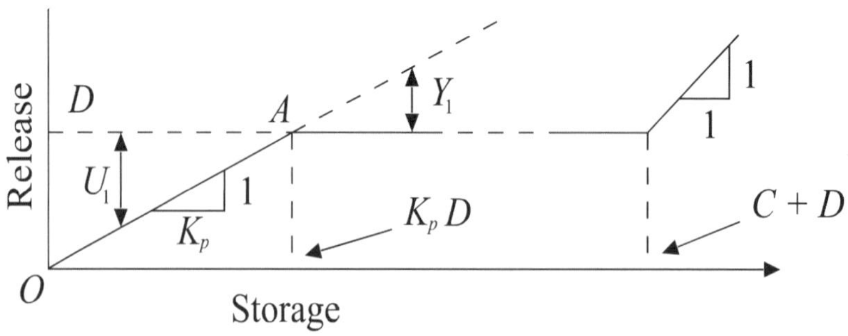

Figure 3.6 Diagrammatic representation of the Hedging Rule for drought management (Shih and ReVelle, 1994)

A separate problem is restricting the consumption to the desired release. This can be achieved by pressure, control, limited time of availability, high tariffs, fines, appeals, or mechanical control valves or orifices. Methods of controlling the consumption by means of tariffs are discussed later. A problem of coordinating consumption with supply availability also exists; because it takes a long time (months) for rationing or fines to become effective and operating rules need to be changed rapidly in times of critical drought. Emergency measures may therefore be needed, e.g. pressure reduction, cutoff supplies, or backup tankers. The practice of demand management is discussed in more detail in Chapter 9.

3.4 PROBABILITY MATRIX METHODS

The graphical mass flow method is used to size reservoirs. Computer simulation can also be used to size reservoirs by trial and error. A lengthy record or time series of inflows is needed to obtain a comprehensive picture of the operating state. Alternative operating rules can also be studied by computer simulation. The frequency of spill or running low can be tabulated. By plotting the reservoir storage fluctuation over a long period one is able to visualise risk of running dry and take corrective steps to minimize the risk. An alternative method of studying the probability of operating in various states (level of storage) is developed below.

3.4.1 Mutually Exclusive Model

Provided the statistics of a flow record are available, the expected operation of a reservoir can be predicted using probability theory. Choose the inflows, draft and storage capacity as integer multiples of some arbitrary volume unit. The following simplistic example demonstrates the technique (McMahon and Mein, 1978).

Figure 3.7 Distribution of reservoir inflows

Assume reservoir capacity is 2 units and there is a constant draft of 1 unit per time period. Inflows are discrete and independent and distributed as in Figure 3.7. Note that the sum of the probabilities of any inflow equals unity. It will be assumed here that the inflows occur at the beginning of each time period and draft takes place at a constant rate over or at the end of the full time period (one year). The example is in yearly increments. A more complex model in monthly increments would require varying monthly probabilities. An alternative model could assume inflow occurs uniformly over the year (the simultaneous model), in which case different results would be obtained. The probability of operating in various storage states is calculated from the probabilities of different inflows.

The constraints may be expressed algebraically and for the mutually exclusive model:

Insufficient water	$Z_{t+1} = 0$	if $Z_t + X_t \leq M$	(3.2)
Normal conditions	$Z_{t+1} = (Z_t + X_t) - M$	if $M < Z_t + X_t < K$	(3.3)
Overflow	$Z_{t+1} = K - M$	if $K \leq Z_t + X_t$	(3.4)

(Assumes first inflow then release)

Solving for inflow, $X_t = Z_{t+1} - Z_t + M$ for normal conditions. Inflow can be anything less for $Z_{t+1} = 0$, and anything more if Z_t gets full. Then the probability of that inflow is obtained from Figure 3.7.

Here Z_t = stored water at the beginning of the tth period
Z_{t+1} = stored water at the end of the tth period or at the beginning of the $(t+1)^{th}$ period
K = capacity of reservoir
X_t = inflow during t^{th} period
M = constant volume released at the end of the time unit or during it.

The first step is to set up the "transition matrix" of the storage contents. A transition matrix shows the probability of the storage finishing in any particular state at the end of a time period for each possible initial state at the beginning of that period. The transition matrix for the above example is a (2 x 2) matrix representing an empty condition and a half full condition as follows:

Finishing state Z_{t+1}	Initial State z_t: Empty 0	1	Full 2	
Empty 0	1/5 + 2/5	1/5	0	
1	1/5 + 1/5	2/5 + 1/5 + 1/5	1	
Full 2			0	
Σ =	1	1	1	(always check)

(3.5)

Each element of the transition matrix is found by applying Eqs. 3.2 to 3.4 to determine the inflows (and hence probability) for the storage beginning and ending in the state corresponding to that element. In the computations, the boundary conditions (empty and full) must be considered and, for the mutually exclusive model, the inflows must be considered separately and prior to the outflows.

Consider the element in Eq. 3.5 which represents a reservoir starting empty and finishing empty. This can happen if there are no inflows for the period (probability 1/5) or if there is one unit of inflow (probability 2/5). In the latter case, the release of one unit reduces the reservoir contents back to zero. Hence, if the reservoir starts empty, there is a probability of 0.6 that it will still be empty at the end of the time period.

Consider now the element (1.0) which represents a reservoir starting empty and finishing half full. If there are two units of inflow (probability 1/5) followed by one unit of release, the reservoir will end the period half full. If there are three units of inflow (probability also 1/5), the reservoir will spill because its capacity is only two units, then after one unit of release, it will again finish half full. Thus the probability of going from empty to half full is 2/5 or 0.4.

Note that the reservoir can never finish (and hence start) in the full condition because of the mutually exclusive assumption about inflows and outflows. Note also that the reservoir must finish in some condition; thus the sum of the probabilities in any column must be unity.

Let us now assume that the time unit is equal to one year and that the reservoir of capacity two units is empty at the beginning of the year one, that is, the initial probability distribution of storage content is:

$$\text{Storage State} \quad \begin{matrix} 0 \\ 1 \\ 2 \end{matrix} \quad \begin{bmatrix} 1 \\ 0 \\ - \end{bmatrix} \quad \Sigma = 1 \tag{3.6}$$

Since the transition matrix expresses the conditional probability of final storage contents given the various values of initial contents, the probability of final contents can be found by the matrix product of the transition matrix and the probability distribution of initial contents. Therefore, at the end of year one (the beginning of year two), the probability of storage is:

$$\underbrace{\begin{bmatrix}0.6 & 0.2\\ 0.4 & 0.8\end{bmatrix}}_{\text{Transition matrix}} \underbrace{\begin{bmatrix}1\\ 0\end{bmatrix}}_{\substack{\text{State of storage at}\\ \text{beginning of year one}\\ \text{(given)}}} = \underbrace{\begin{bmatrix}0.6 \text{ x } 1 + 0.2 \text{ x } 0\\ 0.4 \text{ x } 1 + 0.8 \text{ x } 0\end{bmatrix}}_{\substack{\text{State of storage at}\\ \text{end of year one}}} = \underbrace{\begin{bmatrix}0.6\\ 0.4\end{bmatrix}}_{\Sigma\ 1.0} \qquad (3.7)$$

The conversion process in Eq. 3.7 may be described as follows. The transition matrix shows the probability of the reservoir finishing in a specific state, given an initial state. If the initial state is known in terms of probability, then the joint probability will indicate the likelihood of the storage ending in a specific state. In Eq. 3.7, the transition matrix shows the probability of going from 0 to state 0 as 0.6, and the probability of being in state 0 at the beginning of year one is 1. Thus the probability of ending in state 0 is 0.6 x 1 = 0.6. But it is also possible to arrive at state 0 from state 1 which from the transition matrix has a probability of 0.2 The likelihood of being in state 1 at the beginning of year one is 0. Thus the probability of ending in state 0 but beginning in state 1 is 0.2 x 0 = 0. Hence the combined probability of ending in state 0 at the end of the first year is 0.6 + 0 = 0.6. A similar argument holds for state 1.

The process can now be repeated, using the state vector as the new starting condition. Therefore, at the end of the second year, the probability of storage content will be:

$$\underbrace{\begin{bmatrix}0.6 & 0.2\\ 0.4 & 0.8\end{bmatrix}}_{\text{Transition matrix}} \underbrace{\begin{bmatrix}0.6\\ 0.4\end{bmatrix}}_{\substack{\text{State of storage at}\\ \text{end of year one or}\\ \text{beginning of year}\\ \text{two}}} = \underbrace{\begin{bmatrix}0.6 \text{ x } 0.6 + 0.2 \text{ x } 0.4\\ 0.4 \text{ x } 0.6 + 0.8 \text{ x } 0.4\end{bmatrix}}_{\substack{\text{State of storage at}\\ \text{end of year two}}} = \underbrace{\begin{bmatrix}0.44\\ 0.56\end{bmatrix}}_{\Sigma\ 1.00} \qquad (3.8)$$

$$\begin{bmatrix}0.6 & 0.2\\ 0.4 & 0.8\end{bmatrix} \begin{bmatrix}0.44\\ 0.56\end{bmatrix} = \begin{bmatrix}0.6 \text{ x } 0.44 + 0.2 \text{ x } 0.56\\ 0.4 \text{ x } 0.44 + 0.8 \text{ x } 0.56\end{bmatrix} = \underbrace{\begin{bmatrix}0.38\\ 0.62\end{bmatrix}}_{\Sigma\ 1.00} \qquad (3.9)$$

At the end of the fourth year, the probability of storage content will be:

$$\begin{bmatrix}0.6 & 0.2\\ 0.4 & 0.8\end{bmatrix} \begin{bmatrix}0.38\\ 0.62\end{bmatrix} = \begin{bmatrix}0.6 \text{ x } 0.38 + 0.2 \text{ x } 0.62\\ 0.4 \text{ x } 0.38 + 0.8 \text{ x } 0.62\end{bmatrix} = \underbrace{\begin{bmatrix}0.35\\ 0.65\end{bmatrix}}_{\Sigma = 1.00} \qquad (3.10)$$

At the end of the eighth year the probability of the storage content will be:

$$\begin{bmatrix}0.33\\ 0.67\end{bmatrix} \qquad (3.11)$$

At the end of the ninth period it will be:

$$\begin{bmatrix}0.33\\ 0.67\end{bmatrix} \qquad (3.12)$$

It will be noticed that as successive years are considered, the probability vector of storage content becomes less affected by the initial starting conditions (in this example, the reservoir was assumed empty) and approaches a constant or steady state situation, which is independent of the initial conditions. From the steady state vector, it is seen that there is a 1/3 chance that the reservoir will be empty at the end of any year.

3.5 QUEUING THEORY

For the steady state case, a direct solution of the probability matrix is feasible. Queuing theory (Langbein, 1958) enables fluctuations in reservoir storage level to be correlated with statistical streamflow variations. The theory is so named because inflow to a reservoir is analogous to the random arrival of people to join a queue which is being served according to a prescribed rule. The arrivals (inflows) will conform to some statistical function of time, and the rate of serving (release of water from storage) can likewise be expressed as some mathematical function. The length of queue (volume of water in storage) must therefore be some function of inflow.

A simple numerical method of determining the likelihood of operating at various levels of storage will serve to explain the concepts.

Operation of a reservoir according to a variable draft pattern will be studied. Figure 3.8 indicates a specified operating rule, viz. normal draft is 60% of mean annual river flow (MAR) and once storage falls below 40% MAR, draft is dropped to 40% MAR. If storage should drop to zero, it will be necessary to lower the draft to 20% MAR, i.e. minimum river flow. A decision as to the rate of release is to be taken only once a year, say on the first day of October, and the draft is held constant until that day the following year, when a new decision may be made. In the example, storage capacity equals the MAR and evaporation loss is neglected.

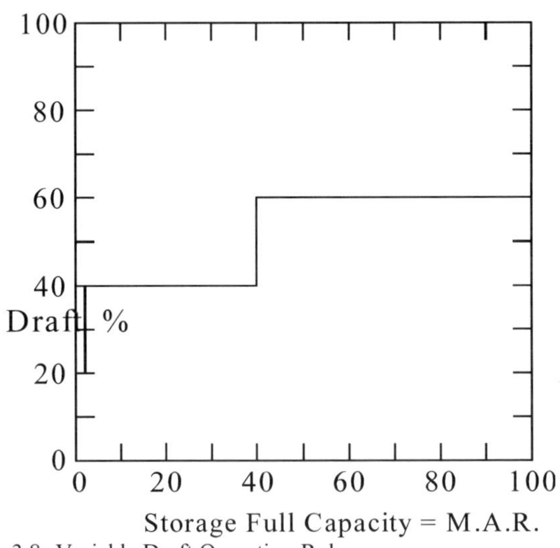

Figure 3.8 Variable Draft Operating Rule

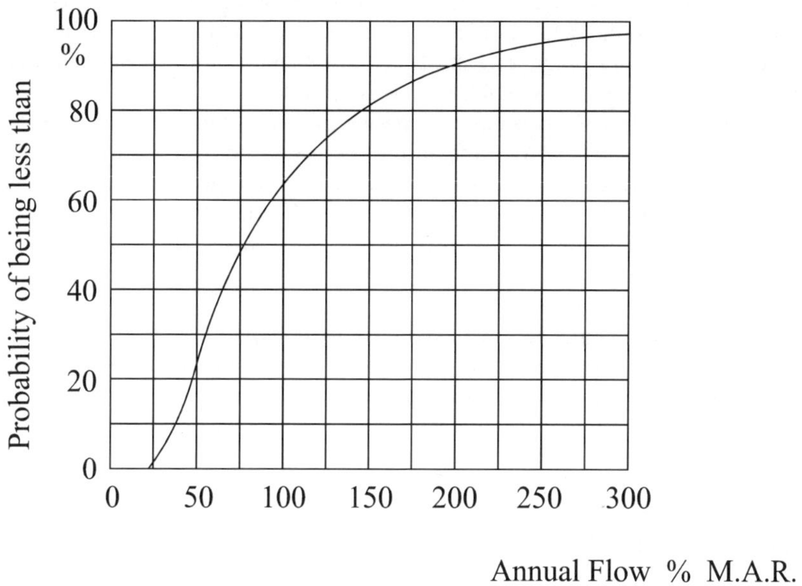

Figure 3.9 Annual Flow-Frequency Curve

Figure 3.9 indicates the frequency with which the annual flow of the river is less than specified values. The graph was prepared from the annual flow records of the Vaal river. From the curve, it can be seen, for instance, that the probability that the annual flow will be between 50% and 100% MAR is 64 – 25 = 39%.

Divide the storage capacity into intervals 0–20%, 20–40%, 40–60%, 60–80%, 80–100%, of the total capacity and denote the probability of commencing a hydrographic year in the range 80–100% full by $P_{0.8\text{-}1.0}$, etc.

The following storage equation is used:

$$1 = S_f - S_i + O \tag{3.13}$$

Where: S_f = final storage
S_i = initial storage
O = outflow

Several combinations of inflow and initial storage could result in the same range of storage at the end of a year. Thus storage state 80–100% could exist at the end of a year if any of the following occurred:

1. If storage stage at the beginning of the year was 80–100% MAR and inflow equalled outflow: The mid-point of the range 80–100% is 90%, which will be taken as the initial storage. The draft for a storage of 90% MAR is, according to Figure 3.8, 60% MAR. Inflow should exceed outflow by an amount equal to the draft less initial storage plus final minimum storage: 0.6 – 0.9 + 0.8 = 0.5 MAR. According to Figure 3.9, the probability of this flow being exceeded is 1.0 – 0.24 =

0.76. The probability of two events occurring simultaneously, however, is equal to the product of the partial probabilities. It follows that the total probability of commencing a year with storage state 80–100% MAR and of inflow being such as to result in a final storage state 80–100% MAR is equal to 0.76 x $P_{0.8-1.0}$.

2. A storage state 80–100% MAR could exist if storage at the commencement of the previous year was 60–80% MAR, and inflow exceeded outflow by an amount equal to the draft less initial storage plus final minimum storage: 0.6 – 0.7 + 0.8 = 0.7 MAR. The probability that this flow will be exceeded is 1.0 – 0.46 = 0.54, so the probability of commencing a year with 60–80% MAR and ending with 80–100% MAR is 0.54 x $P_{0.6-0.8}$.

The total chance of commencing a year with storage 80–100% MAR equals the sum of the probabilities of arriving at that state, from all possible storage states, at the beginning of the previous year. A calculation similar to those in steps (1) and (2) is performed for each possible initial storage state. The coefficients thus calculated are summarized in Table 3.1 in the row coinciding with $P_{0.8-1.0}$. The numbers in that row are in fact the coefficients in the equation:

$$P_{0.8-1.0} = 0.37P_0 + 0.33P_{0-0.2} + 0.41P_{0.2-0.4} + 0.41P_{0.4-0.6} + 0.54P_{0.6-0.8} + 0.76P_{0.8-1.0} \tag{3.14}$$

An equation may be derived for each final storage state. For example, computation of the probability of starting with storage 0–20% MAR and ending with 60–80% MAR proceeds as follows:

According to Figure 3.8, draft for initial storage of 0–20% MAR is 40% MAR. The lower limit to the inflow is that which will raise the storage from 10% MAR (average initial value) to 60% MAR. Draft minus initial storage plus final storage equals 0.4 – 0.1 + 0.6 = 0.9 MAR. The upper limit to the inflow is 0.4 – 0.1 + 0.8 = 1.1 MAR. According to Figure 3.8, the probability that inflow will be in the range 0.9 to 1.1 MAR is 0.67 – 0.59 = 0.8. Similar coefficients are derived for different initial storages and assembled in the appropriate line of Table 3.1.

The matrix in Table 3.1 represents a set of six equations in six unknowns. The simplest method of solving for the six probabilities manually is by successive approximation. A first set of probabilities is arbitrarily selected such that their total equals unity. For example, the set of equations was solved by initially selecting the values 0.01, 0.03, 0.03, 0.08, 0.15, 0.7, for the columns of Table 3.1.

The final probabilities appear at the bottom of Table 3.1. An interpretation of the answer in the last column is that there is a 66.8% chance of commencing the year with storage in the range 80–100% full if the reservoir is operated according to the rule depicted in Figure 3.8. Note that it is assumed a steady state has been reached, i.e. a number of years have passed since the reservoir was built.

Table 3.1 Coefficients in the probability equations

Probability at the end of the year	Probability at beginning of year					
	P_0	$P_{0\text{-}0.2}$	$P_{0.2\text{-}0.4}$	$P_{0.4\text{-}0.6}$	$P_{0.6\text{-}0.8}$	$P_{0.8\text{-}1}$
P_0	0	0.04	0	0	0	0
$P_{0\text{-}0.2}$	0.13	0.20	0.04	0.4	0	0
$P_{0.2\text{-}0.4}$	0.24	0.22	0.20	0.20	0.04	0
$P_{0.4\text{-}0.6}$	0.16	0.13	0.22	0.22	0.20	0.04
$P_{0.6\text{-}0.8}$	0.10	0.08	0.13	0.13	0.22	0.20
$P_{0.8\text{-}1}$	0.37	0.33	0.41	0.41	0.54	0.76
Answer	0.0003	0.006	0.036	0.096	0.194	0.668

The technique described could be used to derive an optimum variable-draft operating rule. If monetary values were assigned to the different levels of draft, the average annual economic loss contingent upon water rationing would be the sum of the probabilities of drawing water at various rates multiplied by the economic loss associated with the corresponding draft. The optimum rule would be selected from a number of computations using alternative operating rules (Morrice and Allan, 1959).

A more direct method of deriving an operating rule than the foregoing would be to treat the draft at each storage state as a variable. The queuing equations would then include a term for each possible storage state and draft. The optimum draft at each storage level could then be selected by linear programming.

The method could be extended to permit a number of seasons to be examined instead of one whole year. The probability of starting any season with a certain storage level would equal the probability of ending the preceding season at that storage level. Recognition of a number of times of the year at which decisions might be made would enable the draft to be varied at more frequent intervals thereby rendering the system more flexible and possibly reducing the incidence of prolonged water rationings. In practice, however, it is unlikely that the effects of sequential correlation of monthly flows can be readily accounted for and the additional computations entailed in breaking the year into more than two seasons would probably not be justified.

Evaporation effects were omitted in the example, but could readily be taken into account in the release function in Figure 3.8.

3.6 CONJUNCTIVE USE OF ALTERNATIVE SOURCES

A way of reducing the risk of running out of water could be to conjunctively use different sources. Different sources of water could be reservoirs on different rivers, or in different hydrological regions. Different types of sources may be used, e.g. groundwater, reclaimed waste water or saline water, rainfall stimulation or snowmelt. Langenbaugh (1970) determines operating policies for conjunctive use of surface and groundwater, and Noel and Howitt (1982) look at multibasin management.

The advantages of conjunctive use of different sources include the following:

1. It is unlikely both sources will suffer shortage simultaneously.
2. Water can be drawn at a greater rate from surface reservoirs knowing there is a backup source in times of drought. So the yield of surface reservoirs can increase.

3. Backup sources, which may be expensive, need only be used in an emergency. Thus operating intensive sources would tend to be used as backup, whereas capital intensive sources would be drawn on more regularly.
4. Standby sources can be slowly recharged, not requiring a high rate of inflow, but drawdown could be at a higher rate.
5. Groundwater aquifers could be recharged with waste water. Slow filtration may even occur to purify the water.

On the other hand, there are a number of options open for utilizing existing resources which include:

1. Optimizing the operation of a single reservoir;
2. Providing different categories of assurance (e.g. Brandehoft and Young, 1983);
3. Restricting the use of water during times of drought and implementing long-term policies;
4. Utilizing multiple reservoirs or groundwater for water transfers;
5. Recycling waste and industrial water.

All the above need to be dealt with optimally and each option considered carefully as the damage due to drought will often depend on the water management policy utilized during the drought (Booker, 1995).

An advantage of utilizing groundwater storage may be that it creates less of an environmental impact, there are smaller losses due to evaporation and seepage, fewer topographical limitations, increased reliability and lower cost. The 'lower cost' is debatable as groundwater is often more expensive than surface water to extract and should then only be used in times of emergency. However, it is clear that use of groundwater reserves needs to be strictly controlled and monitored, as badly managed use often results in problems such as land subsidence or intrusion of inferior quality water (see Chapter 6).

Other disadvantages of groundwater use are the difficulty of accurate modelling of the source, the large area affected due to mobility of groundwater, interdependency of pump sites (which may traverse regional or local authority boundaries); and replenishment of the source, which takes longer than for a surface reservoir. Coupled to this is the interdependency of groundwater and surface water. Groundwater may be recharged by nearby surface water, in which case, drawing supplies from the groundwater will merely deplete the reservoir further. This will create an initial false sense of security in the amount of available water followed by management problems later. Some authors eliminate this problem by assuming that there is no link between the two, e.g. Loucks et al. (1981) assume the groundwater table to be below the stream bed. Gupta and Goodman (1985) present a case where surface water and groundwater are operated cyclically. When the surface water supplies are low, groundwater is utilized. When there are excess surface water supplies, the excess is used to replenish the groundwater. The yield of the system is then considered a joint yield.

Coe (1990) states "Conjunctive use of surface and groundwater can be defined as the management of surface and groundwater resources in a coordinated operation to the end that the total yield of such a system over a period of years exceeds the sum of the yields of the separate components of the system resulting from an uncoordinated operations." Conjunctive use of groundwater poses some additional problems to use of surface waters, e.g. recharge of the source is slower; water quality is more easily affected; effects of over-utilization extend further than the reservoir drying out, switching between surface and groundwater is more costly, ground and surface water are often under the jurisdiction of different government departments or ministries, the question of who owns the water once it has been put into storage, who is responsible if use of groundwater and recharge results in damage to property, etc. Their model minimizes the costs of developing and managing the conjunctive use system whilst taking aspects such as the amount of combined water available, aquifer storage levels and demand of water into account.

Paling (1985) details a conjunctive use system, using groundwater, investigated for the Witwatersrand area in South Africa. He notes that a properly organized recharge system will result in a constant capacity for longer than can be achieved by independent dams as dams silt up over a period of time. As the river water that is being supplemented cannot be used to recharge groundwater, alternative sources need to be considered. In this study, wastewater is considered as a source of recharge (Guyman and Welch, 1990). However, precautions need to be taken to ensure that the water is chemically and bacteriologically safe when abstracted from the ground for conjunctive use. Two methods of recharge are investigated: infiltration and injection through wells. Wells are expensive to construct and operate but may be more feasible under certain constraints such as impermeable surface layers and lack of land. However, wells clog up quickly even with regular maintenance. Surface infiltration can result in a high level of pollutant removal and could result in a decrease in pre-treatment costs of sewage.

The greatest cause for concern when utilizing an aquifer for wastewater storage is the lack of control in the spread of the water. Careful consideration of the physical limits of the aquifer and pre-treatment of the water can overcome this problem. Filtration ponds can contribute to the purification of the water. Pathogenic bacteria and viruses die after travelling approximately 100m in a fine or medium grained aquifer. However, this does not happen in a fissured aquifer. It is important to ensure that wastewater stored in aquifers does not necessarily endanger drinking water sources. Exceeding the capacity of the aquifer may lead to the following:

- Nutrient rich ground water discharged to surface waters which may promote accelerated algal and plant growth in the water;
- Waterlogging of soils can affect agriculture and cause damage to buildings; and
- Overflow from the percolation ponds may flow into surface waters in extreme cases.

3.7 ARTIFICIAL RECHARGE

Artificial groundwater recharge may serve different objectives such as conservation of a certain groundwater level, prevention of saltwater intrusion, water purification, supplementation of groundwater reservoir storage, wastewater disposal or a combination of these. River water is the most obvious source for artificial groundwater recharge, but convenient nearby rivers with adequate surplus flow may be limited.

A largely undeveloped water source can be found in sewage works effluent. Less than 50% of water supplied is used consumptively and this effluent is often used for pasture irrigation and as cooling water for power station and industrial usage. Artificial recharge with sewage works effluent will require greater precautions to guarantee chemically and bacteriologically safe water, but promising results have been seen in

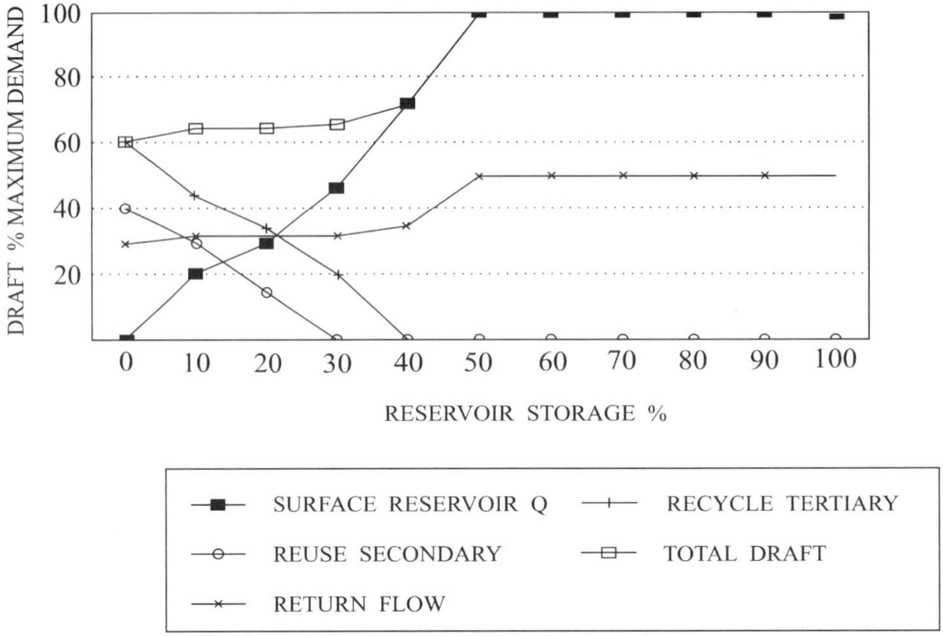

Figure 3.10 Operating Rule: Surface reservoir and wastewater

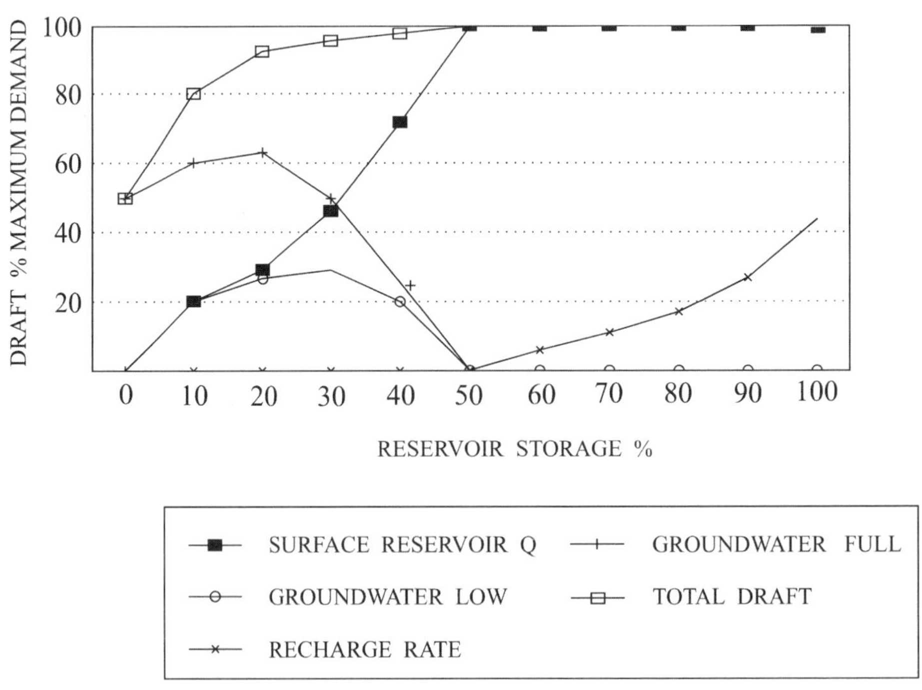

Figure 3.11 Operating Rule: Surface reservoir and groundwater

similar experimental and full-scale projects in the U.S.A. (Bouwer et al., 1980), Israel (Idelovitch and Michael, 1984), Australia (Mathew et al., 1982) and Saudi Arabia (Ishaq and Khan, 1997).

The main causes of clogging of injection wells according to Huisman and Oltshoorn (1981) are:

1. The presence of air bubbles in the recharge water;
2. The presence of suspended matter in the recharge water;
3. The growth of bacteria in gravel pack and surrounding formation;
4. Mechanical jamming;
5. Reactions between the recharge water on the one hand, and the native ground water and aquifer material present in the formation on the other hand.

Except in the case of impermeable surfaces or saturated ground without groundwater flow, every form of surface water will induce infiltration. Although every contact of sewage works effluent with unsaturated ground could therefore be described as recharge with treated wastewater, no further attention will be given to land treatment of wastewater or spray irrigation and flooding as supplementation of groundwater storage is not the main objective in these cases. The common method for groundwater recharge by infiltration is inundation of ditches, ponds or artificial wetlands, for which design recommendations can be found in publications by e.g. Bouwer (1970), Huisman and Oltshoorn (1981) and Gersberg et al. (1983). The initial infiltration rate or a recharge unit is primarily a function of the permeability of the surface layer, soil permeability, moisture content of the underlying beds, basin slope, depth of water in the basin, water temperature and the amount of suspended solids. Although the maximum infiltration rate can be modified somewhat in basin design, it is essentially a fixed property of local geological conditions. It can be decreased though, by clogging of the surface layer and by rising of the groundwater mound beneath the surface to the bottom of the basin. The clogging may be of physical or biological origin, but cheaper ways to prevent clogging and easier remedies to restore the initial infiltration rate are available than is the case with injection wells. Algae growth can be controlled by the addition of chemicals or by intermittent drying of the basin surface. As the clogging is essentially a surface sealing phenomenon, the restoration of the infiltration rate can be achieved by scraping off the top centimeters of the soil and if necessary after a number of treatments, replacing it by a new soil cover.

However, the most important advantage of surface infiltration over injection is the higher efficiency of pollutant removal. After percolating downward through the soil to the water table and moving laterally as groundwater for some distance, the wastewater will have lost its suspended solids, biodegradable materials, micro-organisms, almost all of its phosphorus, and, with proper management of the spreading facility, most of its nitrogen. This may result in a reduction of the pre-treatment costs of the sewage before recharge. The efficiency of this removal process depends to a large extent on the local conditions and process design and should be studied and tested for each individual case.

Optimal infiltration rates vary between 0.03 and 30 m/day (Fetter and Holzmacher, 1983) and flooding schedules from three days flooding – one day drying to eighteen days flooding – thirty days drying were reported (Mathew et al., 1982).

3.8 CASE STUDY

The Min-Der reservoir, in Taiwan, was used for a case study to optimize a reservoir operating rule (Stevens et al, 1998). Taiwan is particularly prone to drought. The majority of river flow occurs during typhoons frequently followed by lengthy periods of drought. Rivers are steep so storage is limited and reservoirs fill rapidly with sediment. Storage is therefore limited and new storage is expensive. Reservoir management is therefore a viable and economic alternative to new supply schemes.

Long-term time series of monthly inflows into the reservoir were generated. River flow records are only available since 1970 and were therefore of limited using in obtaining 100 year or other extreme flows. Time series could be generated synthetically by stochastic means or deterministically. A physically based module, RAFLER (Stephenson and Paling, 1992), was used with monthly rain data from surrounding rain gauges. Acceptable agreement was obtained with the existing streamflow records giving confidence in the longer term projections.

The monthly streamflow series was analyzed to obtain risk of running the reservoir down to different levels with various combinations of starting storage and draft, over the winter and summer seasons respectively. A matrix relating probably end to starting storage and draft was then set up for solution by linear programming. Each draft was associated with a cost coefficient for the optimization exercise.

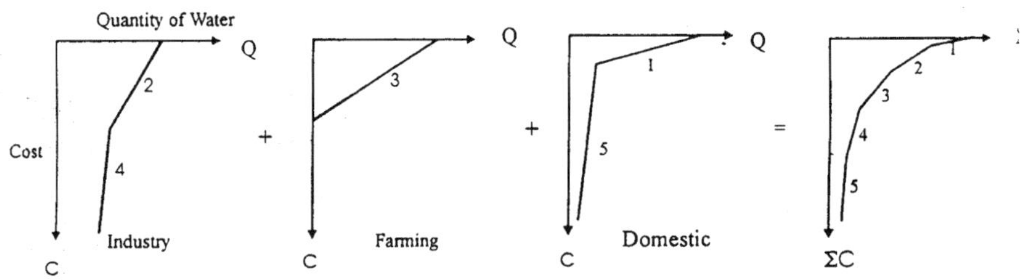

Figure 3.12 Economic loss due to water shortage

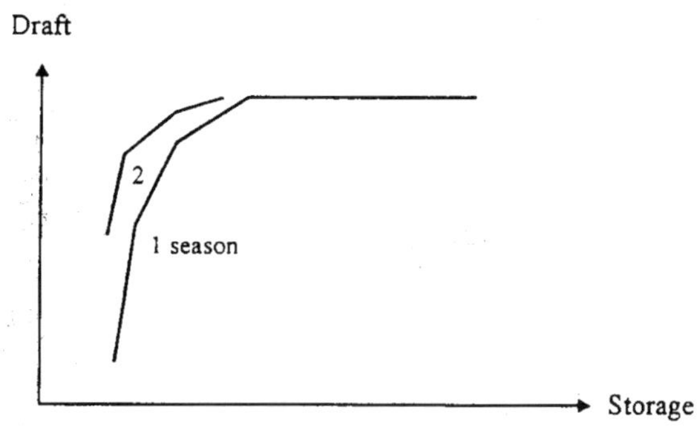

Figure 3.13 Derived operating rule

The economic costs of various levels of water restriction were obtained by questionnaires sent out to various consumers. Figure 3.12 shows the resulting composition of the objective function (economic cost vs. level of supply of water).

The resulting optimum draft levels associated with various reservoir levels and seasons are depicted in Figure 3.13. Once the desirable level of supply had been established, a method of controlling water usage to the desired supply level had to be developed. This was based on the use of tariffs to control water use.

REFERENCES

Alexander, G.N. (1962). The use of the Gamma distribution in estimating regulated output from storages. *Civil Engineering Trans. Inst. Civil Engineers*, Australia, CE4 (1), 29-34.

Basson, M.S., Allen, R.B., Pegram, G.G.S. and Van Rooyen, J.A. (1994). Probabilistic Management of Water Resources and Hydropower Systems. *Water Res. Pubs.*, Col.

Booker, J.F. (1995). Hydrologic and economic impacts of drought under alternative policy responses. *Water Resources Bulletin*, 31 (5), 889-906.

Bouwer, H. (1980). Ground water recharge for renovating waste water. *ASCE, San. Eng.* p59.

Brandehoft, J.D. and Young, R.A. (1983). Conjunctive use of groundwater and surface water for irrigated agriculture. Risk aversion. *Water Res. Research*, 19 (5), 1111-1121.

Brittan, M.R. (1961). Probability analysis to the development of a synthetic hydrology for the Colorado River, in *Past and Probably Future Variations in Streamflow in the Upper Colorado River, Part IV*, University of Colorado.

Chen, L-S (1972). *Rule curve for the operation of an irrigation reservoir*. Joint Commission on Rural Reconstruction, Taipei, Taiwan.

Coe, J.J. (1990). Conjunctive use – advantages, constraints and examples. *ASCE Jnl. of Irrigation and Drainage Engineering*, 116 (3), 427-443.

Dudley, N. (1990). Alternative institutional arrangements for water supply probabilities and transfers. *Proc. Sem On Transferability of Water Entitlement*. Univ. New England, Armidale.

Fetter, C.W. and Holzmacher, R.G. (1983). Groundwater recharge with treated wastewater. *Jnl. Water Poll. Control Fed.*, 46 (2), p260.

Fiering, M.B. (1967). *Streamflow Synthesis*. Harvard Univ. Press, Cambridge, Mass., 139 pp.

Gersberg, R.M., Elkins, B.V. and Goldman, C.R. (1983). Nitrogen removal in artificial wetlands. *Water Research*, 17 (9), p1009.

Gumbel, E.J. (1954). Statistical theory of droughts. *Proc. ASCE*, 80, p.439.

Gupta, R.S. and Goodman, A.S. (1985). Ground-water reservoir operation for drought management. *Jnl. Water Resources Planning and Management*, 111 (3), 303-320.

Guyman, G.L. and Welch, M.R. (1990). Reclaimed wastewater storage in groundwater basins. *Jnl. of Water Resources Planning and Management*, 116 (3), 305-322.

Hazen, A. (1914). Storage to be provided in impounding reservoirs for municipal water supply. *Trans. ASCE*, 77 No. 1539.

Herold, C.E.(1980) *A model to compute monthly diffuse salt loads assoc. with runoff*. Univ. Wits.

Hufschmidt, M.M. and Fiering, M.B. (1966). *Simulation Techniques for Design of Water Resource Systems*. Harvard Univ. Press, Cambridge, Mass.

Huisman, L. and Oltshoorn, Th.N. (1981). *Artificial Groundwater Recharge*. Delft Univ. of Technology, The Netherlands.

Idelovitch, E. and Michail, M. (1984). Soil-aquifer treatment – a new approach to an old method of wastewater reuse. *Jnl. Water Poll. Control Fed.*, 56 (8), p936.

Ishaq, A.M. and Khan, A. (1997).The uses of aquifers in Saudi Arabia to reclaim and store wastewater. *Water for a changing global commun. Proc.27thIAHR Conf., ASCE*, N.Y. 905-910.

Klemês, V. (1979). Storage mass-curve analysis in a system analytic perspective. *Water Resources Research*, 15 (2), 359-370.

Langbein, W.B. (1958). Queuing theory and water storage. *Proc. Amer. Soc. Civ. Engrs. J. Hydr. Div.*, HY5, Oct., 1811, 1-24.

Langenbaugh, R.A. (1970). Determining optimum operational policies for conjunctive use of groundwater and surface water using LP. *ASCE 18th Annual Spec. Conf.,* Minneapolis,

Loucks, D.P., Stedinger, J.R. and Haith, D.A. (1981). *Water Resource Systems Planning and Analysis,* Prentice Hall, New Jersey.

Lund, J.R. and Reed, R.U. (1995). Drought water rationing and transferable rations. *ASCE, J. Water Ress. Planning Management*, 121 (6), Nov/Dec, 429-437.

Maass, A., Hufschmidt, M.M., Dorfman, R., Thomas, H.A. and Marglin, S.A. (1962). *Design of Water Resource Systems*. Macmillan & Co. Ltd., London, England., Ch. 14.

Mathew, K., Newman, P.W.G. And Ho, G.E. (1982). Groundwater recharge with secondary sewage effluent. *Australian Water Resources Council*, Technical Paper No. 71, Canberra.

McKee, T.B., Doesken, N.J. and Kleist, J. (1993). The relationship of drought frequency and duration to time scales, *8th Conference on Applied Climatology*, pp179-184, Anaheim.

McMahon, T.A. and Mein, R. (1978). *Reservoir Capacity and Yield*.Elsevier,Amsterdam, 213pp.

Midgley, D.C. and Pitman, W.V. (1967). Determination of storage requirements to meet variable drafts. *Trans. S. Afr. Instn. Civil Engrs*, 9, Dec.

Moran, P.A.T. (1959). *The Theory of Storage.* Methuen, London.

Morrice, H.A. and Allan, W.N. (1959). Planning for the ultimate hydro development of the Nile Valley. *Proc. Instn. Civ. Engrs*., London, Vol. 14, p101.

Noel, J.E. and Howitt, R.E. (1982). Conjunctive multibasin management, an optimal control approach. *Water Ress. Research*, 18 (4), 758-763.

Paling, W.A.J. (1985). *Systems analysis of conjunctive use of groundwater, wastewater and surface water for the Witwatersrand.* Water Systems Research Report No. 4/1985.

Palmer, W.C. (1965). *Meteorological drought.* Research paper No.45, U.S. Dept. Commerce, Weather Bureau, Washington D.C.

Riley, J.G. and Scherer, C.R. (1979). Optimal water pricing and storage with cyclical supply and demand. *Water Reources Resch*., 15 (2), 253-259.

Rippl,W.(1883)The capacity of storage-reservoirs for water supply.*Min.Proc.ICE,LXXI*,270-278.

Shih, J.S. and ReVelle, C. (1994). Water supply operations during drought, continuous hedging rule. *ASCE, J. Water Resource Planning Management*, 120 (5), Sept/Oct, 613-629.

Stall,J.B.(1962).Reservoir mass analysis by low flow series.*Proc.ASCE*, 88 (SA5),3283,p21-40.

Stephenson,D.(1995)Factors affecting cost of water supply to Gauteng.*Water S.A.*,21(4),275-280.

Stephenson, D. (1996). Drought management as an alternative to new water schemes. *Water S.A.*, Vol. 22, No. 4, Oct, p. 291-296.

Stephenson, D. and Paling, W. (1992). A hydraulic based model for simulating monthly runoff and erosion. *Water S.A.*, 18 (1), 43-52.

Stephenson D. and Petersen, M.S. (1991). *Water Resources Development in Developing Countries*, Elsevier, Amsterdam, 289p.

Stevens, E.G., Stephenson, D., Chu, S-C, and Huang, W-L. (1998). Management of a reservoir during drought. *Water S.A.*, Vol. 24, Oct.

Sudler, C.E. (1927). Storage required for the regulation of streamflow. *Trans. ASCE*, 91, No. 622.

CHAPTER 4

Flood Management

4.1 FACTORS AFFECTING FLOODING

Floods are generally naturally occurring events endangering ourselves or our properties. They occur infrequently and it is the more remote events that are likely to be of concern, as we automatically adjust our lives to accommodate the more frequent events. The question to be answered is what events cannot be readily accommodated and what should be done about them. We refer here primarily to floods caused by intense rain storms. Floods can also be caused by tides, meteorological activity, geological movements, e.g. landslides or dam break, channel blockage or subsidence (see Wohl, 2000).

Floods are usually caused by excessive rainfall. The intensity of rain and the duration of the storm contribute to the flood in different ways, whereas the catchment area and channel topography affect the concentration of that water. The possibility of flooding is further influenced by the characteristics of the flood plains. Theoretically, a flood can be calculated from rainfall patterns and topography. However, the rainfall amount and pattern is seldom known with accuracy in advance.

The ability to cope with floods should be planned for long before the events themselves. A statistical approach is needed for planning. That is, the probability of flooding needs to be specified. The probability or risk of flooding depends on the statistics of the rainfall, i.e. the probability of it being at a certain level relative to the mean or any other datum, as well as the storm duration and antecedent conditions. Although there are global climate models which can track cloud movement and anticipate rain, they cannot do it with the accuracy required for modelling flood volumes and water levels. So long-term flood prediction is not all that much use in deciding risk levels, etc. It is of more use on a smaller time scale for flood management. It is possible that in future we will be able to predict storms in advance by empirical methods, e.g. the cyclic nature related to sun spots or ocean currents (Tyson, 1986) or by global modeling, which requires large computers to model, on a suitable scale, the temperatures, advection of oceans and atmosphere at various levels. These may be combined with topographical data maps and soil and vegetation maps to calculate infiltration and runoff.

4.2 FLOOD CALCULATIONS

Methods of calculating flood peaks and hydrographs from rainfall abound. Which to use depends on the user and the required answers. It also depends on the availability of rainfall statistics and catchment characteristics. It is no use using a sophisticated catchment model if there is a lack of data. The purpose of the calculations may also influence the decision on which method to employ. There is little purpose in generating a complete hydrograph if only the peak flow is required for example to size a bridge opening. If flood routing is to be investigated the full hydrograph is required as well as the flows initially and after the storm, as the total volume is needed. In general a detailed catchment model is required for any management study. The following list indicates some of the constraints and results which can be expected from some flood calculation methods.

Methodology for flood calculation has evolved over years with better understanding of hydrology and more access to computer assistance (see Wilson, 1985, Viessman and Knapp, 1990). Early methods of flood calculation were empirical and based on catchment area only. The recurrence interval or probability only appeared as a variable later.

Table 4.1 Comparison of flood calculation methods

Method	Data requirements	Ease of use	Accuracy	Limitations, Remarks
Lloyd Davies	Catchment area	Simple	Limited but standardized	Applies only to specific rainfall intensity
Rational	Area, Rain intensity, Runoff coefficient C	Two steps	Depends on good guess of C	Assumes storm duration equals catchment concentration time
Time-area	As above and time of entry	More sub-catchments	Better due to more catchment types	Still assumes linear rainfall-runoff
Unit hydrographs	Unit hydrograph and rainfall losses	Standardized hydrograph combinations	Limited	Produces full hydrograph not just peak
Kinematic	Topographic, rainfall, losses	Simultaneous solution of rainfall-intensity and runoff equations	Better	Accounts for non-linearity, non-equilibrium
Conceptual models	Areas and rainfall	Easier than deterministic	Depends on assumptions, cannot extrapolate	Empirical, requires calibration
Deterministic models	Sub-catchment topography, losses, channels, storage, rainfall and pattern	Needs learning time	Good	Best for routing, management, sensitivity studies
Statistical or empirical	Area	Easy	Low	Based on previous flows

Over the years more variables were accounted for, i.e. losses, initially as a fraction of rainfall and later as a function of soil permeability and initial moisture content. Rainfall intensity is related to storm duration so it was assumed that the critical storm duration was equal to the time of concentration of the catchment. Now it is realized that shorter storms may produce higher runoff peaks, and also that there is not a unique concentration time as it is a function of the runoff depth of flow. The rainfall pattern in time and space also influences the runoff peak and volume, so more detailed rainfall data is needed. Often this detail is not available and regional or assumed storm patterns must be considered in a sensitivity study. However, seldom is there both rainfall and runoff data available to verify the model.

The Rational method is the simplest method of calculating flood peak, but it should be confined to small catchments, e.g. less than 100 square kilometres. Its advantage is that it enables a quick mass balance to be made, i.e. comparison of rainfall and runoff. The method assumes a storm duration to calculate runoff rate. The runoff is calculated from the equation:

$$Q=CIA \tag{4.1}$$

Where:

Q is peak flow rate in m^3/s;

C is a runoff coefficient less than one and which is the ratio of runoff to abstraction. It may approach 1 for an impermeable catchment, or be as low as 0.1 for a permeable natural catchment. Typical values are 0.3 for rural and 0.6 for suburban catchments;

I is the rainfall intensity in m/s. Since rainfall may be in mm/h it should be converted;

A is the catchment area in square metres. Feet may be used if the units throughout are consistent, i.e. rainfall in ft/s and runoff in ft^3/s.

The method should not be applied to large catchments because it omits routing effects or, areal reduction of rainfall intensity (which have the effect of reducing flood peak) and does not consider storm durations not equal to the catchment concentration time (which could result in a higher peak). The effects of storage can also not be included in the calculation.

Unit hydrographs are used less and less nowadays as computer models generate specific hydrographs without having to assume a linear rainfall-runoff relationship (which is dangerous for extrapolating to extreme floods as the real system is non-linear). Examples of the use of the kinematic method are given in the next chapter. The kinematic method uses a simplified hydrodynamic equation and neglects backwater and acceleration effects. This enables simple analytical solutions for some cases and also provides the basis for simple computer runoff modelling. Examples of computer models are given in Chapter 13.

It is possible to model the hydrological process from rainfall through runoff to river flow and water levels, and this technology is well advanced. There are statistical and deterministic models for doing this, but they are limited in their accuracy by the availability of rainfall and other data.

To simulate water level at any point, not only details of the rainfall throughout the catchment are needed in time and space, but also details of the catchment surface, i.e. catchment use, cover such as vegetation, soil infiltration capacity and moisture content at the start of the storm, ground slopes and roughness from the point of view of runoff, and river characteristics, i.e. river cross-sectional shape, configuration and mobility of the bed.

There are many factors which can affect the flood flow and water levels in addition to the natural factors indicated above. Reservoirs or dams along the river can detain the floods to varying extents and attenuate downstream peak flows. Catchment management also plays a significant role in determining runoff. Thus ploughed fields and method of ploughing affect the volume of runoff by affecting infiltration or surface water retention. The type of vegetation cover affects the above surface retention, and urbanization can increase runoff with impermeable barriers, thus also enabling the runoff to concentrate more rapidly.

The critical storm which causes a flood is not the same for every catchment or position down the catchment. For smaller catchments, such as paved yards, it is the short sharp storm which is critical from the point of view of maximum water levels. This could be the instantaneous or five minute type rainfall which can fall at a rate of 200mm per hour, albeit for a duration of less than an hour. The bigger the catchment, the longer the duration of the critical storm, and for large catchments of the order of tens of thousands of square kilometers, it is the rainfall over a number of days which becomes critical.

4.3 HAZARD ASSESSMENT

It is recognized that higher water levels cause more damage and more danger than lower water levels. This is because the water could rise out of the river channel and flood areas which are commonly used by man and beast. The level to which the water could rise could vary from storm to storm, but often the river can be simplified into the channel and flood plain, and the flood plain will be flooded only at certain intervals or recurrence intervals.

The effect of water in places where it is not anticipated could be any one or more of the following:

- It could wet property which can be damaged by moisture.
- It could wash away soil and crops and property, referring specifically to agricultural property.
- It could deposit sediment on the land or property.
- It could knock over walls or buildings.
- It could wash away property including vehicles.
- It could drown people and animals by increasing the depth of water or washing them over with a high flow velocity.
- It could prevent access to necessary services such as health services or maintenance services.
- The latter could be caused by wash away of bridges or roads or merely inundation because the services have only been designed against a certain risk of flooding.
- It could harm people or damage fauna or flora.

Services are generally designed to resist or be serviceable against the following probabilities:

- Important roads – the 100 year flood, i.e. a 1% chance of being overtopped in any one year.

- Many roads and buildings – the 50 year recurrence interval.
- Less important roads – the 20 year recurrence interval.
- Storm water drainage pipes and culverts – 20 to 2 year recurrence interval, depending on the consequences of overtopping.

If storm water pipes have a flow exceeding their discharge capacity, surplus water can run down the road in gutters and cover the surface of the road which could cause temporary minor inconvenience, but it acts to conduct the major flood and is therefore often called the major drainage system.

The degree to which a structure is designed to resist floods often depends on the costs as well as the probability of the event. If there is little damage caused by overtopping, the design recurrence interval may be reduced. On the other hand, if there are catastrophic results, danger to life and high economic losses, then the design probability or recurrence interval will be extreme, such as the 100 year flood or more.

For example, dam spillways are often designed for regional maximum floods or even maximum probable floods. These events are obtained from regional or international experience diagrams and may not reflect the floods which have actually been observed at the site or in the river in question. The catastrophe of overtopping a dam, i.e. passing a flow greater than the design capacity, could be that the banks of the dam are washed away and the wall collapses, causing a huge rush of water referred to as the dam break problem.

From the point of view of the rural resident, it is the much shorter recurrence interval or more frequent storm which is of concern. Those living from hand to mouth have a much shorter time horizon for which to plan, and their living experiences are relatively confined so that they may settle in a flood plain a few years after it has been flooded without evaluating the risk of another flood. Hence the design recurrence interval must be related to the hazard effects it causes.

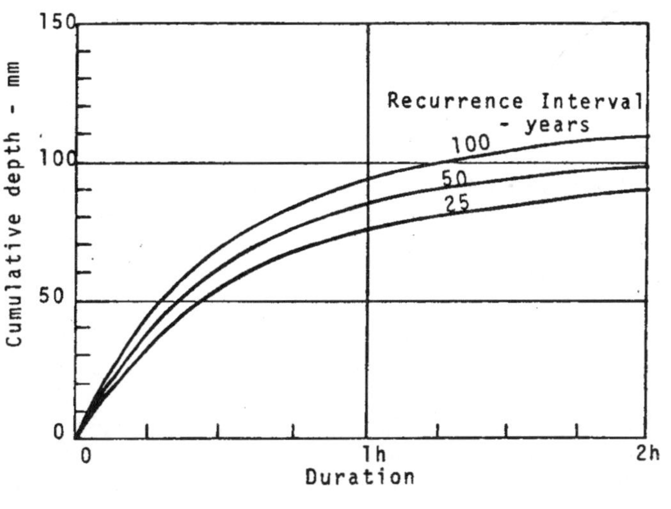

Figure 4.1 Depth-Duration-Frequency relationship for Jan Smuts rain gauge

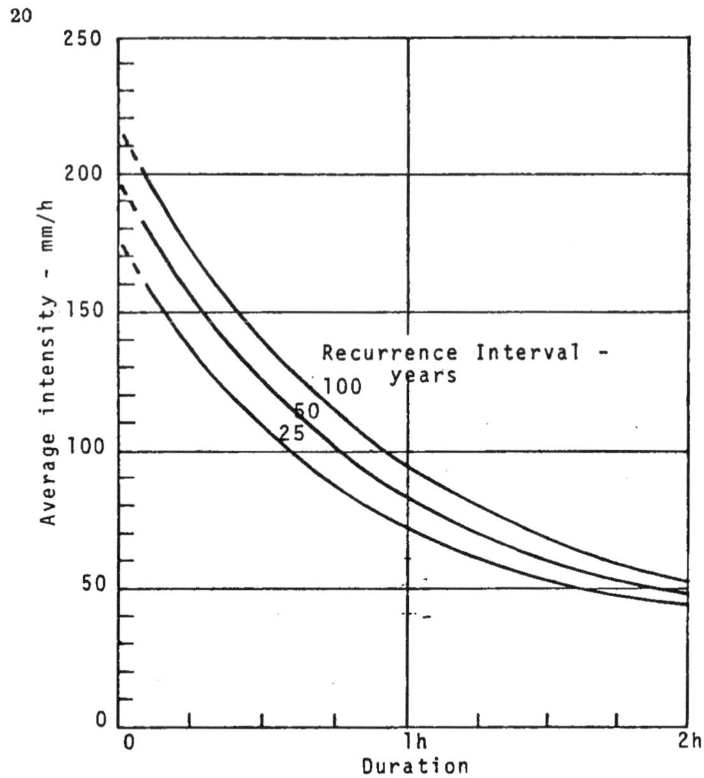

Figure 4.2 Average Rain Intensity-Duration-Frequency relationship for Jan Smuts rain gauge

4.4 ANALYSIS OF RAINFALL DEPTH, DURATION AND FREQUENCY

There have been many studies of short duration rainfall related to floods. It is possible to produce relationships between rainfall depth in millimeters and the duration of the storm and the recurrence interval of the storm. Figure 4.1 is such a curve taken from the Jan Smuts Airport rain gauge, which is a fairly reliable rain gauge taking readings at regular, short time intervals. Sometimes the rainfall rate (depth/duration) is plotted (e.g. Fig. 4.2).

It is not only the depth of rain but also its distribution in time and space which is relevant to the flood magnitude, and time variation of rainfall has been documented by many researchers and typical patterns have been indicated. Spatial variation is somewhat random although the specific storms have been plotted. Storms occur as storm cells which expand and contract as the moisture in the atmosphere builds up or is depleted. The storm cells move with winds at different levels and can travel upstream or downstream, which will give a different flood result. (See Figure 4.3 for a spatial effect.)

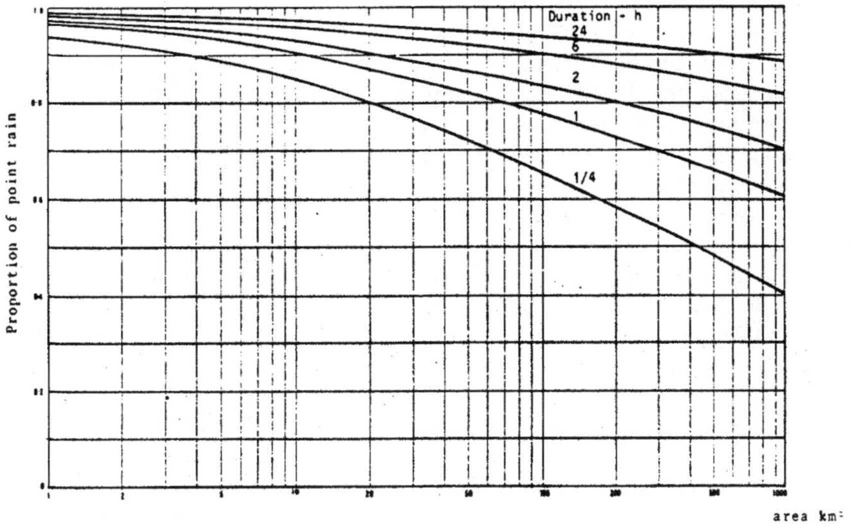

Figure 4.3 Reduction factors for non-point rainfall – Areal reduction factor

Attempts have been made to identify the maximum probable precipitation from the carrying capacity of air and wind speeds and distance from water bodies from where the moisture has been evaporated. The modeling approach was advocated by WMO, but the models are limited by their assumptions and the scantiness of data. Flood models for predicting maximum probable floods are therefore lacking even more because they need to model rainfall as well as the catchment. It is often assumed that the catchment is almost saturated when extreme events occur and similarly reservoirs or dams are assumed full to obtain the worst conditions, i.e. maximum overtopping of dams down the rivers.

Precipitation data can also be collected by satellite and there are data banks of digital terrain models on grids of 1000m x 1000m. This data has been used for modeling runoff but the models are cumbersome and will not necessarily produce the answers required, i.e. what level of flood should be planned for.

4.5 INDICES OF FLOOD MAGNITUDE

4.5.1 Probability

The probability of an event occurring is a simple index of flood severity (Chang, 1989). For example, a 100 year recurrence interval flood sounds an acceptable flood to plan for. The recurrence interval is the average number of years between events equal to or exceeding the event. Thus the 100 year flood is equaled or exceeded once every 100 years on average, e.g. if there were 1000 years of historical records there would be about 10 events equal to or exceeding the 100 year flood. The probability of the 100 year flood occurring in any one year is 1% or 0.01. The recurrence interval is also called the exceedance interval.

The difficulties with using this recurrence interval are as follows: There is a misconception that the 100 year flood will occur only once in 100 years. However, it can occur more than once in any 100 years. In fact there is a 64% chance that the 100 year flood will occur once or more times in any 100 years, so even a 100 year event may be unacceptable for insuring.

If the recurrence interval of a flood is T years, it has a probability of 1/T of being exceeded in any one year. So the probability of it not being exceeded in any one year is 1-1/T. And the probability of it not being exceeded in n years is $(1-1/T)^n$. Hence the probability of it being exceeded in n years is $1-(1-1/T)^n$.

The recurrence interval of an event which has a desired probability of occurring may be calculated as follows:

$$P = 1 - (1 - 1/T)^n \tag{4.2}$$

where P is the probability of the T year event occurring in n years.

$$\text{Therefore } T = 1/\{1 - (1 - P)^{1/n}\} \tag{4.3}$$

e.g. if P = 1%,

$$T = 1/\{1 - (1 - 0.01)^{1/100}\} = 9950 \text{ years Recurrence Interval.} \tag{4.4}$$

i.e. for a 1% chance of a flood occurring in 100 years, the design recurrence interval must be nearly 10000 years.

Ideally, a very long record of events is needed, e.g. hundreds of years to obtain good statistics. It is seldom the case that records exist for longer than 1 century, or even 50 years. So statistically, analysis of whatever data exists is made and various mathematical population distributions are used to extend the prediction of probabilities. Common mathematical probability distributions used are Normal, Weibull, Pearson, Gumbel and Extreme Value (Viessman and Knapp, 1990), and each produces a slightly different figure for extrapolated values.

In fact, rainfall and runoff are not random variables. They are affected by a multitude of parameters. For instance, rainfall can be affected by ocean currents, temperatures, winds, vegetation, ozone layer, carbon dioxide layer in the atmosphere, and runoff is also affected by antecedent soil moisture (introducing a serial correlation), catchment management and development. It would require a complex model to include all the factors, and present Global Climate models are nowhere near predicting all the effects. Digital terrain models are perhaps more advanced, so the hydrologist is largely dependent on the meteorologist for the accuracy of his flood predictions. The parameters are also continuously changing with time, viz. catchment development (such as urbanization), channelization, dams, climate change and even the method of measurement or observation (remote sensing, rain gauges, stream gauges). So historical records cannot readily be extrapolated into the future, even statistically.

The question also arises as to what to do if two or more significant events occur in one year. A frequent approach is to only consider the worst event in any one year (a partial series), or else each event is used (a complete series).

A way of avoiding expressing probability was derived by Francou and Rodier (Rodier and Roche, 1984). Here the measured regional maximum floods are plotted on a single diagram versus catchment area. The upper envelope will gradually move up over years as more extreme floods occur, and a higher F-R index becomes applicable.

4.5.2 Rainfall

Although rainfall is the most important factor causing floods, it is not the flood itself, and is an indirect index, i.e. a number of other factors influence the flood, so it would be better to measure the flood itself. However, rainfall records abound, they often cover a long period of time, they are transferable from one locality to another in the same hydrographic region, and rain occurs in advance of floods.

Rainfall can be measured in terms of:

- Total for 24h at a rain gauge
- Total for other durations, e.g. 7 days, 1 month.
- Average of a number of rain gauges

The rain can be measured in mm depth, or rate of precipitation, or integrated over the catchment area to indicate cubic metres or cubic metres per second. Or it can be expressed as a percentage of the average monthly or annual precipitation. Each of the events indicated above can be ranked to give a recurrence interval or probability of exceedance.

It should be noted that the probability of a rainfall event is not necessarily the same as the probability of the flood event it causes. There are intermediate factors, each of which have probabilities of occurring, e.g. antecedent moisture conditions of the soil. Dry soil can absorb 90% or more of rain. Saturated soil in the catchment will absorb less rain, but puddling (surface retention) and other factors still attenuates the flood. The condition of reservoirs (full/empty, or with operational flood control gates) can affect floods. The pattern of the rainfall is also a major factor in producing floods. In arid catchments, short sharp storms are critical, and in wet catchments long duration storms (up to the concentration time of the catchment) are most significant with regard to flood magnitudes. Reservoirs can absorb shorter flood hyetographs more easily as they have less volume than long duration storms. The bigger the flood due to larger catchment area or more extreme and longer rain, the less effect reservoirs and flood management measures have.

If rainfall is used as an index, modeling of runoff to incorporate all significant factors is necessary. Then a spectrum of floods can be generated from storms. Not only the storm duration but also its time variation and spatial distribution and movement affect the flood peak.

4.5.3 Flood parameters

The factor which is of most concern with respect to flooding is water depth. This could be the water depth in a river or over a flood plain, or reservoir level, or ground water level. Connected with the water level may be a flow of water, and the water velocity is usually a function of the water depth. According to the laws of hydraulics:

$$V = \frac{1}{n}\left(\frac{A}{P}\right)^{2/3} S^{1/2}$$
$$\approx \frac{\sqrt{S}}{n} y^{2/3}$$

where V is the water velocity in m/s, S is the gradient of the water surface, y is the water depth in m, A is the cross sectional area in m^2 and P is the length of wetted perimeter in m. Anything above a velocity of 1 m/s and a depth of 1m are regarded as catastrophic but there are increasing degrees of hazard depending on water depth and velocity. Property damage functions are more related to water depth (Du Plessis, 2000).

4.6 FLOOD MANAGEMENT

We often assume that the floods which occur in nature cannot be changed. That is, they result from extreme storms, which have associated probabilities for recurrence intervals. Gradually our understanding of nature and climate is improving and we are now able to predict, with reasonable confidence, weather patterns in the future due to global wind and ocean patterns. We will probably be able to predict with greater certainty and further into the future as our ability to understand the complex processes associated with rainfall improve.

Engineers and hydrologists are still prone to estimating extreme floods on a probabilistic basis. That is, flood magnitudes are allocated a more and more remote probability for greater and greater floods. Flood magnitudes are allocated a recurrent interval based on experience over past years. Recurrence interval is the average period between events with a magnitude greater than or equal to the stated magnitude.

It is generally the peak of the flood which is of interest as that is the flow associated with the worst flooding along river banks. However, it is not only the peak of the hydrograph but also the volume of runoff over a length of time that affects the flood levels. When we look at the effects of storage in attenuating or routing floods, the volume of water under the hydrograph is of as much concern as the peak of the hydrograph, for it is the volume of water which fills the reservoir prior to it overtopping. In fact the discharge over the spillway or through an outlet from a reservoir is more a function of the storage volume in the reservoir than it is of the peak rate of flow into the reservoir.

With regard to storm drainage design, i.e. the sizing of man-made conduits which affects the water depth in natural channels, the original policy was to remove the storm water as quickly as possible, thereby alleviating local flooding problems. This could be taken to an extreme in many instances as the drainage channels, pipes, culverts and gutters were all designed to be as hydraulically efficient as possible. That is, the water velocities were increased over natural runoff velocities. Impervious cover due to urbanization not only increased the rate of runoff but also the volume of runoff. There was therefore less infiltration and the water resources, in particular the ground water regime of the catchment, were affected. Nevertheless by removing the stormwater as rapidly as possible, there was less flooding on roads and the cross sectional area of man-made conduits was minimized, hence the construction cost was minimized, particularly in the upper reaches of an urban catchment. However, downstream the flood flow rates increased over the natural flows as a result of the rapid concentration of water in man-made conduits.

It is generally the case that for the rapid concentration of water due to urbanization, the time to equilibrium between storm input and discharge is reduced, and therefore it is a shorter duration storm which becomes more critical. Shorter duration storms are associated with higher intensities, so in fact it is a higher intensity storm, and therefore a higher rate of discharge, which is associated with the rapid drainage of the system. This effect is compounded with the increased runoff creating even bigger floods downstream of the man-made drainage system and therefore higher rates of flow and deeper waters or greater flooding once the floods reach natural waterways. The waterways may then be unable to cope with the increased floods, with the result that the streams overflow onto their banks and cause damage previously rarely encountered.

With the realization that floods are not absolute and can be affected by the catchment and man, one can investigate which factors affect the magnitudes of floods. These include:

- *Rainfall*: intensity, duration, movement, spatial and time variation
- *Other sources of water*: snowmelt, dew, hail, groundwater
- *Man induced floods*: dam break, pumping, conduit burst
- *Topography*: slopes, overland, flow lengths
- *Catchment*: cover, roughness, reservoirs, temperature, shape
- *Ground*: antecedent moisture, permeability, capillarity, absorption, porosity, depth
- *Channel*: network, lengths, depths, cross sections, gradients
- *Vegetation*: foliage, height, root system
- *Man made*: cover, buildings, usage, drains

Many of the factors are inherent or cannot be affected by man. These could include topography and soil type. Others can be affected to differing degrees. For even rainfall can be affected by thermal currents from cities, or denudation of forests. The greatest effects of man are probably on the surface system, thereby speeding up runoff and creating bigger floods. Reduction of vegetation, construction of pavements and roofs, add to the volume of runoff. Increased runoff in turn creates erosion, deepens channels, reduces groundwater penetration and thereby accelerates catchment deterioration until the exaggeration effect is irreversible, even if planning policies change.

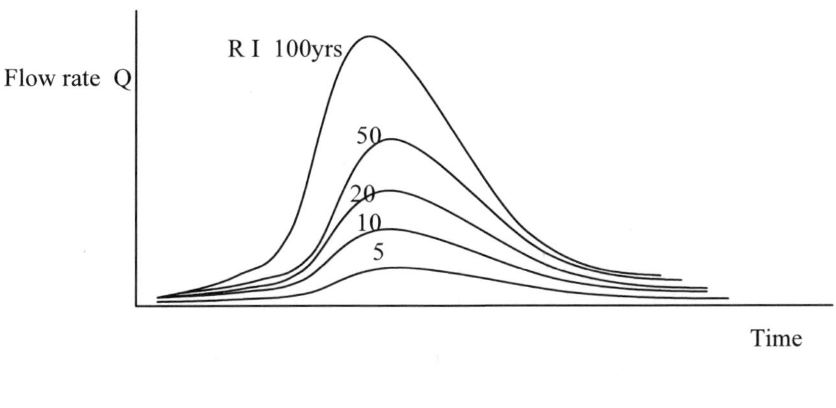

Figure 4.4 Hydrographs for critical storm peak

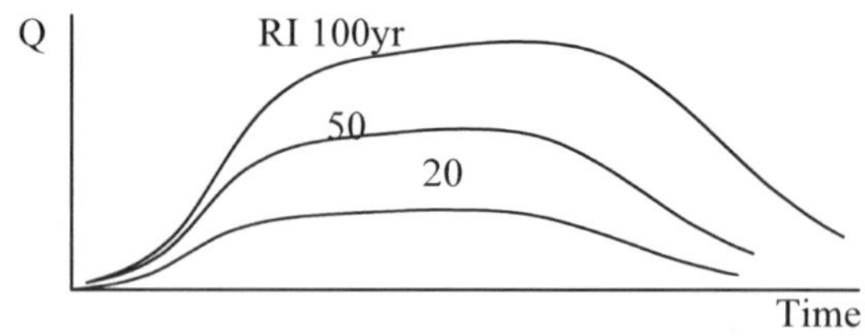

Figure4.5 Hydrographs producing maximum volume

With the realisation of the effects of rapid drainage, the possibility of stormwater detention in order to attenuate the flood became popular. That is, storage basins were built along waterways in order to retard the flood and therefore reduce downstream flow rates, if not to original virgin catchment conditions then at least to an economic effect. Detention storage involves a temporary retaining of the water by means of a dam wall which would subsequently spill and release water downstream, i.e. after the reservoir behind the dam had filled, therefore lopping off a proportion of the peak and also attenuating the hydrograph due to routing.

There could be dangers due to higher floods than the design flood, as the detention storage would tend to catch the initial flows and once full would have a reduced routing effect. The same or even more so would be the case for retention storage, which is the abstraction of water from the river channel and dissipation by means of infiltration or evaporation. Again, once the reservoir is full, the routing effect on the hydrograph becomes negligible.

The same effect can be obtained by encouraging flood plain storage or channel storage, that is, by roughening the channels by means of rocks, gabions or vegetation, the water is retarded and will tend to be stored in the channel and on the banks and that also has a routing effect.

However, the more water which is stored along the waterway, the greater the area which will be inundated, and the average flood band would be wider than for an unattenuated flood. Thus, whereas the storage may act advantageously for smaller and more frequent storms, when extreme floods occur, the storage merely retards the flow to spread it over a greater width and cause greater flooding. Therefore, a proper flood management policy must go with detention storage construction.

The construction of dams and other obstructions such as culverts and bridges should have environmental studies to be aware of the flooding problems and other effects due to the backing up of water. There are a number of ways of reducing this detrimental effect including:

Water volume management in reservoirs. By releasing water from the reservoir prior to the flood hydrograph arriving, it is possible to minimize flood backwater upstream while maintaining the largest possible volume to absorb the flood hydrograph. However, some form of prediction of the inflowing hydrograph is required in order to release the correct volume of water. Too great a rate of release could cause a bigger flood downstream than actually would occur into the reservoir and the operator of the reservoir could be liable for damages. On the other hand, too small a release would minimize the routing effect. It is possible to predict rainfall in order to estimate the release and time available for releasing water from the reservoir using climate models. In the case of larger catchments, the rainfall may occur days or weeks ahead of the flood peak. That is, there is enough time to model the rainfall/runoff process and predict the flows and water levels in the river ahead of the flood. This would be real-time operation rather than projection into the future as the flood flows and water levels can be predicted as accurately as the hydraulic models are able to operate. Alternative management practices include warning systems to minimize damage and danger to life.

Flood release gates, i.e. gates on dams, also offer a way of attenuating floods but to a limited extent because in the long run the more storage the greater the backup and the lesser the effect on large floods. In fact, the net effect for abnormal floods is to increase the flood width upstream of the dam. Such gates can be operated automatically by floats, electrical sensors or by PLC (Programmed Logic Control) or manually. Radial gates which require a small lifting force, and collapsible/inflatable rubber cylinders have been used.

As a last resort, *fuse plug-type spillways* can be used. These are designed to collapse under extreme flows in order to prevent damage to the dam. However, in the extreme they are unlikely to reduce downstream flooding much and in fact may increase the flood due to a rapid increase in outflow from the dam. This type of mechanism includes gravitational structures such as blocks or collapsing walls and spring loaded gates.

Figure 4.6 illustrates simplistically how detention storage volume can be calculated from inflow hydrograph and outflow. The outflow peak occurs on the receding limb of the inflow hydrograph because after the lines cross outflow exceeds inflow. And the shape of the outflow graph depends on the spillway characteristics and stored volume.

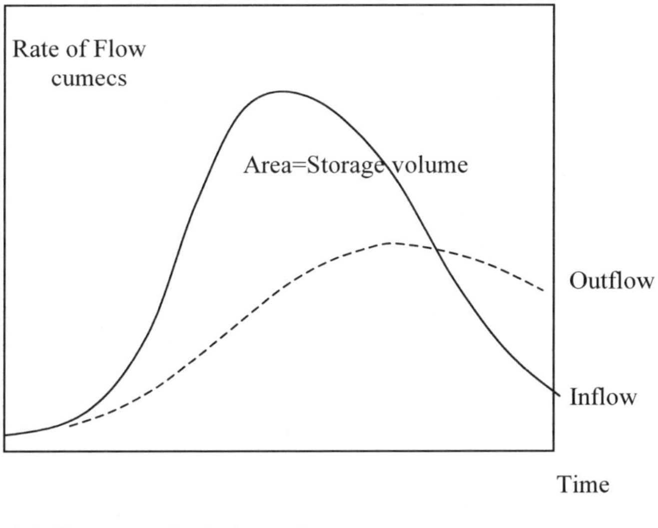

Figure 4.6 Storage routing hydrographs

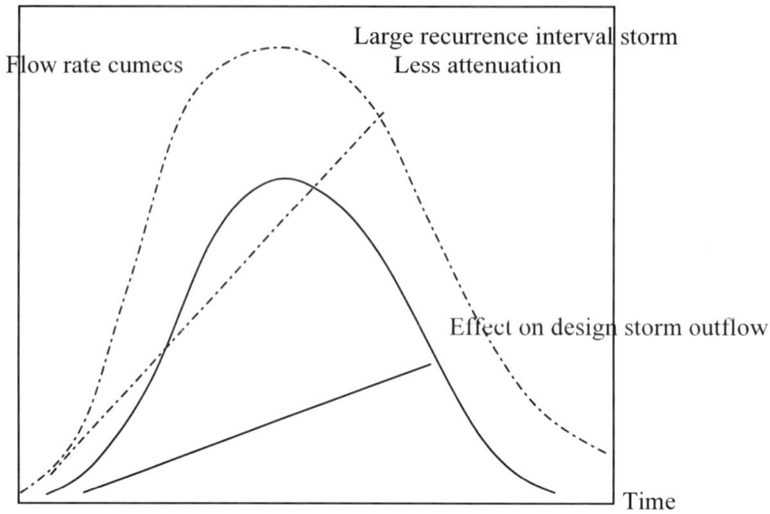

Figure 4.7 Comparison of routing effect of storage on a small inflow and a big inflow

It will be observed from Figure 4.7 that the attenuation effect of a reservoir with limited storage can be small, if not negligible, if a bigger flood occurs than designed for. This would be the case particularly if the outlet were an orifice with low outflow flexibility. An overflow spillway would be more inclined to act proportionally better for larger inflow.

The use of retention storage (abstraction of water) could be problematic for extreme floods. Once the storage was full it would pass the full flow rate. So it may only be the front of the hydrograph which is caught for big floods, and the peak may flow past unhindered. A wide spectrum of floods needs to be considered in designing flood management storage.

One can reach the optimum relationship between storage cost and outflow conduit cost by comparing a number of storage capacities. This could be facilitated using a generalized relationship such as Figure 4.8. This graph was prepared using generalized regional storm depth-recurrence interval data and assuming different outflow rates. All data is rendered dimensionless by dividing by the catchment effective area. Outflow rate was assumed constant during storm.

To use such a chart for any selected Recurrence Interval storm, one selects an acceptable discharge, divides by the catchment area times runoff proportion and reads off the required storage ratio. Multiply this by C x A where A is catchment area in hectares and C is runoff coefficient, to get storage in cubic metres.

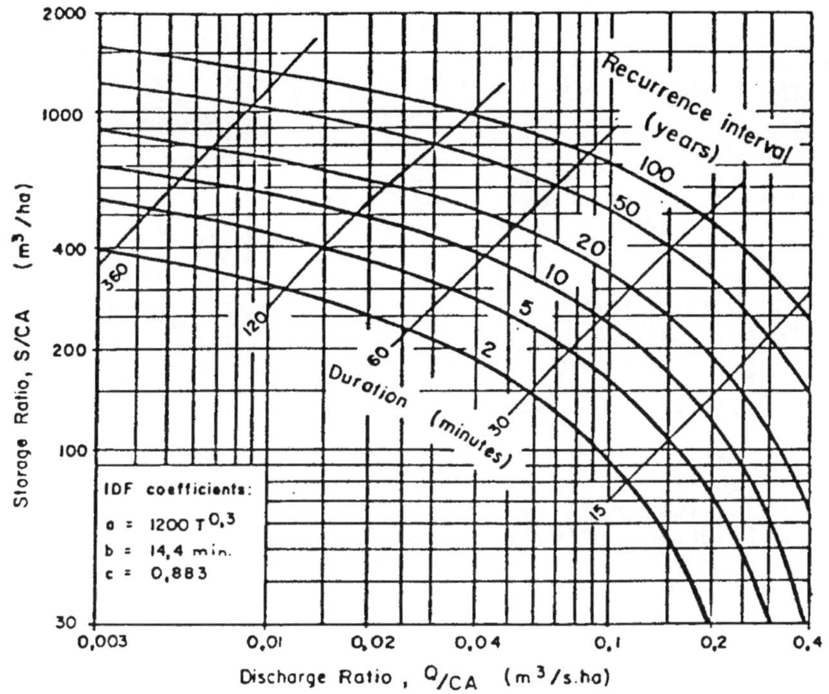

Figure 4.8 Example of curves for the preliminary design of storage ponds

4.7 RESERVOIR ROUTING METHODS

- Routing as a function of storage in the reservoir. This could be an increasing discharge over a spillway (weir flow) or an attenuated outflow such as through an orifice.
- Passive release. This includes spillways where outflow is a function of water level in the reservoir. Automatic gate operation would be included as passive release.
- An active operating rule would be based on the inflow as well as the storage in the reservoir so that there is some anticipatory attenuation. The inflow could be treated in one of three ways:
  - (a) The routing could be done as a function of the measured inflow rate. This could be particularly useful if the inflow were measured a distance upstream of the dam so there was time enough to make a decision as to water to release from the dam.
  - (b) Based on predicted inflow. That is, rainfall data could be telemetered to a computer and the river flows modelled using a deterministic model. Alternatively, climatic models may predict the rainfall rate even further ahead of time.

Method	Example	Limitations	Problems
Absorption	Infiltration	Area total.	Town planning, Directly connected impervious areas
Abstraction	Pumping or bypass	Rate of abstraction	Offchannel storage
Proportion diversion	Weir	Cost	Backup
Retention	Storage	Volume	Longer storms Bigger storms
Detention	Storage dam Channel storage	Volume Depth	High flow Extra flood width
Routing	River reach	Channel storage	Limited efficiency
Flood plains	Wetlands	Area	Pollution
Dual drainage	Pipe and road	Flood	Flooding capacity
Gate operation	Barrages	Mechanical	Over-release
Warning system	Telemetry	Electrical	Human

Hydrographs:

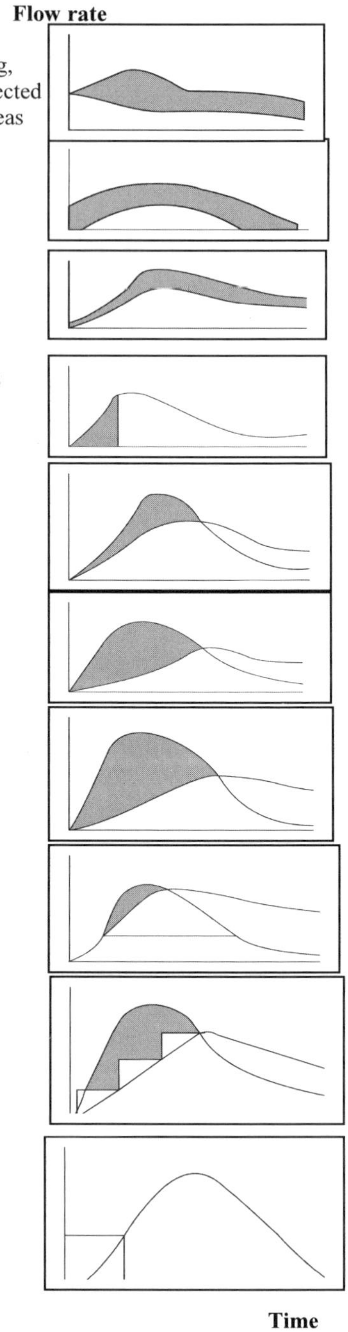

Figure 4.9 Summary of flood routing methods and effects

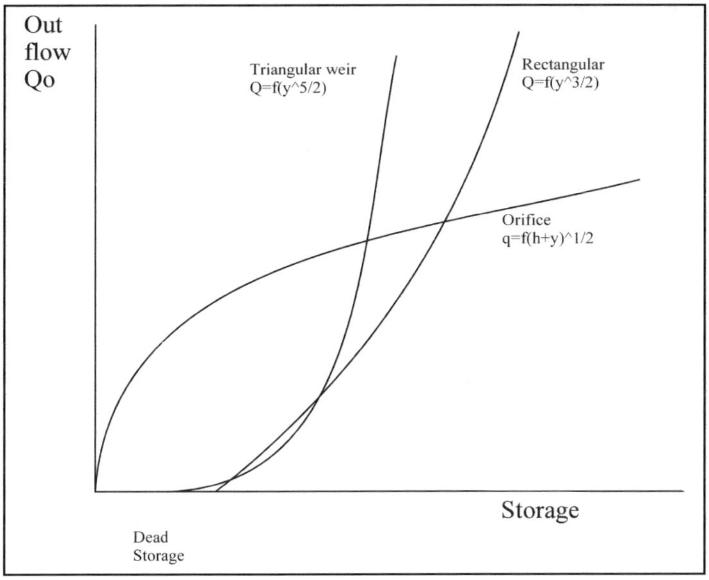

Figure 4.10 The effect of different spillway characteristics on outflow and storage

(c) In either case, a sufficiently long time is made available between the calculated inflow and the actual inflow to enable the release to be optimised. Again, the object would be to release at a rate less than the peak inflow rate.

(d) Probabilistic basis. If a probability analysis of previous inflows is made, then the probability of different rates of inflow and volumes could be made and an optimum operating rule devised based on these. However, there is no anticipatory component in this and individual storms may not be optimised, i.e. it is only over a long period of time that an average damage function can be minimized.

Whatever management practice is adopted, the flood spectrum will change and a new flood frequency probability diagram will result.

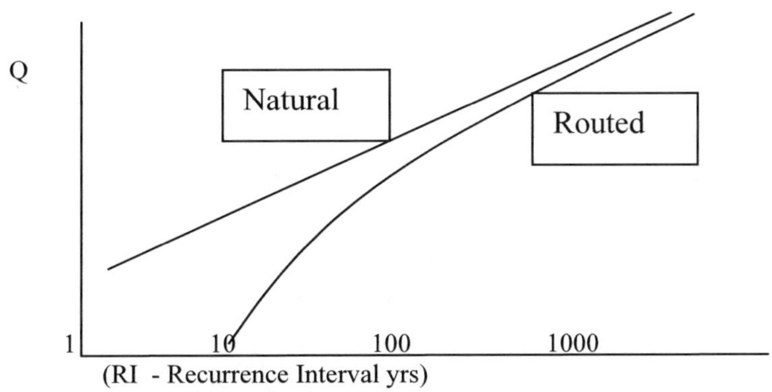

Figure 4.11 Effect of routing on flood spectrum

Table 4.2 Optimization of bridge flood recurrence interval.

A	B	C	D	E	F	G
Recurrence interval yrs	Flood m^3/s	Bridge cost $/an	Damage cost $	Probability = 1/A	Probable damage cost	Total cost C+D+E
2	47	2,700 000	2,000 000	0.5	1,000 000	3,700 000
5	53	2,800 000	2,000 000	0.2	400 000	3,200 000
10	65	2,900 000	2,000 000	0.1	200 000	3,100 000
20	80	3,100 000	2,000 000	0.05	100 000	3,200 000
50	110	3,300 000	2,000 000	0.02	40 000	3,340 000
100	150	3,600 000	2,000 000	0.01	20 000	3,620 000
200	200	4,200 000	2,000 000	0.005	10 000	4,210 000

4.8 FLOOD RISK ANALYSIS

The most appropriate way of selecting a design flood for a structure is with probability analysis. This could apply to a bridge, culvert or spillway. The bigger the recurrence interval of the design flood at a site, the bigger the flood and therefore the higher the cost of the structure. On the other hand the risk of overtopping diminishes and the likely damage is reduced due to lower cost of repairs and danger. The bridge will be damaged if the design flood is exceeded. In Table 4.2 the cost of a bridge over a river is converted to an annual figure, and given for different flood recurrence intervals. The corresponding damage costs are also estimated, i.e. the cost of repairing the bridge and compensating for losses incurred. These are multiplied by the probability of them occurring in any year and added to the annualised cost of the bridge construction. The optimum design flow recurrence interval is where the total cost is a minimum, in this case 10 years.

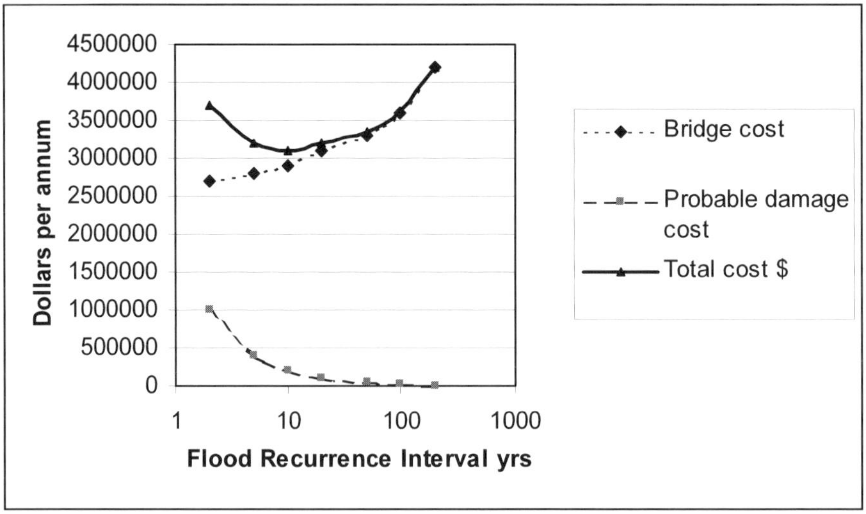

Figure 4.12 Optimum design flood

4.9 FLOOD PLAIN MANAGEMENT

The flood width on the banks of the river is a function of the flood magnitude represented by the recurrence interval as well as the physical factors, such as bank development and land use. There could be regular flooding at successively higher water levels. There is an area which will be inundated regularly, i.e. every few years, and this is referred to as the waterway. No development should be permitted in the waterway as there is a real and frequent danger to life and property.

At a higher level, such as the 1 in 25 year flood, there is still a risk of flooding which could endanger property and life, but it does not occur frequently enough to be a nuisance. The area within the boundaries of this flood would be called the flood zone, which comprises the waterway and the flood plains.

There is an area outside this flood zone at a higher level which could be flooded even less frequently and this area is referred to as the flood fringe. Figure 4.13 illustrates the various zones. Above the 100 year flood line, the possibility of flooding is so remote, it is seldom accounted for in planning, but it should be borne in mind in important developments, e.g. dangerous depositories.

If artificial flood banks are built along the river the flood plain can be confined. Then the question of what recuurence interval to design for is even more important for overtopping could be catastrophic. And it is necessary to detain the side inflow until the main river has subsided.

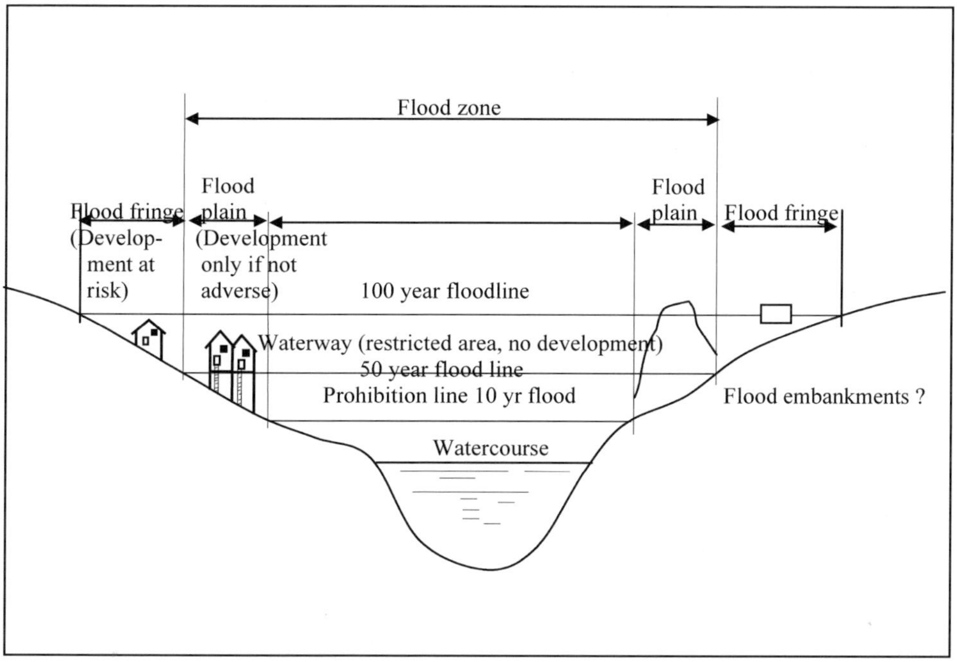

Figure 4.13 Flood plain zones

4.9.1 Hazards associated with flooding

The reason that development of private and public buildings is undesirable within flood areas is not only due to the fact that it endangers lives and property but also because the buildings obstruct the flow and cause backing up of the river. The concern regarding this damage is due to a number of reasons:

If the river is very deep and flows fast, it could wash away buildings and property and also drown people. At successively lower flows corresponding to lower water velocities and shallower depths, the water becomes safer and at a low depth it may not be life endangering, but it could still result in damage to property. However, shallow depths can be countered by protective measures which could be instituted if there was a warning of an impending flood.

In other less dangerous situations, very shallow flooding on roads could be minimized by vehicles avoiding those roads or traveling very slowly. The danger or hazard associated with flow is generally expressed in terms of a diagram such as the flood hazard diagram below. This type of diagram has been adopted by various flood management organizations such as Minnesota in the United States of America (1969) and New South Wales in Australia (1986). The ranking of the hazard was simplified by Stephenson and Furumele (2001) in studies for Eastern Gauteng and a hazard risk index was proposed. Similar studies were done on rural rivers (Du Plessis, 2000; and Adriaans, 2001).

Another point to consider is the risk, or probability, of the hazard occurring. Thus, if the hazard is only likely to occur with a probability of 1% in any one year, it may be tolerable, but if it occurs frequently, e.g. at least once every year, it may be intolerable. This is the risk factor and it can be evaluated in terms of the recurrence interval of the flood. Thus a high risk index could be associated with floods occurring more often than once in 20 years and an intermediate risk index for floods occurring with a recurrence interval between 20 and 100 years, and low risk index for events which occurred less frequently than once in 100 years.

The hazard index and the risk index could be combined by multiplying them to give a hazard risk index which is an indicator of desirability of development within the floodway or flood fringe. The figure below shows the results. A hazard risk index of 2 or less is considered acceptable.

4.10 INTEGRATED FLOOD PLAIN MANAGEMENT

It is thus apparent that a more flexible approach than previously adopted is possible for planning development along rivers. In fact, an integrated management approach is considered at this stage. This implies that a trade off could be made between the implications of flooding and the implications of prohibiting development within the flood zone. The cost of not permitting development in the flood zone will increase over the years as attractive river bank land becomes scarcer. An economic balance between the average hazard cost and the availability of land is needed.

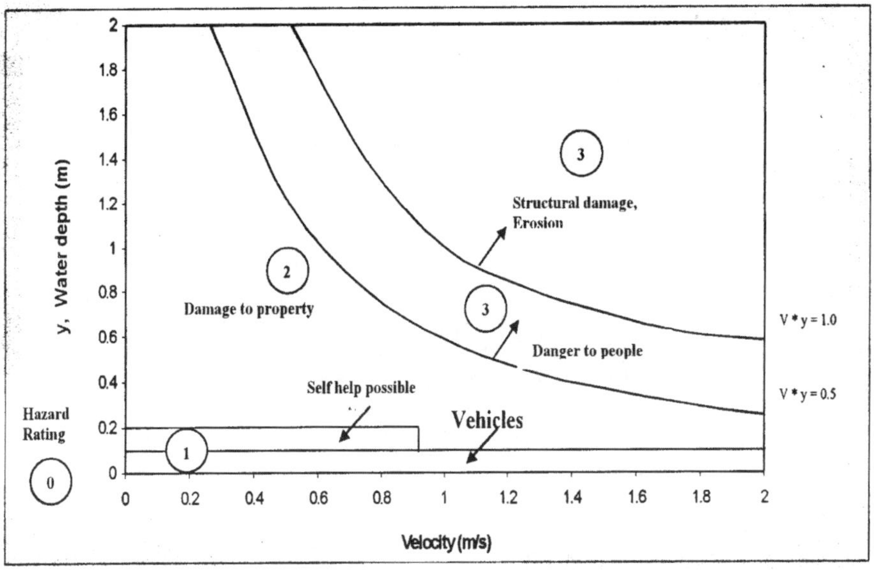

Figure 4.14 Flood hazard diagram

This will require a proactive approach. That is, a flexible approach within guidelines whereby developers may negotiate to develop within the flood fringe or even the flood zone. The developer may be able to physically design his development such that it results in negligible adverse effects but on the other hand has high social benefits. Alternatively, he may be able to contribute economically to the enhancement of the banks in return for being permitted to develop within an approved area. The principles of integrated management can be applied at this stage, and a detailed plan will emerge as more data is gathered. It would also be desirable to benchmark the proposed integrated flood plain management approach with other practices. International attention paid to integrated flood plain management is now producing useful guidelines and an international comparison would be of use.

The decision as to level of development on the banks of a river and how to close the river depends on what the effect of the development will be on the flood, as well as what the effect of the flood is on development. That is, will construction obstruct the flow of water, thereby affecting water levels upstream of the development? Also,

- What the effect of development will be on flood levels
- What a flood will do to the development
- How the value of the new and existing development is affected
- What the damage cost would be
- What pollution threat there is to the water
- The risk of adverse effects

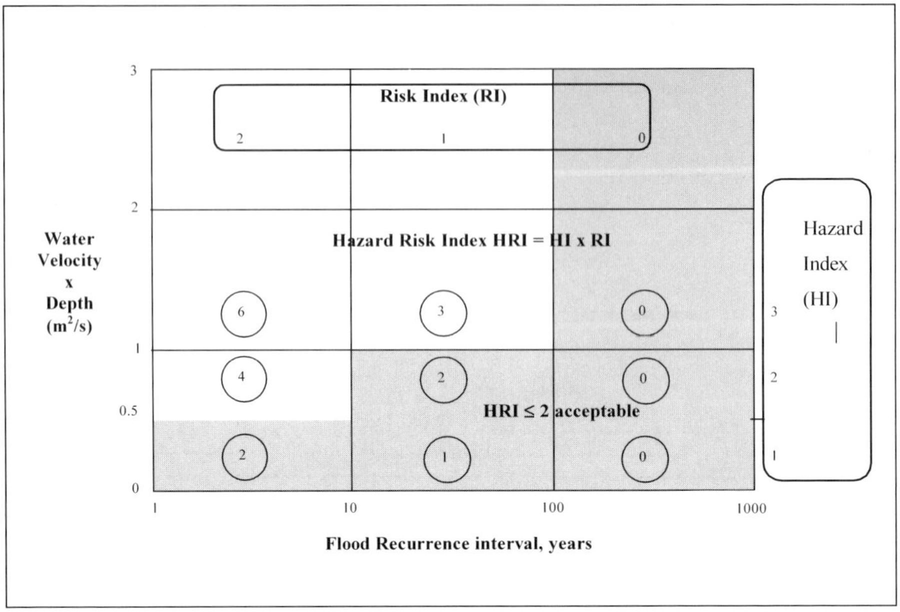

Figure 4.15 Hazard Risk Indices

4.10.1 Alternative planning zones

American practice in Minnesota is to plot the 100-year flood line and divide the waterway into floodway (main channel and edges) and flood fringe (largely dead water). Only pastoral and recreational/sporting activities are permitted in the floodway, and development in the flood fringe is subject to negotiation. Australian policy is more flexible, the objective being to minimize the impact and liabilities of flooding. Social, economic, ecological and hydrological factors are considered. New South Wales derived a valuable criterion in terms of water depth and velocity for deciding whether to permit development.

It therefore appears from experience that the method to adopt is to define different zones, for example as follows:

(a) *Waterway*. No development permitted, except possible channel improvements
(b) *Flood plain*. Development only if it is advantageous and proved not adverse
(c) *Flood fringe*. Development at developer's risk

Delimitations of the different zones could be based on the following:

(a) *Waterway*: either
  (i) Covered by a common flow, for example 100 years or less, or
  (ii) 90% of the 50-year flood, or
  (iii) velocity greater than 1 m/s, or
  (iv) depth greater than 1 m, or
  (v) a combination of (i), (iii) and (iv) giving flood hazard risk index of 2 or less.

(b) *Flood plain*: outside of (a) but bounded by either
- (i) The 25-year frequency flood, or
- (ii) 95% of the 50-year flood, or
- (iii) velocity greater than 0,5 m/s, or
- (iv) depth greater than 0,5 m, or
- (v) a combination of (i), (ii) and (iv), e.g. a flood hazard index of 1 or less.

(c) *Flood fringe*: outside of flood plain but bounded by either
- (i) 50-year or 100-year flood line, or
- (ii) no pollution source.

The designated waterway should be the zone within which obstructive or dangerous development should be prohibited. The suggested prohibition line below which development should be prohibited could be based on a hazard-risk index of above 2. The boundary corresponds to the 10-year flood line. To use a more frequent (e.g. 1-year) flood line would expose developers to a greater risk. The value of the additional land thus available can be evaluated. It can be shown that the value of the additional land thus made available is greater than the cost of additional flooding if building is on stilts.

Developers may propose altering the flood lines on the flood plain, for example by replacing exotic gum trees with indigenous shrubs or reshaping the ground, but the onus should be on them to prove this is beneficial from an economic, social, environmental, water quality and hydrological point of view. This may enable them to make even more area available for development. For the upper limit of the flood plain, the 50 year flood line is suggested. It would not be wise for floor levels of buildings to be below the 50-year flood line as the cost of flooding could be excessive and the backup effect would be grater. Hence, apart from surface development, e.g. roads, recreation areas, on the ground, buildings will have elevated floors, by up to 3m, or more if the consequences of flooding area great.

The flood fringe (between the 50-year and 100-year flood lines) may also be developed to an even greater density and development need not be banned in the flood fringe as long as the developers and owners are aware of and responsible for the risk of flooding and indicate this in writing. The flood lines should be indicated on all development plans, as well as floor levels. The hazard risk index is low and backup effect also negligible due to development on this fringe. In all cases, the responsibility needs legal opinion, and insurance against hazards should be considered.

4.10.2 Warning systems

Flood prediction is a complicated science. The ability to estimate floods in advance becomes less reliable the greater the time horizon. Local floods can be gauged accurately upstream and transposed to the site in question. Rainfall measurement can be converted to flow a few hours or days ahead. Rainfall forecasts can be obtained from climate models months ahead but with some uncertainty. Beyond that we have to resort to statistical projections with a probability range. Koussis et al (2003) describe a model for flood prediction.

The warning system may not only involve hydrological parameters. Structural failures, geological movement and downstream access all influence the vulnerability to flooding. Dam break problems have been modeled particularly with regard to dam safety studies (Zoppoa and Roberts, 2003).

REFERENCES

Adriaans, R. (2001). Development control in the floodplain, a key feature of river management. *IMIESA*, 26 (1), Jan, 22-23.

Chang, T.J. (1989). Characteristics of extreme precipitation. *Water Resources Bulletin*, 25 (5), 1037-1040.

Du Plessis, L.A. (2000). A new and unique approach of flood disaster management. *S.A. Water Bulletin*, 26 (5), Sept., 16-19.

Koussis, A.D.,Lagouvardis, K., Mazi, K., Kotroni, V., Sitzmann, D., Lang, J., Zaiss, T. and Malguzzi, P. (2003). Flood forecasting for urban basin with integrated hydro-meteorological model. *J. Hydrologic Engg*., Am. Soc. Civil Engrs., 8, (1), 1-11.

Lloyd-Davies, D.E. (1905). The elimination of storm water from sewerage systems. Min. *Proc. Inst. Civ. Engrs*., 164 (20), 41-67.

Minnesota (1969). *Floodplain Management Act*, Minneapolis.

New South Wales (1986). *Flood Development Manual*, Sydney.

Rodier, J.A. and Roche, M. (1984). *World Catalogue of Maximum Observed Floods.* IAHS/ Unesco, Wallingford.

Stephenson, D. (1981). *Stormwater Hydrology and Drainage.* Elsevier, Amsterdam.

Stephenson, D. (2002). Integrated flood plain management strategy for the Vaal. *Urban Water,* 4 (4). 423-428,

Stephenson, D. and Furumele, M. (2001). A hazard-risk index for urban flooding. *IAHR Congress*, Theme A: Development planning and management of surface and ground water resources, Beijing, Sept., 413-418.

Stephenson, D. and Paling, W.A.J. (1992). An hydraulic based model for simulating monthly runoff and erosion. *Water S.A.*, 18 (1), 43-52.

Stephenson, D., Van Zantwijk, A. and Hoge, P. (1997). A flood development policy for the Vaal Barrage. *Proc. Instn. Civil Engrs., Mun. Engr*., London, 121, Dec., 199-205.

Tyson, P.D. (1986). *Climatic Change and Variability in Southern Africa*, Oxford University Press.

Van Bladeren, D. (1992). *Historical Flood Documentation, Natal and Transkei*, 1848-1989.

Viessman, J and Knapp, A. (1990). *Applied Hydrology*, Wiley, N.Y.

Wilson, E.M. (1985). *Engineering Hydrology*, Macmillan..

Wohl, E.E. (2000). *Inland Flood Hazards*, Cambridge Univ. Press.

Zoppoa, C., and Roberts S., (2003). Explicit schemes for dam break simulations, *J. Hydraulic Engg.*, Amer. Soc. Civil Engrs. 129 (1), Jan.

CHAPTER 5

Effects of Catchment Development on Runoff

5.1 EFFECTS OF URBANISATION

By constructing buildings and roads on a catchment, we reduce the average permeability and therefore increase the runoff. This simple effect gives rise to a number of other effects which magnify the storm runoff rate. In addition there are changes to the hydrological regime and therefore the ecosystem of the catchment. The more rapid and less planned the development is, the greater the effects on peak storm runoff as well as low flows. Water quality also becomes of concern particularly with the construction of sewers and waste handling facilities which interact with the rest of the catchment.

There is also an interaction between development and micro climate. Thermal currents from the artificially heated buildings create plumes and affect precipitation and cloud pattern. Changes in ground cover and vegetation change the evaporation rates. The gradual sealing of the surface decreases groundwater and therefore affects vegetation and surface evaporation.

On a larger scale, the emission of gasses at different temperatures, releases of pollutants ranging from sulphur to ozone attacking hydrofluorocarbons and releases of carbon dioxide from burning fuel all affect the global climate. These implications are of national and international importance, whereas this chapter is largely confined to the catchment scale effects.

Man needs weather-proofed buildings. Roofs discharge rainwater via gutters and down pipes which could lead to pavements, roads and stormwater drains. The direct connection of roof drainage to the stormwater drains is to be discouraged in town planning as a discharge onto lawns or gardens or permeable surfaces can have a considerable ameliorating affect on the otherwise increased discharges to the stormwater system.

The effect of impervious cover is the easiest to quantify if one considers the runoff process. A 10% impervious cover could be expected to increase runoff by increasing the runoff coefficient for that proportion to unity. However, the disconnection of the impervious area can effectively eliminate that direct runoff by enabling the collected water to permeate elsewhere.

Of less obvious effect is the diversion of stormwater into lined gutters, stormwater pipes and canals. This results in a faster runoff which causes less ponding in the catchment. However, it passes the problem onto the downstream channels and in fact throughout the catchment itself the peak flows are increased considerably.

The smoother surfaces and the deeper flows result in considerably higher water velocities and more particularly, faster concentration times for the catchment. This results in shorter duration storms become the critical ones. Rainfall intensity-duration relationships indicate higher storm intensities for shorter storms so that flood peaks increase just due to faster concentration in urban drainage systems.

The net effect of urbanization can be to increase stormwater peaks by a factor or 10 or more. This does not however apply in the case of the total volume of the hydrograph because the shorter duration hydrographs are associated with smaller volumes as a rule.

The results of actual observations on experimental catchments and the modelling of the runoff are used to illustrate the effect below. Note however that the simpler type of hydrological calculations such as the rational method and next stage linearized methods such as the time-area method, or the unit hydrograph method, omit the non-linearity in the effects described above. At the most, the increased impervious area is accounted for by increasing the rational coefficient 'C'.

It should be noted however that the effects of man-made systems are to considerably change the runoff pattern over time and not just individual hydrographs. When calculating runoff using the simplistic method, i.e. rational method, it is normal to assume that the runoff recurrence interval is the same as that of the storm which causes it. However, the real situation is that the antecedent soil moisture and other effects have an equal if not more significant effect on the peak runoff and therefore the relationship between the series is much more complicated. The system can only really be studied by modelling a number of events over a period of time and re-plotting the ranked series.

Urban development affects the rainfall pattern and statistics as well as the runoff pattern. It has been alleged that blanketing effects due to solar shields affect evaporation and hence the resultant precipitation. The blanket of smog, dust, fumes, etc., may also affect the place in which the clouds release their moisture. So the effect of urbanization on rainfall is difficult to estimate and the statistical properties of rainfall records (e.g. the mean, coefficient of variance, frequency and distribution) will to some extent be affected as well. Rainfall is reputed to fall more on the leeward side of cities due to the heating up of the air over the city and up to 15% more precipitation has been attributed to this effect (Huff and Changnon, 1972; Colyer, 1982). Apart from this, the relationship between rainfall and runoff is affected.

5.2 STORMWATER MANAGEMENT

This section describes some stormwater management systems and comments on their effectiveness. The development of stormwater management practices in any country is affected by history, technology, hydrology, the environment and finance. Whitlaw and Brooker (1995) described the evolution of a stream through Johannesburg, as it became built up over a century. The study was of the Braamfontein stream which suffered erosion and pollution due to urbanization. Beard and Chang (1979) looked at the effects of urbanization from the hydrological point of view. The Australian approach (Bakele et al., 1993) is pragmatic in zoning urban streams and making developers aware of risks of establishing along streams while installing limited management facilities.

Best management practices have lagged behind in many developing countries (Miles, 1979) probably owing to the rapid pace of development. However, the advent of low cost peri-urban housing and associated flooding and pollution problems warrant urgent attention and novel ideas, e.g. dual purpose detention (Wipple, 1979). Low cost solutions are also becoming imperative, e.g. dual drainage (Otterpohl et al., 1997). Progress on treatment of dilute runoff viz. stormwater and combined runoff, has not progressed as far as low cost sewage treatment (and reuse) e.g. Rose, 1999). However, alternative management of urban pollution by social programmes and management of impoundments (Weichers et al., 1996) have been investigated.

The management of stormwater runoff has evolved over the last quarter century from direct runoff facilitation to detention or retention and in some cases even diversion for treatment or semi-treatment of the runoff. The original stormwater drainage approach was to calculate the design runoff using empirical equations such as the Lloyd-Davies (1905) equation or more recently the Rational equation. Stormwater modelling then enabled more accurate calculations to be done, or more specifically to accommodate more complex systems and their effects on runoff volumes and rates. At the same time, with the development of modelling methods, there became an awareness of the advantages of many stormwater management structures.

The attenuation of the hydrograph by means of temporary storage is called the detention of runoff and many detention basins have proved their worth in ameliorating the peak runoff and thereby reducing the cost of the downstream drainage pipes and channels (see e.g. Brooker, 1997; Watson and Miles, 1982). It is possible to size ponds to achieve specified reduction in peaks.

Detention storage can have a disadvantage in that the effect is reduced the greater storm either in terms of volume or peak flow rate occurs, it may have a lesser effect and in fact may have hardly any effect on the more severe storms.

It would be preferable to attenuate extreme storms as well as design storms. Therefore, dual drainage systems have received attention. That is, the minor system copes with recurrence interval storms of the order of 2 to 5 years and sometimes up to 20 years where major flooding is a problem. For higher flow rates, the major system comes into action. That could be the thoroughfare in the way of roads or pavements or overland flow. The latter are more effective in flood routing than conduits as they have a greater resistance due to the shallow flow and therefore perhaps require more consideration. More particularly, a risk analysis is required to consider the possibility of even the major system's capacity being exceeded. And again the severity of this depends on the cost of flooding, both in terms of property damage and danger to life and the environment.

Retention storage, i.e. the permanent capturing of runoff has theoretically even more advantages. That is, the water is retained in the catchment which would previously have been lost to the catchment. The effects of urbanization are well-known, that is an increase in development on the catchment, particularly in the way surfaced roads and roofs, leads to increased volumes as well as rates of runoff. The effect of various storage structures on storm runoff is depicted in Chapter 4.

One should distinguish between in-stream and off channel retention. In-stream retention is only active up to a certain volume and then has a negligible effect except routing. Off channel storage, on the other hand, can have its characteristics controlled by the inlet structure. (In the same way that the outlet structure from a detention pond controls the discharge hydrograph.)

An object of retention storage is to commit water retained longer on the catchment due to man-made obstruction, to infiltrate the ground. This to some extent restores the groundwater moisture content to a more natural condition. Thus vegetation can become established which further retains moisture. Higher water tables or even perched water tables result meaning that less water is required for gardens which has proved an expensive luxury in semi-arid and temperate climatic regions.

The incorporation of knowledge gained from urban stormwater catchment and modelling studies has enabled town planners to control the water regime to a greater extent. The omission of gutters enables roof runoff to be spread over a greater area resulting in more likely infiltration but also less easy diversion to sewerage systems. In fact, the disconnection of impermeable areas has been proved by modelling to be a most important factor in urban areas for improving infiltration and thereby reducing runoff rates and volumes.

While the method of calculation of design floods has improved over the years, there is still a great diversity in the approach. The older empirical formula based on experienced runoff rates in temperate rainfall areas has gradually been replaced by more sophisticated formulae. Even the rational method has been expanded to include various effects such as slope, catchment usage and of more suspect value, the effect of recurrence interval (e.g. Rossmiller, 1980). That is, a proportionally greater runoff generally is associated with a higher recurrence interval. However, this raises the problem of whether it is the 50 year rainfall storm, for example, which produces the 50 year flood. A study by Ridgard (1994) shows that the direct relationship between runoff and rainfall becomes distorted as urbanization occurs (Fig. 5.1).

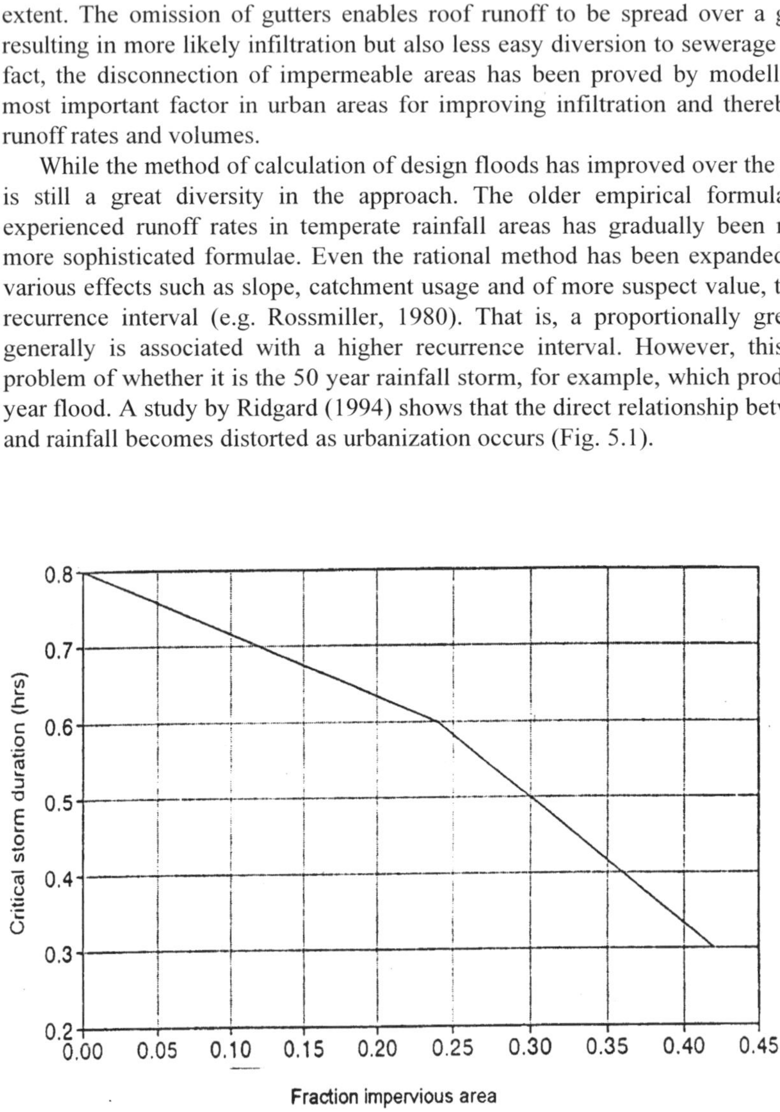

Figure 5.1 Response time versus fraction of impervious area

This is not only because it is a different storm duration which causes the runoff from urbanized areas compared with the natural catchment, but also because other factors need to be included, in particular initial moisture content of the catchment. Many proponents of the rational formula will say that these have been considered in selecting rational coefficients. However, there are many people who still assume that the rational coefficient is the ratio of impermeable to total catchment area and no allowance is made for disconnection of impermeable areas. Computer modelling, even by time area methods, attempts to overcome this problem. However, the time area method still suffers two main disadvantages as follows.

One is that there is not a direct relationship between runoff and rainfall, i.e. the so-called rational 'C' is not a constant but depends on initial moisture content as well as the permeability of the catchment surface. A more logical approach would be to use an infiltration capacity of the soil irrespective of the impending storm intensity with the result that a greater proportion of higher storms would run off than for smaller storms. This may be one reason that more intense storms are often under-estimated, particularly where there is a wide range between extreme and frequent storm intensities.

Another factor which has been highlighted by the introduction of hydraulic methods, in particular the kinematic equations, is the introduction of the sensitivity of the response of the catchment to the rainfall intensity. That is, the higher the rainfall rate the more likely the water depths are to be greater and therefore the runoff rates greater. The response time is thus, according to the hydrodynamic laws, much faster and therefore in fact, a shorter duration storm is more likely to be the cause of an extreme event flood than for more common lower intensity storms (Stephenson, 1981). However, such sophisticated models are not commonly used for minor structures such as road drains and culverts, particularly where catchment areas are less than a few square kilometres. Figure 5.2 ranks some of the effects of management for a series of storms as modelled.

5.2.1 Effect on recurrence interval

A study by Sutherland (1983) indicates little correlation between rainfall recurrence interval and the recurrence interval of the flood when assessed in terms of the peak flow rate. He proposed that antecedent moisture conditions, measured in terms of the total precipitation in preceding days, should be a parameter in runoff-duration-frequency relationships. His contention is that the probability of a certain runoff intensity is more related to the probability of the soil being at a certain saturation than the rainfall intensity.

Table 5.1 Effect of alternative management practices

	System	2 year peak m^3/s	10 year peak m^3/s	Cost implication
1	As is	2.2	6.6	Nil
2	Disconnect impervious areas	1.6	5.0	Low
3	Re-arrange road plan	1.3	3.9	High
4	Flood plain & channel storage	1.5	5.1	Low
5	Detention storage 1000 m^3	1.2	4.4	Medium
6	Dual drainage	1.9	5.3	Medium

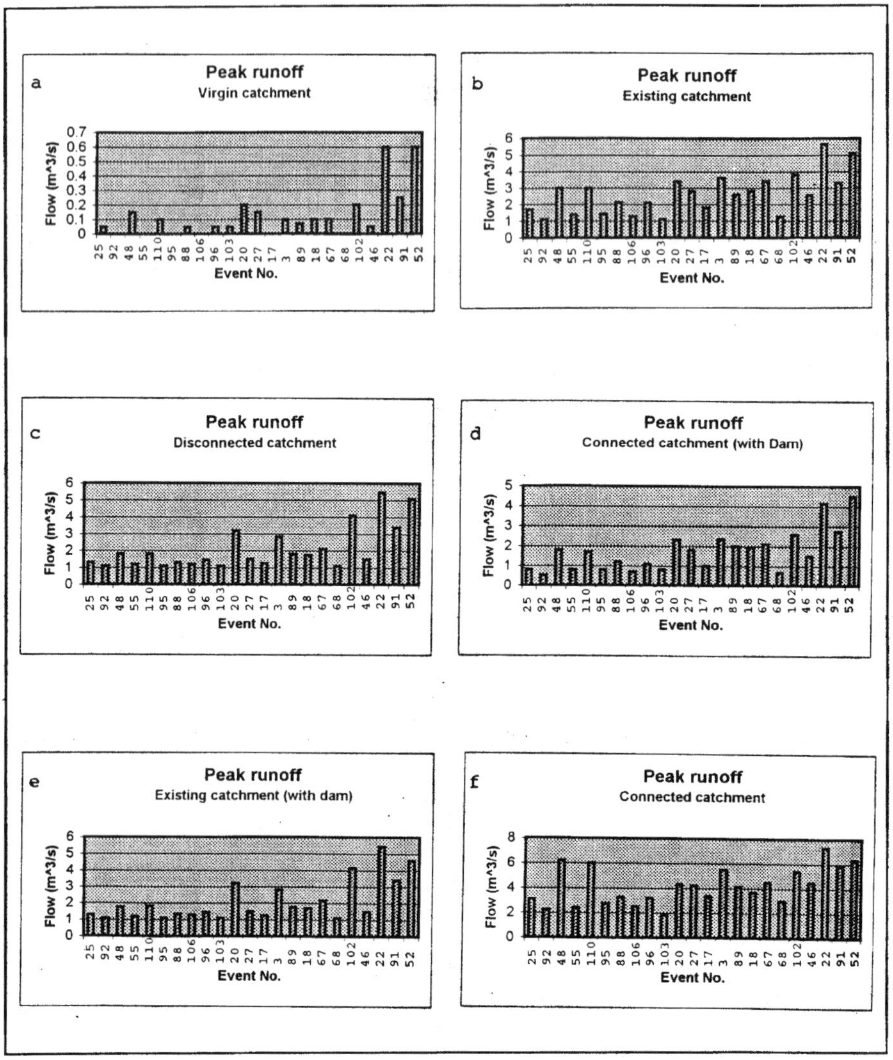

Figure 5.2 (a-f) Runoff spectrum for all catchment scenarios as obtained from modeling studies

How does urbanization affect the argument? In fact, it counters these ideas. The more the natural surface cover is replaced by impermeable surfaces the more runoff becomes a direct response function to rainfall. In the limit for 100% runoff, soil does not feature and the recurrence interval of runoff is equal to that of the storm causing it.

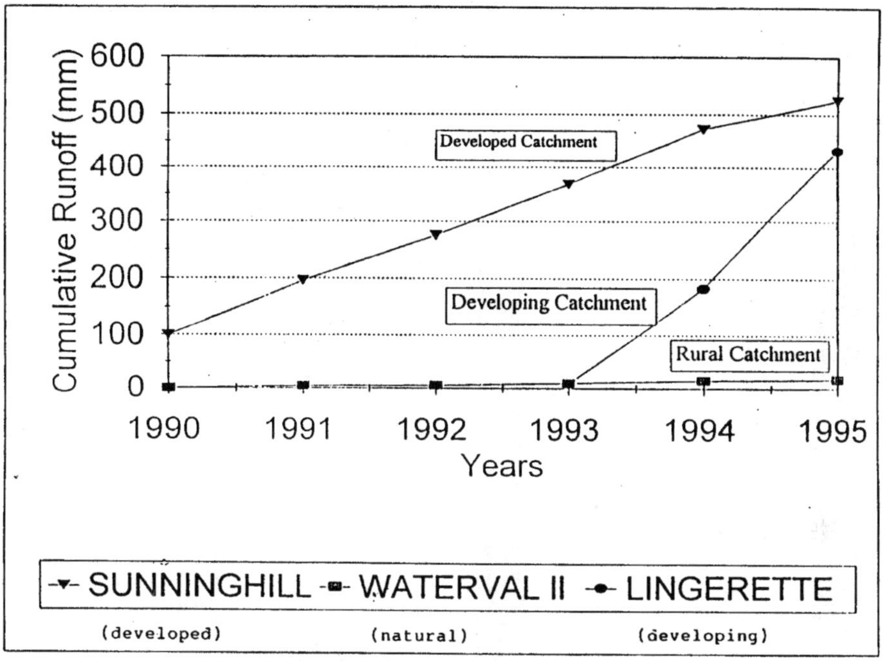

Figure 5.3 Triple mass curve relationship for 3 catchments

5.3 CASE STUDIES

The relative effect of various management strategies was compared by modelling (Stephenson, 1989). The results for a 75 ha urban catchment north of Johannesburg (Sunninghill) are summarized in Table 5.1. This gives the peak runoff rate for two storms, i.e. 2 year and 10 year storm, resulting from the introduction of alternative stormwater management practices.

Three similar catchments north of Johannesburg, South Africa, were selected for study (Stephenson, 1997). These were similar sized catchments with similar slope and in fact adjacent to each other. One of the catchments is described at 'developed' and it comprised 75 hectares of upper class housing with some townhouses and limited commercial development in the catchment. The second catchment was a naturally covered catchment, i.e. grazing land with some trees but also no defined runoff channels. This enabled one to determine the overland flow characteristics which in fact indicated peak storm runoffs of the order of a quarter of the magnitude of those from the urbanized catchment. The various reasons contributing to this include obviously the fact that there is greater infiltration into the natural catchment, but also due to the fact that runoff was slower owing to the lack of channelization and a rougher surface.

The third catchment was one which was developed over a period of four years during the experiment. Here, the peak runoffs were found to increase by up to 16 times compared with conditions prior to development of the catchment. Figure 5.3 shows a water balance comparison of the undeveloped and developed catchments and Figure 5.4 shows a breakdown of the component in the mass balance.

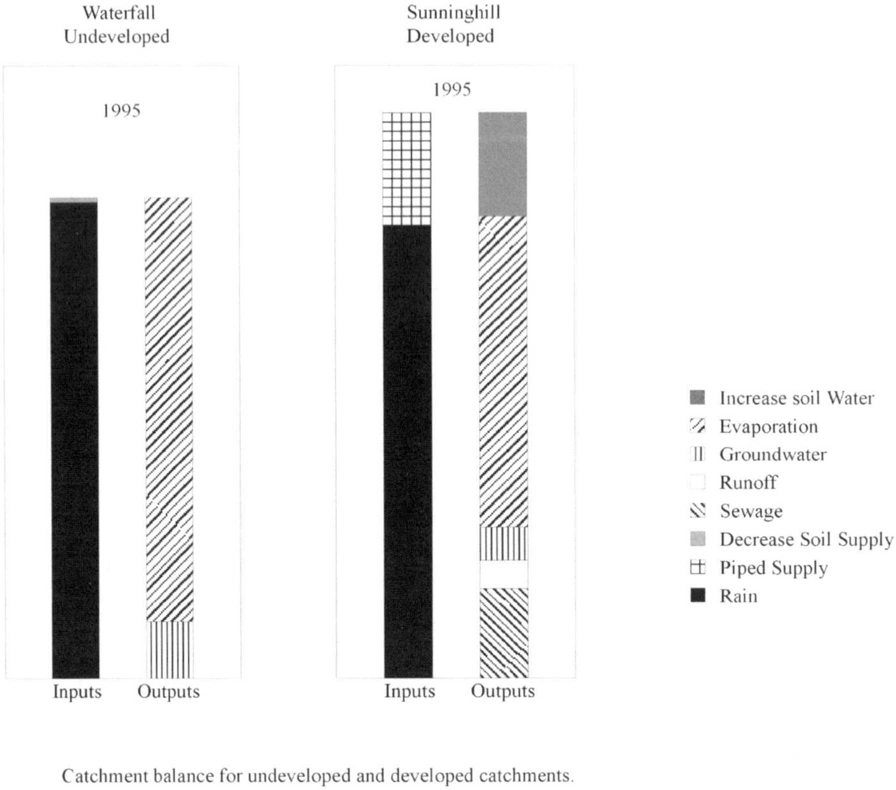

Figure 5.4 Catchment balance for undeveloped and developed catchments (Stephenson, 1997)

An example of a detention basin which reduced the danger of flooding in an industrialized area is the Booysens Road intersection in Johannesburg. A basin 10000m^3 in volume abstracted water by means of a siphon spillway from an overloaded culvert once the water level reached the priming level. The rate of outflow was controlled by a restricted weir. The capacity of the system was thereby increased from 50 to 88 m^3/s. The basin also served as a settling basin and screens were constructed to remove the large amounts of litter from the water. A problem which has been experienced in Johannesburg is the dirty water which limits the use of the detention basin after storms. That is, the high dirt load means that the areas cannot be readily used for playing fields or even parks. Nevertheless, the clean-up process had benefits all the way downstream.

In Pretoria, the Apies river flows through the centre of the town and again the urbanization of the surrounding suburbs increased the runoff manyfold since construction of the channel. The channel runs under road bridges but also around essential services such as major communications cables and electrical supply cables.

Figure 5.5 The effect of urbanization on runoff

The result is that the channel could not be deepened notwithstanding the fact that it was concrete lined.

The channel needed careful remodelling to take advantage of its steep slope so that where the height restrictions applied, the water was shot through at supercritical flow and by minimizing the reversion to subcritical flow downstream, it was contained within the original channel confines once more. However, upstream catchment re-configuration to attenuate peak flow was not possible owing to the dense road network.

In general, the application of retention storage and infiltration have been limited in the interior of South Africa owing to the high intensity storms and the impermeable soils. The only area where infiltration has proved successful and in fact has been incorporated in town planning ordinances is in the Western Cape where areas such as the Cape Flats and Saldanha Bay have sandy aquifers of considerable depth over the bedrock (SAICE, 1988). There, the infiltration has proved so successful and the management of groundwater flow equally successful that recharge and groundwater abstraction are being considered with modelling studies.

5.3.1 Example: Calculation of peak runoff for various conditions

The effect of urbanization on runoff can be illustrated with the following example. In particular, it will be seen that the peak flows increase (as well as the volume of runoff) due to various effects, namely decreased permeability, decrease in roughness and channelization. The relative effects are evaluated and the effect on the rational coefficient "C" is indicated.

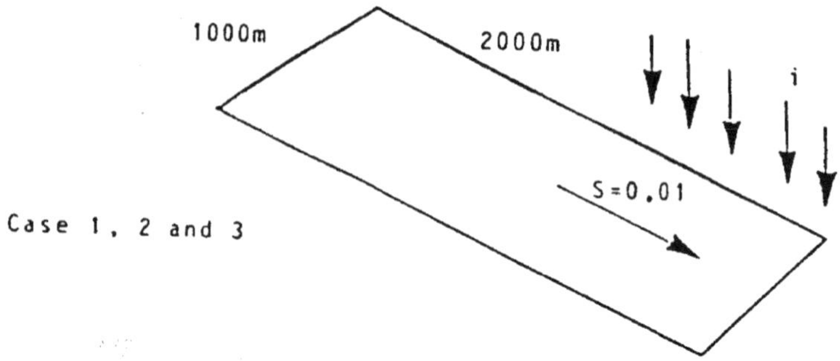

Figure 5.6 Simple catchment analyzed

(i) Virgin catchment

The simple rectangular catchment depicted in Figure 5.6 will be studied to indicate the various effects of urban development on the storm runoff peak. The effects computed are reduced roughness, impermeable cover and channelization. A constant frequency, uniform rainfall intensity duration relationship is used as follows:

$$i(mm/h) = \frac{a}{(0.24 + t_d)^{0.89}} \tag{5.1}$$

Where t_d is the storm duration in hours.

This is typical of a temperate area, and the value of 'a' for this region is estimated to be 70 mm/h for storms with a 20 year recurrence interval of exceedance.

The catchment is assumed to have a constant slope of 0.01 and initially the cover is grass. The representative Manning roughness for overland flow is estimated to be 0.1. The initial abstraction (surface retention and moisture deficit make up) is 30mm and subsequent mean infiltration rate over a storm is 10mm./hr.

Thus $\alpha = \sqrt{S}/n = \sqrt{0.01} / 0.1 = 1.0$

Infiltration ratio F = f/a = 10/70 = 0.143

Initial loss ratio U = u/a = 30/70 = 0.429

Length factor in SI units $LF = L/36\alpha a^{2/3} = 2000/36 \times a \times 70^{2/3} = 3.27$

From Figure 5.7 (for U = 0.40) read equilibrium $t_e > 4h$ (off the graph) but the peak runoff factor for this F is QF = 0.23 which corresponds to a storm duration of t_d = 2.2h. The peak runoff rate is

$$Q_p = 0.23B\alpha a^{5/3}/10^5 = 0.23 \times 1000 \times 1 \times 70^{5/3}/10^5 = 2.74 \text{ m}^3/\text{s} \tag{5.2}$$

The total precipitation rate over the catchment of area A for the same storm duration is

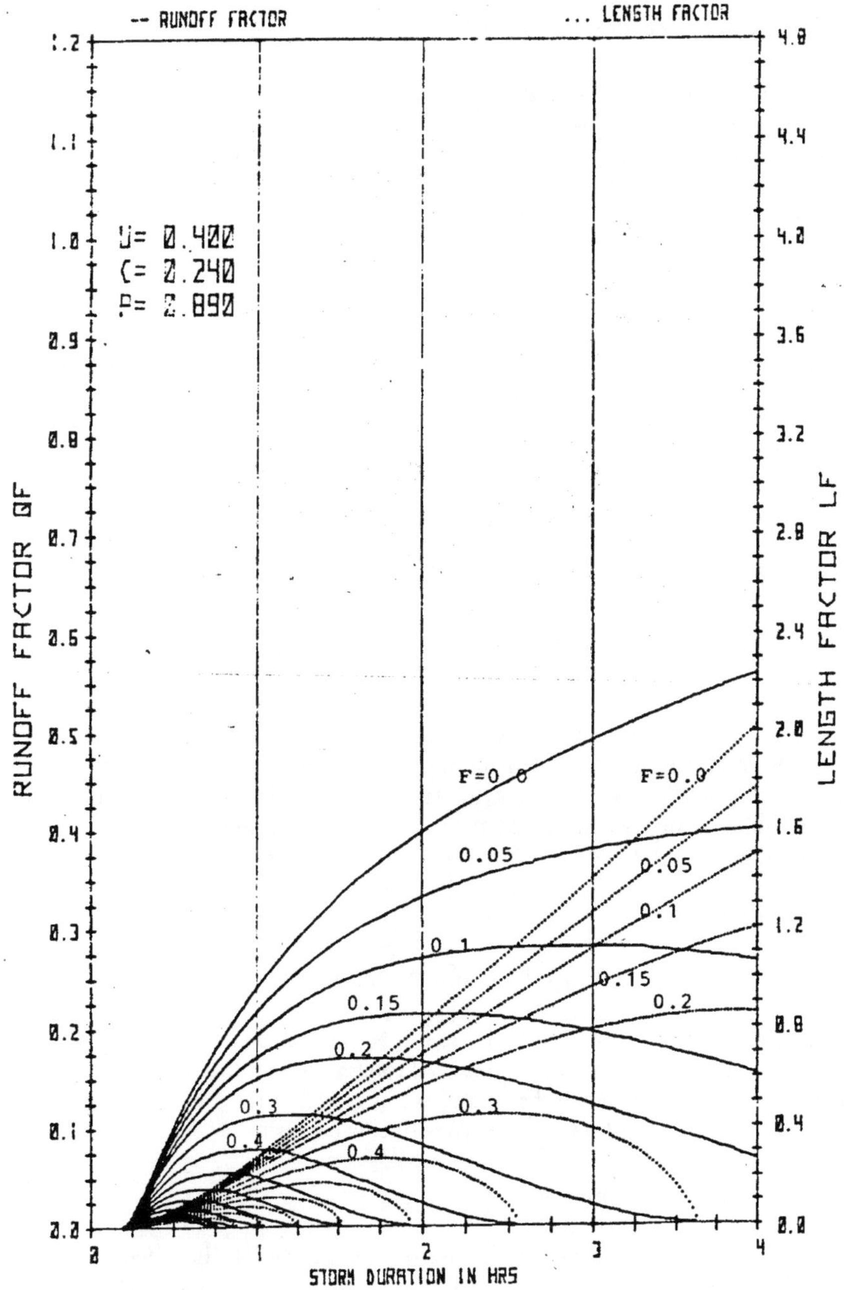

Figure 5.7 Peak runoff factors for U = 0.40 (Stephenson and Meadows, 1986)

$$\text{Ai} = \frac{70\text{x}1000\text{x}2000}{(0.24+2.2)^{0.89}\text{x}3600\text{x}1000} = 17.6\text{m}^3/\text{s} \tag{5.3}$$

So the rational coefficient C = 2.74/17.6 = 0.16. (5.4)

Note however that the full catchment is not contributing at the time of peak runoff for the design storm, so C does not only represent the reduction in runoff due to losses, it also accounts for only part of the catchment contributing. The runoff for the full catchment would be less as the storm duration would be longer than 2.2h so the intensity would be less and the losses relatively higher.

(ii) Reduction in infiltration

If the infiltration and initial abstractions are reduced by urbanization, the peak runoff increases. The construction of buildings and roads could reduce infiltration rate to 7mm/h and initial abstraction to 14mm. For F = 7/70 = 0.1 and u = 14/70 = 0.20 (Fig. 5.8) then the LF = 3.27 as for case (i). The time to equilibrium is off the chart but the critical storm has a duration of 2.2 hours and the corresponding peak flow is

$$Q_p = 0.44 \times 1000 \times 1.0 \times 70^{5/3}/10^5 = 5.24 \text{ m}^3/\text{s}. \quad (5.5)$$

The corresponding rational coefficient C works out to be 0.30

(iii) Effect of reduced roughness due to paving

With the construction of roads, pavements and building the natural retardation of the surface runoff is eliminated and concentration time reduces. That is, the system response is faster and as a result shorter, sharper showers are the worst from the point of view of runoff peak. For the sample catchment, the effective Manning roughness could quite easily be reduced to 0.03. Then α = 3.33 and LF = 0.98. This time to equilibrium would therefore be 3h, but the peak intensity storm has a duration of 2.2h as before. In this case, the extent of the storm over the catchment is greater however, and the peak runoff is

$$Q_p = 0.23 \times 1000 \times 3.33 \times 70^{5/3}/10^5 = 9.12\text{m}^3/\text{s} \quad (5.6)$$

The corresponding increase in C is from 0.16 to 0.52, an appreciable increase, if it is borne in mind this is only due to reduced roughness and does not account for reduced infiltration. It will be noted that the effect of reducing roughness is even greater than decreasing infiltration for this case. The effect is magnified in the following example.

(iv) Effect of canalization

The effect of a stream down the centre of a catchment is illustrated in the following example. The same surface roughness (n = 0.1) and permeability (f = 10mm/h, u = 30mm) as for case (i) are assumed. The cross slope is taken as 0.04 for overland flow and 0.01 for an 8m wide channel down the catchment. The dimensionless hydrographs in Figure 5.10 are used.

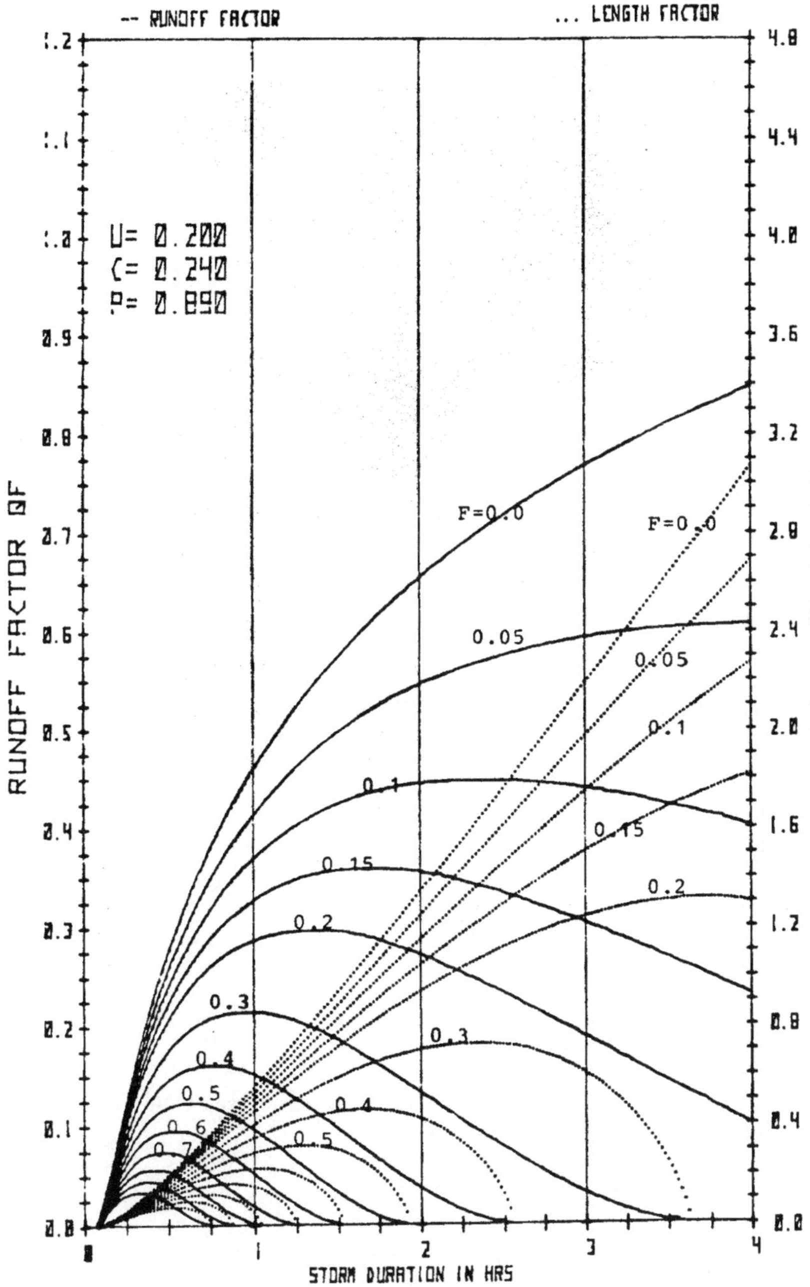

Figure 5.8 Peak runoff factors for U = 0.20 (Stephenson and Meadows, 1986)

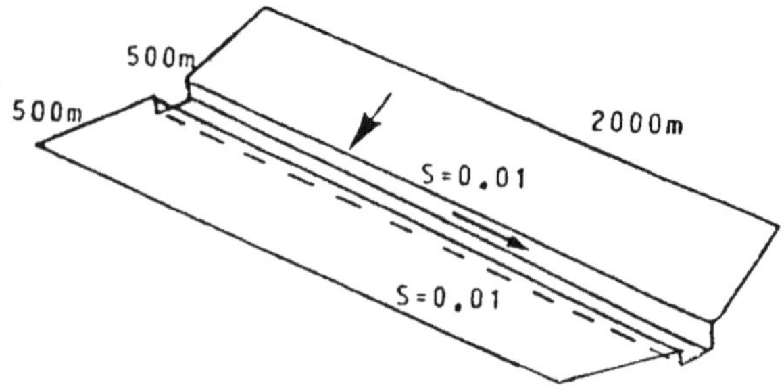

Figure 5.9 Catchment with channel

$$\text{The stream catchment ratio } G = \left(\frac{2L_S{}^{0.6}}{b\alpha_S}\right)\frac{b\alpha_o{}^{0.6}}{2L_o} = \left(\frac{2x2000}{8x1}\right)^{0.6}\frac{8x2^{0.6}}{2x500} = 0.50 \tag{5.7}$$

By trial, guess storm duration resulting in peak runoff of 1.5h, then

$$i_e = \frac{70}{(0.24+1.5)^{0.89}} - 10 = 42.7 - 10 = 32.7\text{mm/h} \tag{5.8}$$

$$t_{ed} = t_d - t_u = 1.5 - 30/42.7 = 0.80\text{h} \tag{5.9}$$

$$F = 10/32.7 = 0.31 \tag{5.10}$$

$$t_{co} = \left(\frac{L_o}{\alpha i_e{}^{m-1}}\right)^{1/m} = \frac{500}{2x(32.7/3600000)^{2/3}}^{3/5} = 2860\text{s} = 0.80\text{h} \tag{5.11}$$

$$T_D = (5/3)t_{ed}/t_{co} = (5/3)0.8/0.8 = 1.67 \tag{5.12}$$

Therefore $t_d = t_{ed} + t_u = 0.8 + 30/42.7 = 1.50$h, which agrees with the initial guess. Using Figure 5.10 again, the peak factor Q = 0.85.

Peak flow $Q_S = Qai_e = 0.85 \text{ x } 2 \text{ x } 10^6 \text{ x } 32.7/3.6 \text{ x } 10^6 = 15.4\text{m}^3/\text{s}$

Rational coefficient C = 15.4/(42.7 x 2/3.6) = 0.65

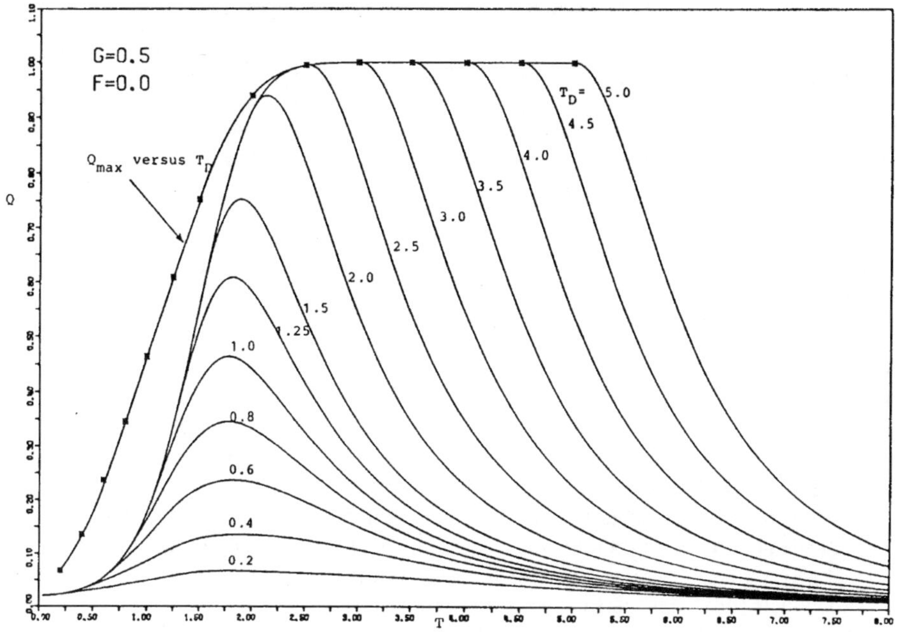

Figure 5.10 Dimensionless runoff hydrographs for the V-shaped catchment with stream (G = 0.5, F = 0.0)

(v) Combined reduced roughness and reduced losses

If roughness is reduced by paving to 0.03, then $\alpha = 3.33$ and LF = 0.98 as for case (iii). The reduced loss factors become F = 0.1 and U = 0.2 as for case (ii). From Fig. 5.8, t_c = 1.7h and the corresponding QF = 0.43.

Hence the peak flow Q = 0.43 x 1000 x 3.33 x $70^{5/3}$ = 17.0m^3/s. The rainfall rate for a storm of this duration is

$$\frac{70\text{x}1000\text{x}2000}{(0.24+1.7)^{0.89}\text{x}36000\text{x}1000} = 21.6\text{m}^3/\text{s}, \qquad (5.13)$$

So C = 0.79.

The relative effect of each variable on peak runoff can be compared with the aid of Table 5.2. The effect of reducing infiltration 30% and initial abstraction 40% is to double the peak runoff. The critical storm duration was not affected but the effective area contributing increased slightly. The effect of reducing surface roughness is even more remarkable however. Even maintaining the same losses (both initial and abstraction and infiltration) as for the natural catchment the runoff peak increased by a factor of 4.

Table 5.2 Showing effect of different surface configurations on peak runoff from a 2000m long by 1000m wide catchment $S_o = 0.01$, $i = 70mm/h/(0.24h + t_d)^{0.89}$

	Case	n	f mm/h	u mm	t_c h	t_d h	i mm/h	$Q_p m^3/s$	C
1	Virgin catchment	0.1	10	30	5	2.2	36.7	2.74	0.16
2	Reduced losses	0.1	7	14	4	2.2	36.7	5.24	0.30
3	Reduced roughness	0.03	10	30	3	2.2	36.7	9.12	0.52
4	Canalization (stream width 3m)	0.1	10	30	0.8	1.5	42.7	15.4	0.65
5	Reduced losses and roughness	0.03	7	14	1.7	1.7	38.8	17.0	0.79

The area contributing increased notably although the critical storm duration was not affected. Reducing roughness even more would not necessarily increase runoff much as practically the entire catchment contributes for case (iii), whereas the area contributing in case (i) was much less. Only for case (v) with reduced roughness and losses is the concentration time equal to the critical storm duration.

The effect of canalization is somewhat similar to reducing roughness – water velocities and concentration rates are faster. This is due to the greater depth in channels ($Q = B\sqrt{S}y^{2/3}/n$). Consequently, a greater area contributes to the peak.

Not much sense can be made out of comparing the resulting rational coefficients (ratio of peak runoff rate to rainfall rate times catchment area). That is because the time of concentration for each case is different due to different roughness, rainfall rate, etc. In any case, it is irrelevant when it is compared with critical storm duration which is shorter than the time to equilibrium.

5.4 DETENTION STORAGE

The following analysis reveals the relative significance of type of control and channel roughness on catchment storage and consequently flood peaks (Stephenson, 1984). An introductory solution for detention storage volume was made by McEnroe (1992).

Although the kinematic equations as presented previously cannot accommodate reservoir storage, they may be rearranged to illustrate the storage components in them. The St. Venant equations which include terms for storage when water surface is not parallel to the bed, are

$$\frac{\partial A}{\partial t} = -\frac{\partial Q}{\partial x} \tag{5.14}$$

$$\frac{\partial v}{\partial t} = g(S_o - S_f) - g\frac{\partial y}{\partial x} - v\frac{\partial v}{\partial x} \tag{5.15}$$

The first equation is the continuity equation and the second the so-called dynamic equation. The first equation does not give the total storage in the reach, it represents the rate of change in cross sectional area of flow as a function of inflow and outflow. The second equation contains more about the distribution of storage. The last two terms represent the wedge component of storage, which are absent in the kinematic equations.

The kinematic equations therefore treat storage as a prism, with storage in blocks and no allowance for difference in slope between bed and water surface is made. Since the second equation is replaced by a friction equation, and $S_o = S_f$ in the kinematic equations, only the first equation in the case of the kinematic equations can be used to calculate storage changes.

The continuity equation may be written as

$$\frac{O-I}{\Delta x}+\frac{A_2-A_1}{\Delta t}=0 \tag{5.16}$$

where O is outflow, I is inflow over a reach of length Δx, and A_1 and A_2 are the cross sectional areas before and after Δt respectively. If $0 = (O_1 + O_2)/2$ and $I = (I_1 + I_2)/2$ and $A\Delta x$ is replaced by S, and storage which is a function of A_I and A_O, which in turn are functions of flow rate, e.g. $S = [XI + (1 - X)O]$, then equation (5.6) becomes the one frequently used for open channel routing.

$$O_2 = c_1I_1 + c_2I_2 + c_3O_1 \tag{5.17}$$

Where c_1, c_2 and c_3 are functions of Δx and Δt. The latter equation is referred to as Muskingum's equation used in routing floods along channels. If $X = 0$, the routing equation corresponds to level pool or reservoir routing, The more general equation with $X = ½$ represents a 4-point numerical solution of the continuity equation as employed in kinematic models (Brakensiek, 1967).

5.4.1 Channel storage

Channel storage performs a similar function to pond storage in regard to flow, and there are many analogies which can be drawn between the two. Channel storage is a function of friction resistance and channel shape and can be controlled in various ways.

The form of friction equation, as well as the friction factor, affect the reaction speed of a catchment and the volume stored on the catchment. The excess rain stored on the catchment, whether in channels or on planes, is a form of detention storage, and as such, affects the concentration time and consequently the peak rate of runoff. Some friction formulae used in stormwater drainage practice are listed below.

	S.I. units	English units	
Darcy-Weisbach	$Q = (8/f)^{1/2}A(RSg)^{1/2}$	$Q = (8/f)^{1/2}A(RSg)^{1/2}$	(5.18)
Chezy	$Q = 0.55CA(RS)^{1/2}$	$Q = CA(RS)^{1/2}$	(5.19)
Manning	$Q = AR^{2/3}S^{1/2}/n$	$Q = 1.486AR^{2/3}S^{1/2}/n$	(5.20)
Strickler	$Q = 7.7A(R/k)^{1/6}(RSg)^{1/2}$	$Q = 7.7A(R/k)^{1/6}(RSg)^{1/2}$	(5.21)

R is the hydraulic radius A/P where A is the area of flow and P the wetted perimeter. R can be approximated by depth y for wide rectangular channels. S is the energy gradient, f is the friction factor and k is a linear measure of roughness analogous to the Nikuradse roughness coefficient.

Both the roughness coefficient α and the exponent m of R or y in the general flow equation (5.24) affect the peak flow off a catchment. This is largely due to the attenuating effect of friction resulting in a larger time to equilibrium. A rainfall excess intensity-duration relationship is required to evaluate the effect of each coefficient on peak runoff rate and maximum catchment storage. The following expression for excess rainfall intensity is assumed:

$$i_e = \frac{a}{(c + t_d)^p} \qquad (5.22)$$

In this equation, it is customary to express i_e and a in mm/h or inches per hour and b and t_d in hours, where t_d is the storm duration assumed equal to time of concentration t_c for maximum peak runoff of a simple catchment.

Starting with the kinematic equation for continuity:

$$\frac{\partial y}{\partial t} + \frac{\partial q}{\partial x} = i_e \qquad (5.23)$$

and a general flow resistance equation:

$$q = \alpha y^m \qquad (5.24)$$

then it may be show that $t_c = (L/\alpha i_e^{m-1})^{1/m}$, where q is the runoff rate per unit width of the catchment and y is the flow depth. The rising limb of the hydrograph is given by the equation:

$$Q = \alpha(i_e t)^m \qquad (5.25)$$

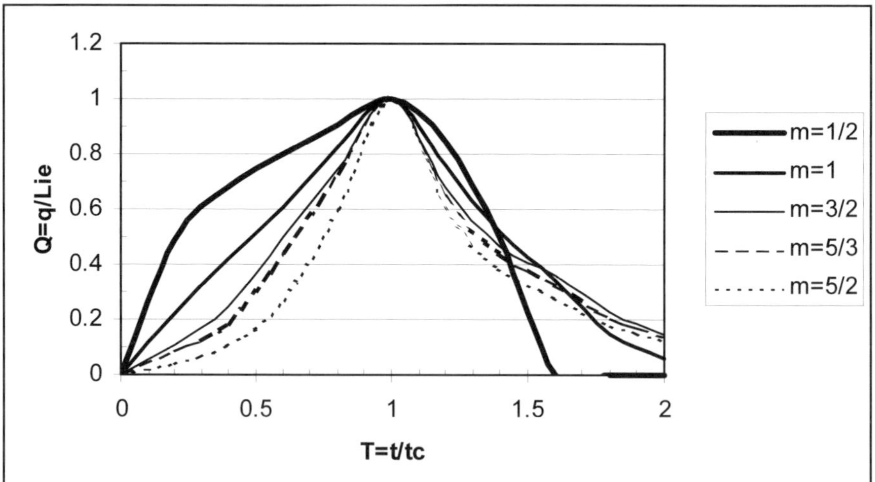

Figure 5.11 Hydrograph shapes for different values of m in $q = \alpha y^m$

Another expression may be derived from the falling limb. In Figure 5.11, dimensionless hydrographs are plotted to illustrate the effect of m on the shape of the hydrograph. The graphs are rendered dimensionless by plotting $Q = q/i_eL$ against $T = t/t_c$. m is used as a parameter. Thus m = ½ represents closed conduit or orifice flow, m = 1 represents a deep vertical sided channel, m = 3.2 represents a wide rectangular channel according to Darcy or a rectangular weir, m = 5/3 represents a wide rectangular channel if Manning's equation is employed, and m = 5/2 represents a triangular weir. The graphs immediately indicate the effect of m on catchment detention storage since the area under the graph represents storage.

The smaller m, the greater the storage. Thus, provided storage is economical, then by throttling outflow one may increase storage and increase concentration time thereby reducing discharge rate (which is not immediately apparent from these graphs as they are plotted relative to excess rainfall intensity). In practice, as the concentration time increases the greater the storage, so that the lower intensity storms become the design storms. This has a compound effect in reducing flow rates since total losses increases and it is possible that the entire catchment will not contribute at the time of peak flow .

A general solution of peak flow and storage in terms of intensity-duration relationships is derived below. Solving (5.22) with $t_d = t_c$ for maximum rate of runoff per unit area and generalizing by dividing by a,

$$q_m / aL = i_e a = \frac{1}{\left\{c + \frac{\left(L/\alpha i_e^{\,m-1}\right)^{1/m}}{3600}\right\}^p} = \frac{1}{\left\{c + \frac{\left|L/\alpha(a/360000)^{m-1}\right|^{1/m}}{3600(i_e/a)^{1-1/m}}\right\}^p} \qquad (5.26)$$

The term $L/\alpha a^{m-1}$ is referred to as the length factor. Constants are introduced for a in mm/h, and time of concentration in hour units. The maximum peak flow factor i_e/a is plotted against length factor in Figure 5.12, since it is not easy to solve (5.26) for i_e/a.

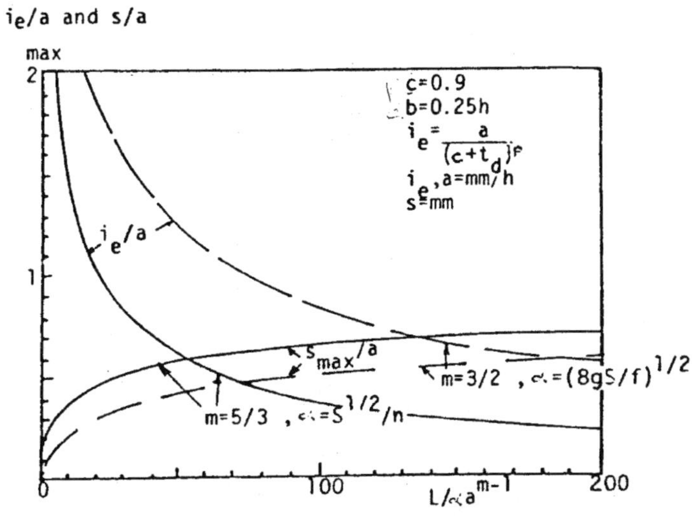

Figure 5.12 Peak flow and storage versus length factor

An expression for the corresponding catchment storage is derived below. At equilibrium the flow per unit width at a distance x down the catchment is

$$q = i_e x$$
$$= \alpha y^m$$
$$\text{therefore } y = (i_e x/\alpha)^{1/m} \tag{5.27}$$

Integrating y with respect to x yields the total volume on the catchment

$$V = \frac{Lm}{m+1}\left(i_e L/\alpha\right)^{1/m} \tag{5.28}$$

Or in terms of the average depth of storage s = V/L

$$s/a = \frac{m}{m+1}\left(\frac{i_e}{a}\right)^{1/m}\left(\frac{L}{\alpha(a/3600000)^{m-1}}\right)^{1/m}\frac{1}{3600} \tag{5.29}$$

Where s is in mm, and i_e and a are in mm/h. s/a is also plotted against length factor in Figure 5.13. It will be observed that the average storage depth does not increase in proportion to $L/\alpha a^{m-1}$. In fact, the rate of increase reduces beyond $L/\alpha a^{m-1} = 50$, and the rate of reduction in peak flow i_e/a also decreases beyond the figure, indicating reducing advantage in increasing channel length or roughness ($\alpha = K_1\sqrt{(S)/n}$). Since total channel cost is a direct function of storage capacity, it would appear to be an optimum at some intermediate value of $L/\alpha a^{m-1}$ if there is a cost associated with peak discharge, e.g. culverts or flooding downstream (see Fig. 5.13).

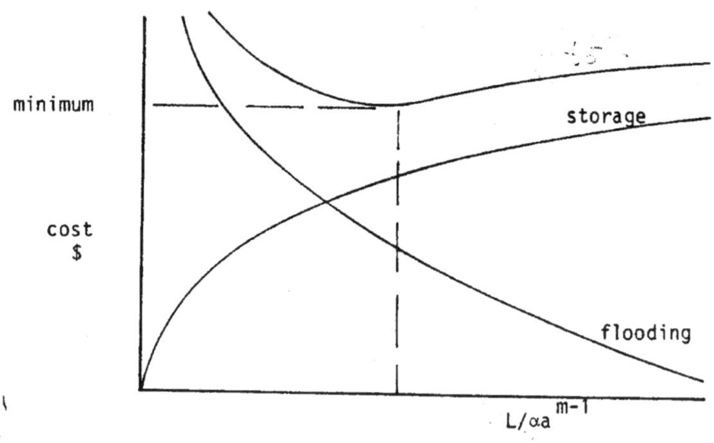

Figure 5.13 Optimum catchment storage volume

Note that infiltration after the rainfall stops is neglected in the above analysis. Inclusion of that effect would lower the i_e/a and s/a lines to the right, implying a larger $L/\alpha a^{m-1}$ is best. The model provides an indication of total storage in the system. The location (and volume) of storage could be further optimized using dynamic programming methods or by detailed modelling. It should be found generally that it is most economical to provide pond storage (m = ½) at the outlet, whereas channel or catchment storage (m = 5/3) is most economical at the head of the system.

5.4.2 Town planning

The detailed method of setting out a town and method of drainage can have a considerable effect on runoff control and retention. Figure 5.14 illustrates some good and bad practices. Some are in the hands of building control, such as ensuring stormwater is not discharged into sewers or even onto paved areas, but should be directed to pervious areas, especially in suburban development. In the case of dense urban areas underground facilities can be provided. Road layouts should avoid flow directly downhill, but care is also needed to avoid catching runoff in dips. Drains should have overflows leading to flood plains to absorb excess flow, due to extreme weather conditions or bad planning. Dual drainage systems assist in this since the major system can usually handle excess flood waters. And where land is available, detention storage or preferably retention basins should be provided. Open spaces therefore need to be planned at township conception stage not only for aesthetic reasons but also for acting as a balance in natural occurrences such as floods and droughts.

The drainage engineer needs to work together with the town planner to ensure a safe, economical infrastructure and a pleasing surrounding. Best management practice therefore goes beyond water quality control and flood control.

5.5 WATER QUALITY

Urbanisation has a detrimental effect on water quality. Wastes including sewage, storm runoff, seepage and leachates from landfills all pollute adjacent waters.

The quality of urban runoff is a particular problem in developing communities and poor towns. Litter, e.g. plastic bags, refuse and even unwanted household items, are washed into conduits and may block them or restrict their capacity. Screens are necessary at key points and these must be serviced regularly.

Overspill of sewage in overpopulated sites, e.g. squatter camps (Wimberley, 1992), is frequent if services have not kept pace with growth. Monitoring systems and awareness campaigns are necessary from the health point of view.

Even dissolved chemical loads, due to atmospheric washout, concentrations at first flush, and illegal chemical discharges reach severe proportions. The ratio of informal businesses to regulated industries results in a higher chemical concentration than is accepted in developed countries. Limitations on capital and inadequate policing make it almost impossible to catch up. It points to higher costs of providing potable water in the long run, since untreated stormwater finds its way to watercourses which are the source of water supplies.

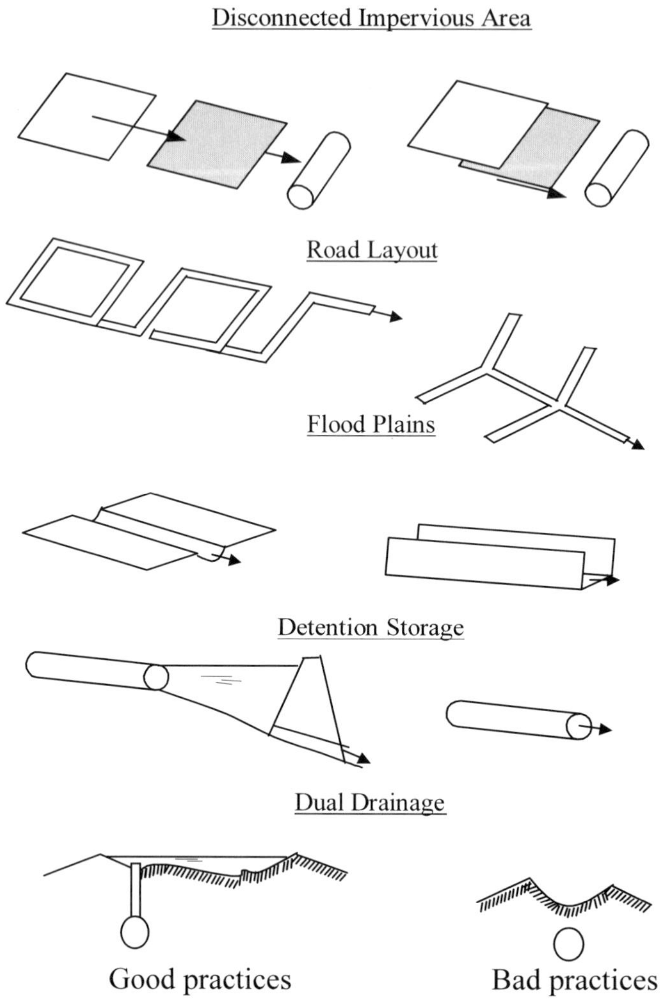

Figure 5.14 Alternative urban drainage practices

An emerging problem is the dense peri-urban development along river banks. This is often uncontrolled or informal housing and the communities may be unaware of flood dangers. Warning systems are not sufficient and unlikely to work years after installation. Channelization adjacent to townships becomes necessary, and to counterbalance this concentrating effect, detention structures are needed in the urban areas. The ideas of Parikh (1999) applied in India would seem most applicable, that is by providing adequate stormwater drainage in the peri-urban areas, flood water is drained more effectively from the city as well.

Further ideas for draining low cost peri-urban areas were suggested by Stephenson (2001). Instead of the fashionable separate sewer systems, it may be expedient to revert to combined sewer-stormwater systems in temperate countries. A third drainage system was suggested for poor areas, that is sullage drains to take the bulk of household drainage, plus tanker emptied toilets, and surface stormwater drainage. Low cost sullage drainage could be treated to a low level, but combining with stormwater which may require limited treatment is a fourth option, again for low rainfall areas.

REFERENCES

Bakele, G., Allen, M.D. and Argue, J. (1993). Recent developments in on-site stormwater management in Adelaide, S. Australia. *Proc. Intl. Conf. Storm Drainage*, Niagara.

Beard, L.R. and Chang, S. (1979). Urbanization impact on streamflow. *Proc. ASCE* 105 (HY6), June.

Brakensiek, D.L. (1967). Kinematic flood routing. *Trans Am. Soc. Agric. Engrs.*, 10 (3), 340-343.

Brooker, C.J. (1997). *Detention storage for control of urban stormwater runoff.* MSc(Eng) thesis, Univ. Witwatersrand, Johannesburg.

Colyer, P.J. (1982). The variation of rainfall over an urban catchment. *Proc. 2nd Intl. Cong. Urban Storm Drainage*, University of Illinois.

Huff, F.A. and Changnon, S.A. (1972). Climatological assessment of urban effects on precipitation at St. Louis. *J. Appl. Meteorology*, 11, 823-842.

Lloyd-Davies, D.E. (1905). The elimination of storm water from sewerage systems. Min. *Proc. Inst. Civ. Engrs.*, 164 (20), 41-67.

McEnroe, B.M. (1992). Preliminary sizing of detention reservoirs to reduce peak inflow. *J. Hydr. Eng.* Am.Soc.Civil Engrs., 118 (11), Aug.

Otterpohl, R., Grottker, M. and Lange. J. (1997). Sustainable water and waste management in urban areas. *Water, Science and Technology*, 3 (9), 12-33.

Parikh, H. (1999). Slum networking – an alternative way to reach the urban poor. Keynote lecture. *Intl. Cong. Urban Storm Drainage*, Sydney.

Ridgard, A.M. (1994). *Investigation into the effects of urbanization on critical design storm.* MSc(Eng) thesis, Univ. Witwatersrand, Johannesburg.

SAICE (1988). Westfleier and Atlantic Industrial stormwater management systems. *Civil Eng. in S.A.*, Feb.

Stephenson, D. (1981). *Stormwater Hydrology and Drainage*. Elsevier, Amsterdam.

Stephenson, D. (1984). Kinematic analysis of detention storage. *Proc. Storm Water Management and Quality users Group Meeting*, USEPA, Detroit.

Stephenson, D. and Meadows, M (1986). *Kinematic Hydrology and Modelling*, Elsevier, Amsterdam, 245pp.

Stephenson, D. (1989). Application of a stormwater management program. *Civil Eng. in S.A.*, June, 181-6.

Stephenson, D. (1997). *Research on monitoring the effects of catchment developments on urban runoff and water balance*. Exec. Summary, WRC report 319/8/97, Pretoria.

Stephenson, D. (2001). Problems of Developing Countries, in *Frontiers in Urban Water Management – Deadlock or Hope*, ed. C. Maksimovic and J.A. Tejada-Guibert, IWA Publishing for UNESCO.

Sutherland, F.R. (1983). *An improved rainfall intensity distribution for hydrograph synthesis.* Water Systems Research Programme, Report 1/1983, University of the Witwatersrand.

Watson, M.D. and Miles, L.C. (1982). Current stormwater drainage practice in S.A. *Civil Eng. in S.A.*, June, 251-7.

Weichers, A.U.S., Freeman, W.J. and Howard, M.R. (1996). *The management of urban impoundments in S.A.*, WRC TT 77/96. Vol. 1.

Whitlaw, R. and Brooker, C.J. (1995). The historical context of urban hydrology in Johannesburg. *Jnl. S.A. Instn. Civil Engrs.*, 37 (3).

Wimberley, F.R. (1992). *The effect of polluted stormwater runoff from Alexander Township on the water quality of the Jukskei river.* MSc(Eng) thesis, Univ. Witwatersrand, Johannesburg.

Wipple, W. (1979). Dual purpose detention basins. *Proc. Amer. Soc. Civil Engrs.*, 105 (WR2), 403-412.

CHAPTER 6

Groundwater

6.1 THE EXTENT OF GROUNDWATER

Of the earth's 10^9 cubic kilometers of water, less than 1% comprises fresh water, and of this, three-quarters is frozen in icecaps and one-quarter occurs in aquifers or groundwater. Only 2% occurs on the surface in rivers and lakes, i.e. the volume in fresh water in the ground is about 20 times the volume accessible on the surface. Groundwater is therefore not a small resource and indeed the fact that it is stored for years in aquifers makes the aquifer a huge storage system. The problems arise when we try to abstract groundwater because the aquifers are frequently not very permeable and many boreholes are needed to abstract large volumes from the ground. The long-term effects of abstraction need consideration, as well as the recharge rate.

A large proportion of rainfall infiltrates into the ground. Up to 90% of rainfall seeps into the upper layers of the ground but a proportion of this evaporates through plants, or by capillary action migrates to the surface and evaporates from the surface of the soil. There is still a big proportion however that reaches the water table. This water then may migrate slowly through the aquifer and some of it emerges in the form of springs or seepage in low lying or steep areas and eventually reaches the sea. There are however many aquifers which are confined or so remote from the sea or watercourses that the water remains there for thousands of years. In fact, there are large bodies of water deep below deserts which are now being tapped as a source of water. Present climates will not replenish this water for thousands of years if ever and presumably there were different climatic conditions when the water collected.

The scale of aquifers is enormous compared with the size of surface reservoirs we see. Aquifers can be hundred of metres deep and hundreds of kilometres wide. Thus both the time scale and space scale are orders of magnitude larger than for surface water.

Many communities rely on groundwater for drinking water and for agriculture. Often rural communities subsist on this water alone. Hand-dug wells or, nowadays, augured boreholes can provide water for villages who are used to living with minimal amounts of water. In other areas, whole cities receive their water from the ground. In karst areas, many cumecs (cubic metres per second) can be pumped from the aquifer and even replenished naturally from rain or melting snow. Perhaps less is written about groundwater because it is not visible. It needs sophisticated works to abstract it and it is expensive to retrieve. It is estimated that about 20% of human water requirements comes from groundwater.

It is difficult to measure groundwater parameters. Intensive borehole investigation and geophysical investigation can still only sample the system beneath the ground. Aquifers can be highly non-homogeneous and non-isotropic. Fissures and dykes may cross aquifers to isolate compartments or to cross-connect compartments. Layers of different permeability may result in confined aquifers and separate aquifers. The movement of water is generally controlled by the less permeable strata or layers whereas the volume of water contained is more influenced by the larger aquifers or the average storativity. Great strides have been made in recent decades by petroleum exploration companies in geophysical exploration. Satellite data collection and other types of remote sensing can provide vast amounts of data.

The majority of groundwater occurs in the pores of soils. Soil has around 20-30% of its volume as pores between the grains. The majority of this volume below the water table is occupied by water. However, there are often air pockets and air layers in aquifers and these can affect the permeability of aquifers considerably, particularly in the upper layers. As water permeates downwards, so air must travel upwards as it is displaced. The contraflow can obstruct the downward seepage of the water and evaporate water and transport it back to the surface. Capillary forces are strong in the ground and water may never reach the permanent water table if deep but may be held in surface layers to be drawn up by plants. Sometimes water is contained in dissolved spaces in the rocks such as in dolomitic rock. Alternatively, the rock may have had some of the minerals dissolved so that it is highly porous. The same often applies to loams. Material with big pores can be used to insulate aquifers, i.e. to prevent moisture evaporating from the surface of the soil.

Groundwater is seldom conspicuous, yet it generally makes up the bulk of water in any catchment. It can extend from the ground surface or from many metres below the surface, down to solid bedrock. Deeper groundwater may have existed there for thousands of years ("fossil water"). Groundwater nearer the water table (water surface) may be fresher and may be in a continual state of flux, the water surface rising with more infiltration from above, or falling due to abstraction by plants, evaporation or lateral migration through the aquifer. Although the rate of movement of groundwater is very slow (typically a few metres per year) the existence of groundwater is fragile. It may not easily be replenished if overdrawn. Pollutants could remain for hundreds of years. It may support plant and animal life in a state of delicate equilibrium and small changes in water table may drastically change the amount of water available at springs, or in the root zone of plants. The base flow of streams is often from aquifers and if streams dry up the surface population will be affected.

Figure 6.1 shows the relative proportions of water in different forms on earth. Groundwater makes up the bulk of the volume of fresh water available to us. However groundwater replenishment rate is low, 25% of precipitation gets to the groundwater table, the balance running off the land via rivers, or evaporating, or being used by plants. The rate of recharge of groundwater is only about 1% of its total volume per year.

More than 50% of the world's drinking water is taken from groundwater, because it is reliable, clean and ubiquitous. Large abstractions are also made for agriculture and industry. Some projects (e.g. The Great Man Made River in Libya) abstract at much greater rates than the replenishment rate, which is called groundwater mining.

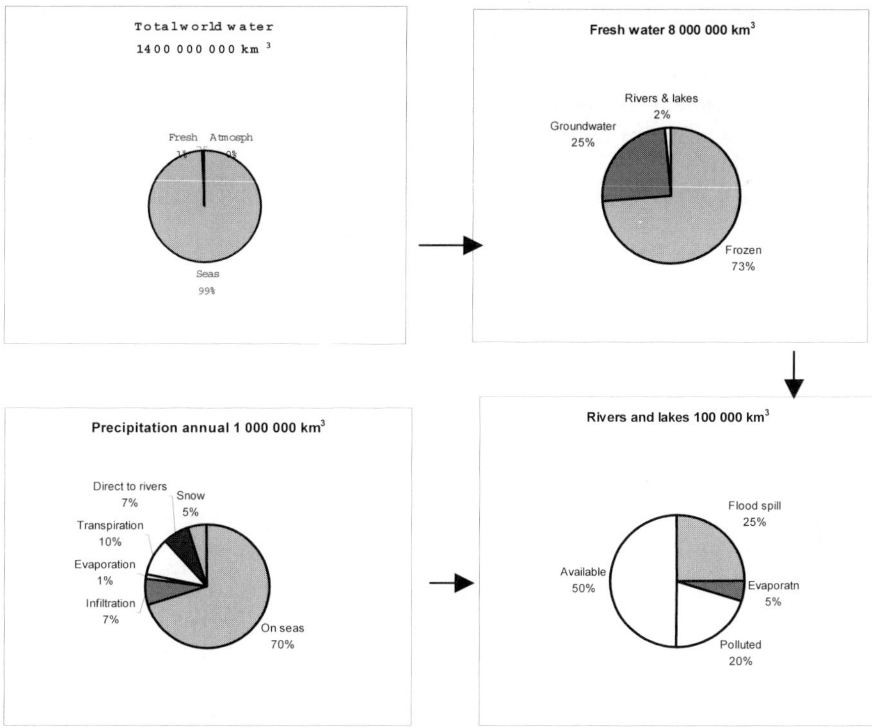

Figure 6.1 Distribution of the Earth's Water Resources

Users seldom consider the consequences of using groundwater, i.e.:

- Lowering of the water table
- Changes in catchment ecology
- Faster soil erosion
- Ground subsidence
- Poor transmissivity could require a lot of boreholes
- Steepening water table resulting in faster lateral flow
- Drawing in saline water, polluted water and other undesirable water.
- Due to the slow movement of groundwater changes are not readily noticed or remedied.

6.2 FLOW OF GROUNDWATER

Two important hydrogeological parameters affect the yield of an aquifer. One is the volume of water released per unit volume of aquifer per unit change in piezometric head, ie the storage coefficient or storativity. This can be as high as 30% for sand but is generally 10 to 20%. It is less than the porosity (ratio of void space to total volume). Not all the water held in the pores of soil is accessible. Some is held by capillary action or molecular attraction or chemical bonding, some is held in closed pores and some flows away. For confined aquifers, the storage coefficient, S, can be as low as 10^{-4}

indicating that large pressure changes are needed to produce water. The pressure could decline in the case of a falling water table in an unconfined aquifer, or be increased to obtain a yield from a confined aquifer.

The specific yield of the aquifer on the other hand is the amount of water released solely due to the influence of gravity. It is lower than the storage coefficient. Specific yield can be as low as 0.01 for an unconfined aquifer.

The second hydrogeological parameter of the aquifer is the transmissivity. This is the rate of flow per unit width of catchment for a unit hydraulic gradient. Expressed mathematically,

$$q=kih \tag{6.1}$$

where q is the flow rate in $m^3/s/m$, h is the depth of aquifer, i is the gradient of the free water table and k is the permeability. k can be as low as 10^{-10}m/s for clays and 10^{-3}m/s for coarse sand.

Alternatively we can write

$$Q=kiA \tag{6.2}$$

where Q is the total flow in m^3/s and A is the cross sectional area. This is known as Darcy's equation. The apparent water velocity is given by ki, but a drop of water actually travels faster than this as the open area between soil particles is less than the total. The water velocity or travel rate is actually ki/n where n is the soil porosity.

In addition, the extent of the aquifer, i.e. the distances to the boundaries, affects the amount of water available, as well as the recharge rates, whether from rain, rivers, neighbouring water bodies or artificially, e.g. from reclaimed wastewater. There may be discontinuities in the aquifer, e.g. dykes, or preferred routes, e.g. fissures. The aquifer may be layered or otherwise non-homogeneous. It may be non-isotropic, e.g. different transmissivities in different directions making abstraction complicated. All these characteristics have to be surveyed before an accurate picture of the aquifer can be built. Extensive exploration by drilling and geophysical methods is needed followed by laboratory tests and mathematical modelling.

A broad catchment water balance should be obtained before detailed modelling is undertaken. To do this, the catchment first needs defining. Remote imagery, topographical plans and geological maps are needed. It is easier to mark the surface watershed than the subsurface watershed. Ground water movement may be against the contour due to dips in the water bearing strata, or via porous zones or fissures.

Once the physical boundaries have been defined, inflows and outflows and changes in storage should be estimated. Precipitation, evaporation, surface runoff and subsurface flow should be measured by whatever means possible. Evaporation and groundwater flow are difficult parameters to measure. Generally an evaporation model is used and groundwater flow can be deduced by performing a mass balance.

It is only after the cross-boundary flows have been estimated that mathematical models can be used for more detailed study of movement within the aquifer. There are a number of groundwater flow and groundwater quality models available from companies, universities and on the Internet.

The cost of modelling is low compared with the cost of obtaining subsurface data. Average parameters are not sufficient. It is the impermeable barriers and permeable fissures or layers which control the subsurface flow rates and stored volumes. So dense borehole logging and testing are needed. Geophysical exploration of an aquifer can take years and even then there may be a degree of uncertainty concerning aquifer properties.

Non-homogeneities, and non isotropic properties are the bugbears. The exploration may take place in stages. Overall impressions may be obtained from sampling, and remote sensing. Satellite images can indicate topography and exposed obstructions, such as dykes. Thermal data will be useful for detecting moisture. Seismic, magnetic and resistivity surveys may indicate discontinuities. These tests are done by traversing the surface. The next stage may be boreholes and testing. In situ pumping tests will yield geohydrology and borehole logs will indicate soil profiles. Laboratory testing of samples taken from boreholes will provide more detailed information, i.e. geochemistry, porosity and soil structure. In situ geotechnical tests may also be carried out for densities and compressibilities.

6.2.1 Groundwater parameters

Haimes and Dreizin (1977) consider stream flow variations important for surface water balance, but precipitation and evapotranspiration as important for groundwater balance. The location and number of wells in relation to the recharge affect the local drawdown pattern. When dealing with groundwater, physical properties of the actual aquifer are very important. Nachtnebel (1994) lists parameters relating to the aquifer that are important for the modelling of complex groundwater systems:

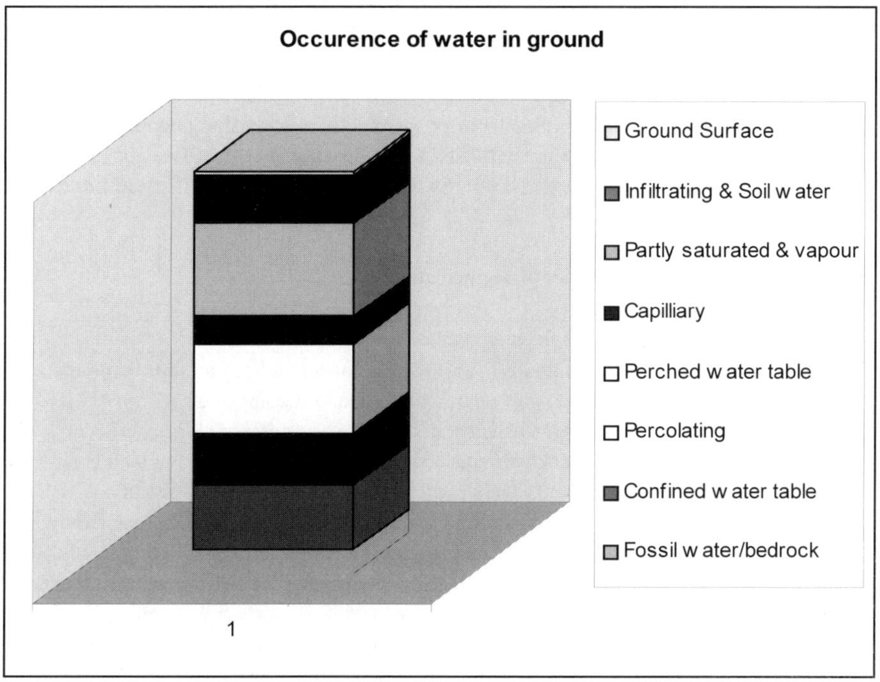

Figure 6.2 Distribution of water in the ground

- Spatial distribution of several layers of the aquifer;
- Permeability or transmissivity values throughout the aquifer;
- Storage coefficient of the confining layers;
- Specific yield of the aquifer;
- Thickness of the top layer of soil;
- Different soil types and their saturated permeabilities;
- Natural groundwater recharge and its spatial pattern (Das Gupta and Paudyal, 1988);
- Location and type of boundaries; and
- Location of pumping wells and the amount of pumping.

Land use connected with the aquifer also has to be considered, i.e.:
- Bottom level of dumping sites, if applicable (because of leaching);
- Possible locations of groundwater pollution from industry, mines, agriculture, etc.; and
- Spatial pattern of land use, e.g. agricultural areas, villages, ecologically valuable regions.

Gupta and Goodman (1985) present a table which classifies the components and variables of an augmentation scheme (Table 6.1).
Water in an aquifer moves under the influence of pressure or hydrostatic head. A gradient in the water surface causes water to flow from the higher head to the lower head. According to Darcy, the water flow rate is $Q = kiA$, where Q is the flow in cubic metres per second, i is the hydraulic gradient in metres per metre, k is the permeability in metres per second, and A is the cross-sectional area through which the water flows. The coefficient of permeability, k, can be as high as 10^{-2} m/s for gravel, changing to 10^{-4} m/s for sand, 10^{-6} m/s for silts and 10^{-8} m/s for clays.

Although the apparent flow velocity through a porous media can be calculated from the equation $v = ki$, in fact the wetted front proceeds faster than this because the porous area is less than the total cross-sectional area. Owing to stratification and non-homogeneity of the aquifer, water will emerge at different points at different times.

Table 6.1 Components and variables of augmentation

Component		Related variables
1.	Streamflow augmentation	Various hydrologic parameters related to time distribution of water supply demand and streamflows
2.	Groundwater pumping	Area extent and thickness of aquifer, storage coefficient, and hydraulic conductivity (transmissivity) of aquifer, water withdrawals, water levels, boundary conditions
3.	Induced infiltration	Thickness, storage coefficient and hydraulic conductivity or stream bed. If cutoff trench to reduce infiltration is incorporated, above parameters are also relevant for trench
4.	Recharge	Parameters of Item 3 and intake capacity of soil. Precipitation and wastewater disposal pattern in the basin are not considered quantitatively in the study

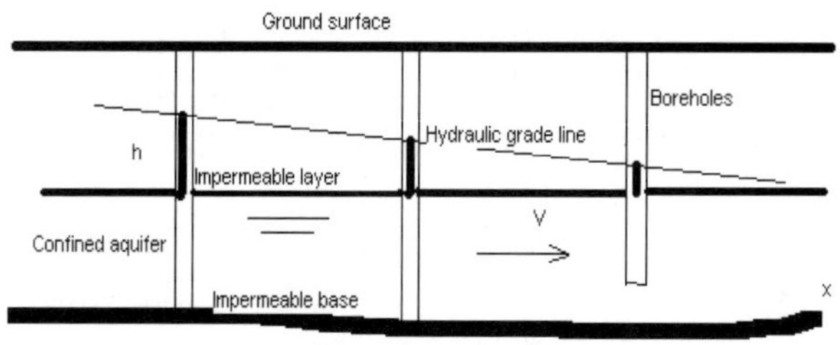

Figure 6.3 Steady unidirectional flow in an aquifer

This gives rise to a type of macro dispersion. The dispersion effect is important for groundwater pollution studies. It should also be noted that unsaturated flow can be much slower than when the aquifer is saturated and the effective dispersion coefficient can be higher.

6.3 GROUND WATER AND WELL HYDRAULICS

6.3.1 Steady radial flow to a well

When a well is pumped, water is removed from the aquifer surrounding the well and the water table or piezometric surface, depending upon the type of aquifer, is lowered. The drawdown at a given point is the distance the water level is lowered. A drawdown curve shows the variation of drawdown with distance from the well (see Fig. 6.3). In three dimensions, the drawdown curve describes a conic shape known as the cone of depression. In addition, the outer limit of the cone of depression defines the area of influence of the well.

Unconfined aquifer:
An equation for steady radial flow to a well in an unconfined aquifer can be derived with the help of the Dupuit assumptions. As shown in Figure 6.5, the well is assumed to completely penetrate the aquifer down to the horizontal impermeable base and a concentric boundary of constant head surrounds the well. The well discharge is given by the equation;

$$Q = 2\pi r K h \frac{dh}{dr} \tag{6.3}$$

which, when integrated between the limits $h = h_w$ at $r = r_w$ and $h = h_0$ at $r = r_0$, yields

$$Q = \pi K \frac{h_0^2 - h_w^2}{\ln(r_0 / r_w)} \tag{6.4}$$

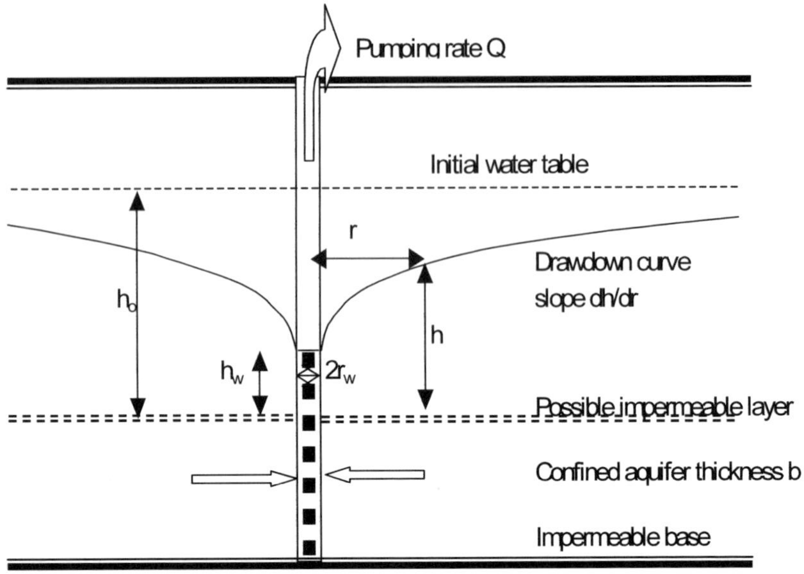

Figure 6.4 Radial flow to a well penetrating a confined aquifer

Because of large vertical flow components, this equation fails to describe accurately the drawdown curve near the well; however, estimates of Q for given heads are good. In practice, the selection of the radius of influence r_0 is approximate and arbitrary, but the variation in Q is small for a wide range of r_0. Suggested values of r_0 fall in the range of 500 to 1000 feet. These distances do not indicate the limits within which drawdown can be observed but, rather, they serve as approximations for practical application of the equation. Any pair of distances and heads also can be substituted in Eq. (6.4).

6.3.2 Unsteady radial flow to a well

When a well penetrating an extensive aquifer is pumped at a constant rate, the influence of the discharge extends outward with time. The rate of decline of head times the storage coefficient summed over the area of influence equals the discharge. Because the water must come from a reduction of storage within the aquifer, the head will continue to decline as long as the aquifer is effectively infinite; therefore, no steady state flow can exist. The rate of decline, however, decreases continuously as the area of influence expands.

From the differential equation which applies to this situation, in plane polar coordinates, becomes

$$\frac{\partial^2 h}{\partial r^2} + \frac{1}{r}\frac{\partial h}{\partial r} = \frac{S}{T}\frac{\partial h}{\partial t} \qquad (6.5)$$

where T is the coefficient of transmissibility (T = Kb, where b is the aquifer thickness) and t is the time since start of pumping

The non-equilibrium equation permits determination of the formation constants S and T by means of pumping tests of wells, or if the constants are known, the drawdown can be computed for a given well discharge. The assumptions required in applying equation (6.4) should be noted: the aquifer is homogeneous and isotropic and is of infinite areal extent; the well penetrates the entire aquifer; the well diameter is infinitesimal; and the water removed from storage is discharged instantaneously with decline of head.

Jacob method of solution:
Jacob (1950) noted that for small values of r and large values of t, u is small, so that the series terms in the expansion of Eq. (6.4) become negligible after the first two terms. As a result, the drawdown can be expressed by the asymptote

$$h_0 - h = \frac{Q}{4\pi T}\left(-0.5772 - \ln\frac{r^2 S}{4Tt}\right) \tag{6.6}$$

Rewriting as

$$h_0 - h = \frac{Q}{4\pi T}\left(\ln\frac{4Tt}{r^2 S} - 0.5772\right) \tag{6.7}$$

This reduces to

$$h_0 - h = \frac{2.30Q}{4\pi T}\log\frac{2.25Tt}{r^2 S} \tag{6.8}$$

Therefore, a plot of drawdown $h_0 - h$ versus the logarithm of t forms a straight line. From drawdown measurements in an observation well during a pumping period, $h_0 - h$ and t are known and can be plotted. The slope of the line fitting the data enables the formation constants to be computed.

6.4 GROUNDWATER MODELLING

Computer models for analysis of flow, water depth and quality have proved immensely valuable.

In the case of groundwater resources studies, a two-level approach for optimizing groundwater systems can be used; firstly optimizing for each local area and then optimizing the whole system. This tends to be an iterative exercise. Yu and Haimes' method (1974) involves a local authority optimizing its own sub-regional problem on the first level, whilst on the second level the regional authority optimizes the variables among the first level subsystems, thereby influencing the optimal solution for the sub-regions.

Haimes and Dreizin (1977) propose a model where only cost objective functions are considered. The cost functions are construction costs, pumping costs (including costs associated with steady state lift, drawdown in a well due to pumping from other cells in that cell, drawdown in a cell due to pumping from other cells), surface water supply costs, artificial recharge costs, and depletion of stream penalty (flow of water from a stream into a cell due to pumping from inside the cell or from other cells, and normal steady state flow from the stream into the aquifer cell). The sum of the cost functions is minimized. The constraints are concerned with minimum water requirements, drawdown, pumping capacity, recharge facility capacity, surface water supply upper limit and infiltration rate limits. Gorelick (1983) lists various modelling methods.

Gupta and Goodman (1985) also utilize the concept of two levels. Level 1 determines the best physical arrangement of wells within each aquifer unit by utilizing a simulation model. Level 2 establishes the policy for the net quantity of augmentation from the different cells in the system by means of a linear programming model.

The 3-dimensional differential equation of motion is;

$$\frac{\partial}{\partial x}\left(T_x \frac{\partial h}{\partial x}\right)+\frac{\partial}{\partial y}\left(T_y \frac{\partial h}{\partial y}\right)+\frac{\partial}{\partial z}\left(T_i \frac{\partial h}{\partial z}\right)=S\frac{\partial h}{\partial t}+bW(x,y,z,t) \tag{6.9}$$

where h is the hydraulic head; T_i are the transmissivity tensors along the x, y and z axes; S is the storage coefficient; b is the thickness of the hydrologic unit; and W(x,y,z,t) is the source/sink term.

The interaction of surface water and groundwater and the effect of abstracting groundwater at one point can be visualized in terms of the equations of flow. Many cases exist of over-exploitation of groundwater at one point in an aquifer resulting in dropping of the groundwater table somewhere else. Other damage can be caused. Water quality can change due to drawing in of water from hitherto confined pollution sinks. Often the hydraulic gradient is sufficient to suppress groundwater contamination.

Seawater intrusion is a classical example. Normally the interface between fresh water and seawater extends a few metres inland. It is maintained in this position by the continuous flow of fresh water from inland towards the sea. If the fresh water aquifer is tapped, it will draw down the hydraulic gradient and seawater can flow inland, with the resulting deterioration in the previously fresh aquifer water.

Not only hydraulic continuity needs to be considered, but also the geotechnical aspects associated with groundwater abstraction. By dropping the water table, the effective stresses in the soils are increased resulting in settlement. This can be damaging to structures and can result in cracking.

The so-called safe yield of a borehole depends on the way it is assessed. Borehole drillers often do a quick single well test and maintain a pumping rate for a certain period of time before quoting that as the yield of the borehole. Even a two or multi-borehole test may not test the entire aquifer. It could take days or months before the extent of the aquifer is reached in terms of drawdown and the yield of a borehole is reduced. It should also be noted that the sum of the yield of a number of boreholes in an aquifer is not the same as the yield of the aquifer in total because each well will affect the other.

By abstracting water from an aquifer, the water table will be lowered and the aquifer will dry out as well as the catchment as a whole, resulting in changes in surface cover. This may mean more water is required for gardening or alternatively there is a greater potential for infiltration when it rains. The result could be decreased surface runoff. It is important to do overall mass balances for the catchment, or preferably the aquifer as a whole, before deciding safe yields and the effect of any abstraction should be modelled and discussed with all those who could be affected by changes in groundwater.

When aquifers are used for water supply, then they could be recharged artificially because of the greater storage capability. Either surplus river water could be used to flood fields or treated wastewater could be discharged into the aquifer. There are many such artificial recharge basins at various stages of testing. While it is recognized that there is some degree of purification as wastewater is discharged into the aquifer, the long-term effects have not been reported on convincingly yet.

6.5 GROUNDWATER POLLUTION

It is often only many years or decades after contamination has occurred that the effects are felt. Groundwater moves so slowly and dispersion occurs so that it is difficult to pinpoint when many contamination effects have been caused. Nevertheless, there is a growing awareness of some of the extreme effects that waste disposal, including toxic wastes, have had on potential water resources. Some aquifers have had to be abandoned and others extensively remedied followed possibly by years of no abstraction. A lot of money being spent on remediation of groundwater in Europe and the United States and this has led to improved assessment of aquifer properties and modelling techniques.

Groundwater pollutants have a number of factors affecting their movement or change or dispersion. These include:

- Advection or lateral transport by moving water. This could be one-directional, two-directional or even three-directional.
- Gravitational movement of contaminant in suspension in water of a different density. There may occur advection with the water or there may be a separation and density current effect.
- Salinity differences causing osmotic and gravitational gradients. Alternatively there may be mixing across interfaces.
- Diffusion or dispersion, i.e. mixing due to molecular, or more likely turbulent motion of the transporting medium, water.
- Differential movement in different layers of aquifer. Slight differences in horizontal layers can divert most flow to the more permeable layers. And Impervious barriers can divert the direction of flow.
- Preferential movement in different directions due to non-isotropic nature of aquifer.
- Lateral movement, i.e. across streamlines due to mixing.
- Alternative phase, i.e. vapour movement. Water vapour may escape with upflowing air displaced by downflowing water.
- Adsorption onto soil particles, by roots or insects.
- Reaction with soil, rock or other material encountered.
- Obstructions such as boulders, dykes or air pockets.
- Separation of pollutants, by vaporization, solidification or gravitation.

In the past there was indiscriminate discharge of wastes into pits, wells waterways and seas. It is largely the pollutants in the groundwater which remain, and until extensive remediation is performed many groundwater sources are rendered unusable. There also exist waste disposal deposits waiting to reach groundwater. Lack or deterioration of linings, or slow movement have delayed contaminants from reaching aquifers. The removal or neutralization of ground deposits is more difficult than for surface deposits.

Types of pollutants include:

- Chemicals especially toxic ones
- Hydrocarbons
- Organic chemicals, e.g. chlorinated hydrocarbon
- Radioactivity
- Heavy metals such as mercury which concentrate in organisms
- Colourants, taste, temperature and odour affecting

Some of these pollutants have lives of hundreds of years. Concentrations of volatile organic compounds or heavy metals of a few parts per billion are dangerous, whereas concentrations of some chemicals such as chloride can reach 100 mg/litre without danger. Pollutants endanger lives in many ways. They may be poisonous, cause cancer, nervous disorders, heart, kidney or liver problems, lung infections, skin ailments, affect sexuality and fertility or induce breathing problems. Although our primary concern is human health, there is also danger to fish, animals, insects, birds and vegetation. The soil and water can become unsuitable for crops or livestock. There may be warning signs from fauna before humans notice. Fish are particularly sensitive to water quality.

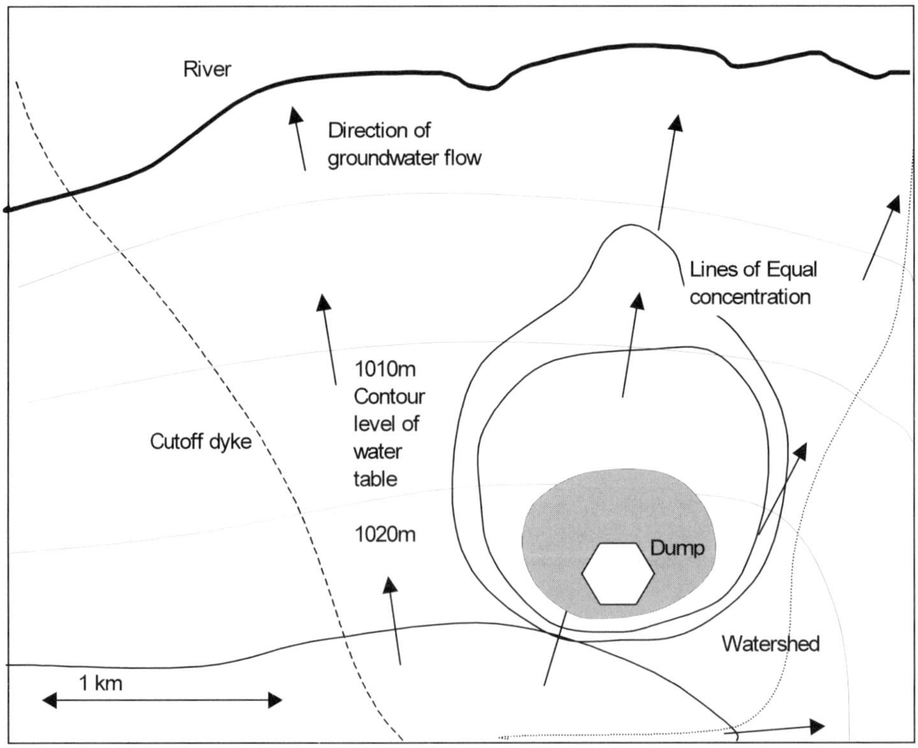

Figure 6.5 Map of groundwater pollution plume

The nature of groundwater pollution from surface sources is complicated and difficult to model. Pollution from the atmosphere or from wastes, both solid and liquid discharged onto the ground surface will partly infiltrate into the ground. Some of this pollutant may seep to the groundwater table or a perched water table, to eventually be abstracted or flow laterally to emerge far away. Some polluted surface water runs off to streams or drains directly. Some of the polluted water evaporates or is taken up by plants to transpire. The latter processes have the effect of concentrating contaminants in the upper ground layer. The contaminant may decompose in the sunlight, or may be adsorbed onto soil particles or be caught in the voids. Much of the previously dissolved salt or other compounds remain in a dry or semi-dry form, until they can be re-dissolved by more water. This may occur during the first rains of the season. There may then occur a concentrated washoff, the first flush effect.

Sources of groundwater pollution include:

- Industrial deposits in landfills, pits and containers
- Municipal landfill
- Septic tanks, pit latrines, soakaways
- Decomposing organic matter on the land surface
- Agricultural pesticides, insecticides, fertilizers
- Illegal dumping and discharges
- Mining and mine waste dumps
- Petroleum and gas industry
- Seepage from polluted rivers
- Seawater intrusion
- Geochemical leaching
- Runoff from cities
- Atmospheric fallout and washout

Pollutants dissolve in water and get transported into the ground. There may be a lag between surface pollution and groundwater pollution, due to evaporation, subsequent re-wetting, time needed to dissolve pollutant and travel time There is some treatment of pollution in the ground provided it is not overloaded. Filtering, aeration, adsorption, reaction, microbial action and retention all act to improve water quality in general.

There are links between groundwater and surface water. Pollutants can emerge or disappear depending on the direction of movement, whether it is infiltration, gravitational or pressure flow, or emergence at springs or into wetlands or watercourses.

Remediation of groundwater quality is difficult and expensive (orders of magnitude greater than abstracting water). The ways of remedying contaminated groundwater or neutralizing the pollution source include:

- Injecting neutralizing chemicals
- Injecting steam
- Changing state e.g. vaporization
- Abstraction , treatment and recharge
- Treatment of only water abstracted for use
- Excavation and treatment
- Isolation by membranes, slurries
- Cutoff trenches
- Capping to stop infiltration
- Solidification

- Diversion by barriers

After research has developed and proved a method it is applied on large-scale pilot tests, then in practice. An intensive monitoring system is installed around the test site. There are at least 3 scales of monitoring. The other two include sampling of water abstraction points and the third is on a regional scale to detect diffuse pollution from the surface. The depth of sampling needs to be established by modelling to consider drawdown effects. The sampling period and interval also need to be established by modelling and initial courser monitoring.

6.5.1 Dispersion

Dispersion is the mixing of a pollutant across an interface. It occurs on a macro scale due to various reasons. On a micro scale molecular diffusion occurs due to Brownian movement of molecules. On a larger scale advection of the fluid creates velocity gradients so that some particles move faster than others, and there is also lateral and vertical mixing between the layers of liquid due to turbulence or preferential flow paths. The largest mixing effect is generally due to different permeabilities between layers through which the water travels. A highly permeable layer will yield some water rapidly and other slow moving water may arrive much later, or not at all.

The disposal of wastes in porous media requires estimation of dispersion in the medium. Disposal or storage of wastes in the ground in arid areas has been undertaken. These include hydrocarbons, radioactive wastes and chemical wastes. Movement of contaminant is less likely in dry soil than wet. Groundwater contamination is also unlikely to be a problem as the water table is usually deep, and the groundwater is used to a limited extent. Arid areas pose less of a danger as population densities are low, there is less likely to be objections, land is cheap and There is likely to be less environmental effect. In addition movement by animals and birds is less likely. Of concern is the exploitation of poorer peoples lands for waste disposal. Often arid areas are poor and the communities may be unaware of the consequences of accepting wastes.

Absence of precipitation and groundwater are not the only requirements with respect to minimizing seepage. A low aquifer permeability is also desirable since in case of local contamination water movement should be minimized. Ideally the permeability should be associated with a high storativity in order to capture any seepage which may occur and minimize its movement laterally and vertically.

The storage capacity of an unsaturated aquifer can be high, in fact nearly equal to the porosity. The voids between the soil particles are initially filled with air, and this is displaced gradually when water pervades the aquifer. The air is displaced slowly as water permeates the aquifer. The upward flow of the air counters the downward flow of water, further reducing the effective permeability. Even as the groundwater pressure increases hydrostatically, some air pockets are trapped between soil particles and by surface tension or downward drag of the water, which diminishes the permeability considerably in comparison with a fully saturated medium. Often the air bubbles will disappear only when dissolved into the water permeating downwards.

The dispersion of an aqueous discharge into an unsaturated or dry aquifer is particularly complex (Stephenson and De Jesus, 1885). Flow of the liquid is influenced by gravity, the directional properties of the aquifer, surface tension and the release of air. There is considerable mixing at the front of the permeating liquid, even more than could be predicted by diffusion or even hydrodynamic dispersion. Figure 6.6 depicts dispersion in a homogeneous medium and Figure 6.7 in a real aquifer. Figure 6.8 shows the theoretical variation at a point with time for one-dimensional flow.

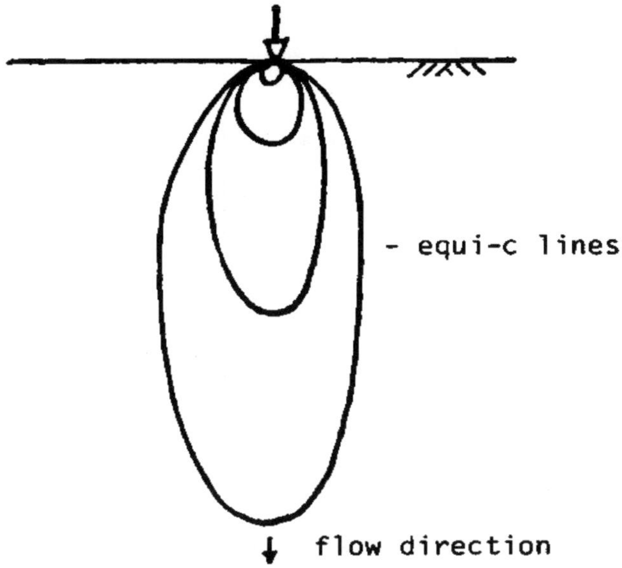

Figure 6.6 Dispersion in a homogeneous medium

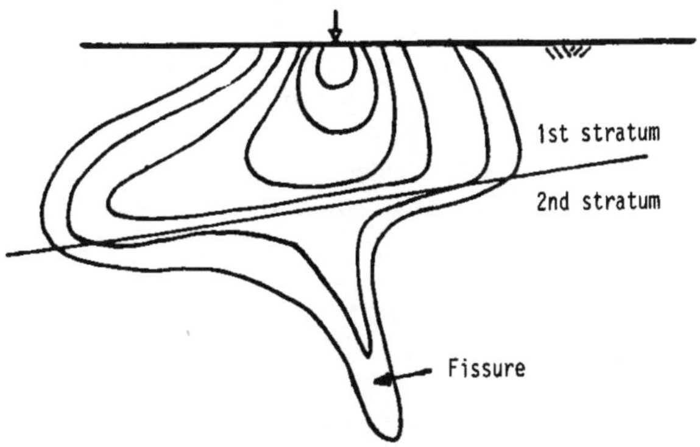

Figure 6.7 Dispersion in a real aquifer

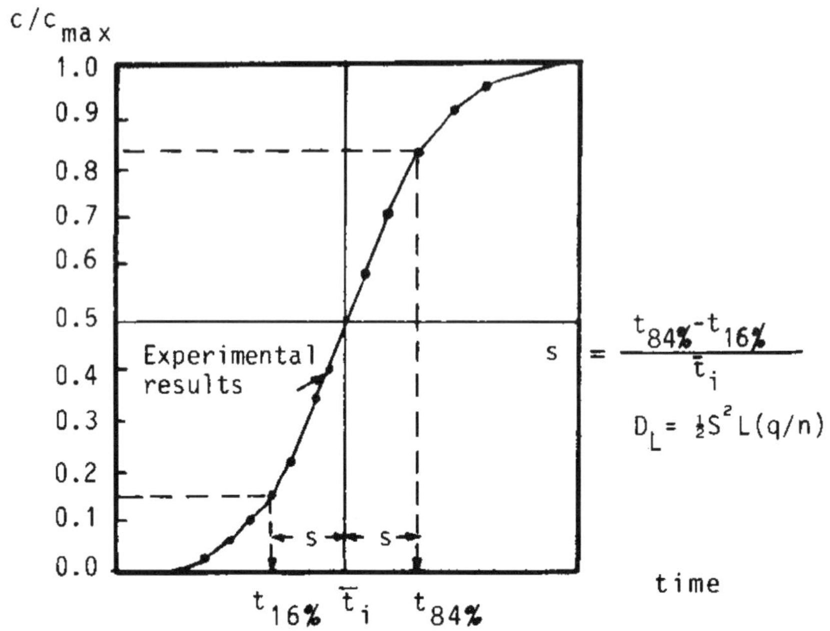

Figure 6.8 Computing dispersion coefficient from a one-dimensional dispersion experiment

It is generally recognized that the dispersion coefficient is a function of the flow velocity. Bear (1972) proposes a coefficient designated the dispersivity which is a ratio of the dispersion coefficient to flow velocity. Even the dispersivity is a function of many other factors and must be determined under site conditions to obtain a realistic estimate of it. Many methods have been developed to assess the dispersivity of saturated aquifers especially in the case of horizontal flow (e.g. Lenda and Zuber, 1970; Meyer et al., 1981). Study of flow in unsaturated aquifers is complex (Stephens and Neuman, 1982), and the mathematical analysis equally problematic (e.g. Narsimhan and Witherspoon, 1977).

Lenda and Zuber (1970) present the following statistical model for concentration of a tracer moving at a velocity v in the x-direction, at any point x, y, z, at time t:

$$C(x, y, z, t) = m(4\pi D_x t)^{-1/2} \exp\left[\frac{(x-vt)^2}{4D_x t}\right](4\pi D_y t)^{-1/2}$$

$$\text{x} \exp\left[\frac{-y^2}{4D_y t}\right](4\pi D_z t)^{-1/2} \exp\left[-\frac{z^2}{4D_z t}\right] \tag{6.10}$$

where m is the mass of tracer injected and D_x, D_y and D_z are the dispersion coefficients in the x, y, z directions respectively. D is assumed to equal av where 'a' is the dispersivity (approximately a constant).

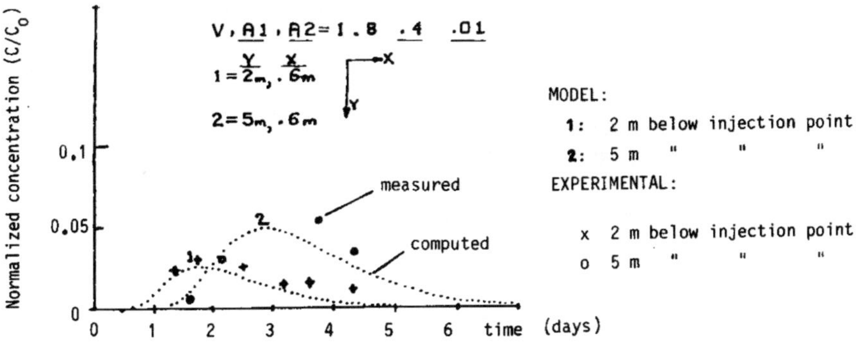

Figure 6.9 Correlation between probe data and model results

When considering continuous injection, if m is proportional to the volume of water injected, which in turn is proportional to x^3 if it expands and saturates in three directions in a homogeneous medium, then:

$$\frac{C}{m/x^2} \propto \exp^3\left(-x^2/4Dt\right)/\left(4\pi Dt/x^2\right)^{3/2} \tag{6.11}$$

This function has a maximum at $x^2/Dt \cong 1.2$. Thus, for D = ac = 0.1m x 1m per day, and t = 10 days, then x ≅ 1m, which is the proposed distance between observation hole and injection hole, assuming vertical flow.

Freeze and Cherry (1979) present equation (6.10) as the analytical solution for the concentration of activity at a point x, y, z in a medium at time t. The equation applies to an instantaneous point-source injection with flow velocity v in the x-direction and no adsorption. x, y, z are orthogonal directions relative to the point of injection. For a partly saturated aquifer an instantaneous injection may not travel far. It would gravitate downwards and spread until the surface tension was sufficient to suspend the cloud of water in the vardose zone (equivalent to adsorption of the liquid).

6.5.2 Finite difference solution

A similar curve for C(x, y, z, t) was obtained from a numerical simulation of the dispersion equation with specific boundary conditions. This method was more tedious than the analytical solution described above but was able to account for the boundary configuration at the well. It was also used to provide a solution under condition of continuous injection of water whereas the analytical solution is for an instantaneous injection.

The general equation for convection-dispersion in a three-dimensional medium, with steady uniform flow (not varying in time or space) with an adsorption term, is (Freeze and Cherry, 1979):

$$\frac{\partial C}{\partial t}=\frac{w}{n}\frac{\partial s}{\partial t}+D_x\frac{\partial^2 C}{\partial x^2}+D_y\frac{\partial^2 C}{\partial y^2}+D_z\frac{\partial^2 C}{\partial z^2}-v_x\frac{\partial C}{\partial x}-v_y\frac{\partial C}{\partial y}-v_z\frac{\partial C}{\partial z} \tag{6.12}$$

where v is the linear velocity q/n, q is the flow per unit area of medium, n is the porosity, w is the bulk density of medium and s the mass of chemical adsorbed per unit mass of medium. For flow along the x direction, with no adsorption and lateral homogeneity ($D_y = D_z$), then on any diameter:

$$\frac{\partial C}{\partial t}=D_x\frac{\partial^2 C}{\partial x^2}+D_r\frac{\partial^2 C}{\partial r^2}+\frac{D_r}{r}\frac{\partial C}{\partial r}-v_x\frac{\partial C}{\partial x} \tag{6.13}$$

For a numerical solution at x_m, r_n, t_i (Fried, 1975):

$$\frac{\partial C}{\partial t}=\frac{C_{m,n,i+1}-C_{m,n,i}}{t_{i+1}-t_i} \tag{6.14}$$

$$\frac{\partial^2 C}{\partial x^2}=\frac{C_{m+1,n,i}+C_{m-1,n,i}-2C_{m,n,i}}{(x_{m+1}-x_m)^2} \tag{6.15}$$

A two-pass solution is preferred, i.e. the advection term is first accounted for and then the two dispersion terms in Eqn. (6.13) are solved. It is necessary that $\Delta x/\Delta t = v$ for true advection.

By comparing measured concentration fluctuations at two depths in the observation holes with those obtained from a numerical solution to Eqn. (6.13) or (6.10), the values of downflow velocity and dispersion coefficients could be obtained by deduction. It was necessary to revise the values for v_x, D_x and D_r successively until a satisfactory fit was obtained.

The value of v_x could be estimated fairly easily by observing the time taken for C to reach 50% of its equilibrium value at any depth. Then v = x/t. The rate of increase over the rising C curves is largely a function of D_x, and the final equilibrium values are a function of D_r. It was also theoretically possible to distinguish between the two orthogonal horizontal directions along which the two sets of observation holes had been drilled

6.6 GROUNDWATER PROTECTION

Discharge of wastes into watercourses and onto the ground is easy and tempting. Until the consequences in the way of penalties are enforceable and high enough it will continue. There are many ways of controlling discharges of pollutants, some of which are more effective than others (Travis and Etnier, 1984):

- Legislation on concentration, amount and location of discharges
- Legal action
- Patrolling and punishment
- Financial contribution by industry towards cleanup
- Insurance

- Remediation
- Standards for drinking water quality
- Monitoring and feedback
- Remote sensing
- Closure of wells
- Zoning and classification
- Zero discharge systems for industries
- Establishment of national water quality agencies and staff
- Fund research in remediation
- Subsidize effluent collection and treatment.

Whatever steps are taken to protect groundwater will cost less than later remediation. And the long-term emergence of the problem means that we may yet realize the consequences of past indiscriminate dumping. Until treatment and zero discharge methods are feasible, the establishment of safe controlled disposal sites must be a reality. The establishment of dangerous waste storage and handling sites requires intensive environmental impact studies, testing and negotiation. Methods of lining ponds to prevent loss include plastic sheets, bentonite clay and concrete. But research into treatment of the wastes should be proceeding at a greater rate. Industry needs incentives to investigate methods of neutralizing, fixing or concentrating chemicals.

REFERENCES

Bear, J. (1972). *Dynamics of Fluids in Porous Media*. Elsevier, New York, N.Y., 764 pp.

Das Gupta, A. and Paudyal, G.N. (1988). Estimating aquifer recharge and parameters from water level observations. *J. Hydrol.*, 99, 103-116.

Freeze, R.A. and Cherry, J.A. (1979). *Groundwater*.Prentice-Hall, Englewood Cliffs, 604 pp.

Fried, J.J. (1975). *Groundwater Pollution*. Elsevier, Amsterdam, 330 pp.

Gorelick, S.M. (1983). A review of distributed parameter groundwater management modelling methods. *Water Res. Research*, 19 (2), 405-419.

Gupta, R.S. and Goodman, A.S. (1985). Ground-water reservoir operation for drought management. *Jnl. Water Resources Planning and Management*, 111 (3), 303-320.

Haimes, Y.Y. and Dreizin, Y.C. (1977). Management of groundwater and surface water via decomposition. *Water Ress. Research*, 13 (1), 69-77.

Jacob, C.E. (1950). Flow of ground water. In *Engineering Hydraulics*, ed. Rouse, H., J. Wiley and Sons, New York, 321-386.

Lenda, A. and Zuber, A. (1970). Tracer techniques in groundwater experiments. Proc. Symp. Isot. Hydrol., Int. Atomic Energy Agency, Vienna, 619-641.

Meyer, B.R., Bain, C., De Jesus, A.S.M. and Stephenson, D. (1981). Radiotracer evaluation of groundwater dispersion in a multi-layered aquifer. *J. Hydrol.*, 50, 259-271.

Nachtnebel, H.P. (1994). A decision support system for iterative groundwater modelling and management. Multicriteria decision analysis in *Water Resources Management*, UNESCO, Paris, 283-302.

Narasimhan, T.N. and Witherspoon, P.A. (1977). Numerical model for saturated-unsaturated flow in deformable porous media, I. Theory. *Water Resour. Res.*, 13 (3), 657-664.

Stephens, D.B. and Neuman, S.P. (1982). Vadose zone permeability tests, I; Summary, II; Steady State Results, III; Unsteady Flow. *Proc. Am. Soc. Civ. Eng. (ASCE)*, 108 (HY5) papers 17058-17060, pp. 623-639, 640-659, 660-677.

Stephenson, D. and De Jesus, A.S.M. (1985). Estimate of dispersion in an unsaturated aquifer. *J. Hydrol.*, Vol. 81, 45-55.

Todd, D.K. (1959). *Ground Water Hydrology.* John Wiley & Sons Inc., USA/Toppan Co. Ltd., Japan.

Travis C.C. and Etnier EL (1984). Groundwater Pollution: Environmental and Legal Problems. Amer. Assn. Advancement of Science, Washington.

Yu, W. and Haimes, Y.Y. (1974). Multilevel optimization for conjunctive use of groundwater and surface water. *Water Ress. Research*, 10 (4), 625-636.

CHAPTER 7

Water Quality and the Environment

7.1 WATER POLLUTION

Poor water quality is the greatest threat to water availability. Were it not for salinity the oceans would be the source of unlimited water. If industrial and domestic wastewater discharges were unpolluted the water could be recycled and this would reduce demands from fresh water sources by 75%, for that is the ratio of wastewater discharge to fresh water use. And in many situations a relatively small discharge of polluted water can contaminate huge volumes of fresh water so the pollution problem multiplies. On the other hand nature has a capacity to purify water to a limited extent and we should head the limits to enable this facility to be used to its optimum.

Pollutants in water can be classified as:

- Suspended solids – floating debris, settleable sediment or buoyant, e.g. algae.
- Dissolved chemicals – many cations, anions, with different effects, e.g. on health, equipment or toxicity.
- A quality unrelated necessarily to dissolved or suspended solids, e.g. temperature, acidity, colour, odour or taste. The latter are often associated with matter in the water, however.
- Bacteria, pathogens, viruses, organic micropollutants.

Pollution can be from many sources:

- Natural, e.g. geochemical, erosion, decay
- Atmospheric, e.g. ionization, windborne, smoke, fumes
- Agricultural, e.g. fertilizers, weed killers, decaying crops, animals, increased erosion
- Industrial and mining, e.g. wastes (solid and liquid), spoil, storage
- Human and domestic e.g. sewage, wastes, litter
- Urban, e.g. vehicles, energy generation, wastes, partly or untreated.

The effects of pollution manifest in degraded environment, animal and human health, industrial corrosion and other problems, increased water treatment costs and aesthetics. Unless developers take heed of pollution possibilities they will be subject to harassment and legal action. There appears an increasing environmental awareness in developed countries, but undeveloped countries often cannot afford to tighten up on pollution control.

7.1.1 Public health

The probability of contracting a water-related disease is statistically related to the concentration of pathogens or viruses. Improved laboratory techniques have resulted in the identification of harmful organic and inorganic matter, These may occur in water sources naturally but also from industry and agriculture. Also of concern to health are heavy metals such as mercury, cadmium and lead as well as pesticides such as DDT and carcinogenic compounds.

Diseases may also be transmitted by water based parasites, e.g. snails, and vectors, e.g. mosquitoes, especially in the tropics.

Pathogenic bacterial related diseases include cholera, typhoid and dysentery, while viruses, which are smaller and require a host, include hepatitis. It is difficult to detect bacteria in water, so evidence of excreta is sought and indices such as *E. coli* are used.

Water related diseases may be grouped as follows:

a) *Waterborne diseases*:
Waterborne diseases are those which are mainly spread through contaminated drinking water. The main infecting organisms are bacterial (*vibrio cholerae, salmonella typhi, shigella*), viruses (*hepatitis A, orbiviruses,* and *enteroviruses*) and protozoa (*giardia lamblia, histolytica*). Contamination of the water occurs through faecal matter entering the water – that is, through poor hygiene and sanitation.

b) *Water-washed diseases*:
These diseases are mainly infections which can be significantly reduced by an improvement in domestic and personal hygiene. They depend on the quantity of water that is available, rather than on the quality. All diseases with faecal-oral transmission fall into this category – such as typhoid and cholera. Others are skin and eye diseases, such as skin sepsis and trachoma, and infections carried by parasites on the skin surface, such as lice. Most of the intestinal worms also belong in this group, including roundworm, threadwork, whipworm and pinworm.

c) *Water-based diseases*:
In the case of water-based diseases, the pathogen has to spend part of its life cycle in the water. The best known of these is *Schistosomiasis* (Bilharzia). It is a water-contact disease that has infected many millions of people in the tropics. It is spread through schistosome eggs in human excreta, which hatch on reaching water. The resultant larvae invade suitable snail hosts and multiply.

d) *Water-related insect vectors:*
Water provides the environment necessary for the development of many insects that transmit diseases. Malaria is a water-habitat vector-borne disease, certain mosquitoes

being the host. Other such diseases are filariasis and elephantiasis (also transmitted by mosquito), and onchocerciasis (transmitted by the black fly).

e) *Water shortage*:
Dehydration, malnutrition and dirty eating conditions are due to shortage of water.

Apart from diseases, many other factors in water can render it unsafe. High dissolved salts can lead to high blood pressure. Alkaline water can affect the arteries and cause skin disorders. Heavy metals and some organic pollutants concentrate in the life chain and can reach lethal proportions. The factors causing cancer are only beginning to be understood.

Then there are physically dangerous factors: flooding, drowning, land slides due to groundwater pressures, pipe bursts, dam break, wave action, temperature and drought.

Water must be acceptable to the users. Aesthetic appeal, odour and taste are largely the basis that individuals use to determine acceptability. Thus limits for colour and turbidity have been included in all standards. The presence of excessive amounts of dissolved salts or water that is too acidic or alkaline influence odour and taste.

7.2 WATER QUALITY STANDARDS

A water quality standard or criterion provides the quantification of indicators of pollution and may further tabulate the method(s) used to detect the indicators. These indictors are used for virological, biological, radiological, physical and chemical examination of water, primarily used for domestic purposes. In addition to appropriate sampling for laboratory determination of quality, a survey of the sanitation facilities associated with the water use is desirable. The World Health Organization (WHO, 1984) endorses sanitary surveys on the basis that a complete knowledge of the conditions, at the sources of supply and in the distribution system, is required to carry out a bacteriological or chemical examination. The standards may be different for non-domestic water, but wherever human contact is possible monitoring and assessment are advisable.

Ability to determine the quality of water or conversely to detect pollution has improved dramatically. Apart from rapid laboratory tests there are available probes for direct readout of many parameters. Bacteriological quality is measured using existing determinands such as total coliforms plus the detection of viruses. Besides the measurement of total dissolved solids (TDS) for aesthetic and industrial reasons, the detection of heavy metals and hydrocarbons, which are carcinogenic, is becoming more significant with the intensification of industrialisation. The detection of toxic organic chemical compounds is assuming greater importance in developed countries because of effluent re-use. The reaction of living organisms in the water also reveals pollutants. Changes in micro organisms, or mutations, or in the extreme, death of fish, have been of great use in detecting changes in water quality. New substances are continually being added to the lists of determinands of concern to humans and ecology. Of greater importance now is the desire to maintain ecosystems in original condition. But even this goal may be transitory. Over centuries there have been changes in nature despite man. So change in itself is not to be shunned, it is the consequences of change which need impassionate evaluation.

Table 7.1 Guideline values for physical/chemical water quality parameters

Determinant	Unit	Recommended	Maximum level for insignificant risk	Maximum level for low risk
Turbidity	NTU	1	5	10
Fluoride	mg F/ℓ	1	1.5	2
Nitrate	mg N/ℓ	6	10	20
Iron	mg Fe/ℓ	0.1	1.0	2.0
Hardness	mg $CaCO_3$/ℓ	20 - 300	650	1300
Conductivity	mS/m 25°C	70	300	400

A comparison of the international approach to establishing water quality shows that there are essentially two approaches, namely enforceable standards and guidelines. The United State of America enacted a Safe Water Drinking Act in 1974 (amended in 1986) which required the USEPA to establish national primary drinking water regulations along with the identification of enforceable maximum contaminant levels and non-enforceable maximum contaminant level goals.

The European Economic Community (EEC) in 1980 adopted a set of standards that address maximum admissible concentrations of water contaminants. The WHO recognized that uniform water quality standards could not practically be applied throughout the world. However, it noted the need for guidance to regulatory agencies on water quality to ensure maintenance of good health. It decided on, and in 1984 published, drinking-water quality guidelines to be used as a basis for the development of standards in each country. The WHO also stated that the judgement of acceptable risk levels would be undertaken by society as a whole. Therefore, the adoption of the proposed guidelines is for each country to decide. The guidelines were developed scientifically assuming lifelong consumption and that specific geographic, socio-economic, dietary and industrial conditions would also have to be considered. Guideline values for the above physical/chemical water quality parameters are given in Table 7.1.

7.3 STORMWATER POLLUTION

Stormwater runoff from urban catchments in particular contains a surprisingly high pollutant concentration even if no sewage or wastewater is discharged into the system. Setting aside combined sewage-stormwater systems, the source of contaminants ranges from material precipitated from the atmosphere (dust or in rain) to seepage from waste tips or industrial zones. There may be discharge of wastes from factories and commercial concerns which, intentionally or not, finds its way to stormwater drains. This type of pollution is referred to as point-source, whereas natural spread inflow is referred to as non-point source pollution. Animal faeces, garden fertilizer, soil erosion, motor vehicles (oils and rubber from tyres) and decaying vegetable matter are some of the known sources. Litter deposited on streets, wastes in contact with water, and runoff or leachate from managed waste sites all cause pollution.

The concentration of sulphates, nitrates and suspended solids in rain, is not inconsiderable. Increasing attention is being focussed on the acidity of rain – caused by fumes from traffic and industry. Runoff from rural catchments is also frequently contaminated. Fertilizers and decaying vegetable matter are frequently the source in agricultural areas, and salts leached from the ground may contaminate water in mineral-rich areas.

The rate and amount of pollution of streams due to incoming stormwater can vary widely. The intensity of rain will affect the rate of transport. The 'first flush' is known to bring down most of the pollutants. In fact, many pollution models assume an exponential decay rate in the pollution washoff through a storm. The Environmental Protection Agency (EPA) (1971) in the USA indicated a storm depth of 12.5mm would remove 90% of road surface particles.

Table 7.2 indicates a reasonable range of measured parameters as obtained from various sources. The values cannot be taken as representative for any particular catchment. They merely indicate that pollution does occur and the degree of pollution can vary widely.

The parameter by which pollution is measured is in terms of concentration. Thus dissolved salts, e.g. chlorides and sulphates, are measured per litre (mg/ℓ) as are suspended solids such as silt. A specific nutrient such as nitrate is measured in terms of mg/ℓ of nitrogen. The total nitrogen content may comprise organic nitrogen, ammonium nitrogen, nitrite and nitrate. The oxidation of ammonia to nitrite then nitrate, and subsequent biological reduction to free nitrogen, is termed the denitrification process and occurs in nature but is also forced at wastewater treatment works. Phosphate is also a nutrient, and in the correct proportions in the presence of nitrate, can support life such as aquatic weeds and algae. Formation of algae in warm climates, in particular, is objectionable, and is termed eutrophication.

Biological matter in waters requires oxidation in order to render it innocuous. This includes decaying vegetable matter, faeces and some industrial wastes, e.g. from paper factories or abattoirs. The oxygen required is measured in mg/ℓ and termed the biochemical oxygen demand (BOD). It is a slow test to determine BOD and frequently the 5-day or 20-day values are taken as indicators of the ultimate BOD. Due to the difficulty in measuring BOD, many researchers prefer to use chemical oxygen demand (COD) or total organic carbon (TOC) as an indicator of oxygen demand.

Table 7.2 Urban runoff quality characteristics

Characteristics	Units	Low	Average	High
BOD	(mg/ℓ)	10	30	500
Suspended solids	(mg/ℓ)	20	200	10 000
Coliform	(No./100mℓ)	50	10 000	100×10^6
Total chlorides	(mg/ℓ)	10	200	10 000
Total dissolved solids	(mg/ℓ)	300	1 000	10 000
pH	(mg/ℓ)	5.3	7	8.7
Nitrogen	(mg/ℓ)	1	3	100
Phosphate	(mg/ℓ)	0.1	1	50
Phenols	(mg/ℓ)	0		0.2
Oils	(mg/ℓ)	0		110
Lead	(mg/ℓ)	0		2

Free oxygen is measured as dissolved oxygen (DO). Other parameters of interest are the conductivity which is often related to the dissolved salts content, turbidity (related to suspended solids), colour and temperature. pH is a measure of acidity, with 7.0 being neutral and lower values indicating acidity. Bacteria are measured in terms of the most probable number (MPN) per 100 mℓ sample. Faecal coliform and faecal streptococci are the significant bacteria.

The water quality has an effect on many uses of water. Thus agriculture can accommodate nutrients but not high salt contents. For domestic water supplies, coliform count, colour and taste are important as well as most other parameters. For recreational purposes, similar criteria are often applied. The standards required vary from country to country and Table 7.3 indicates some of the acceptable limits.

Table 7.3 Accepted limits for selected water quality parameters in streams

Parameter	Designation	Limit	Units	Reason
Dissolved oxygen	DO	5 minimum	mg/*l*	Aquatic life
Temperature	T	30 max	°C	Life
Free hydrogen	pH	6-9		Acid-Alkali
Coliform	MPN	10 000 max	No/100 m*l*	Disease
Total dissolved solids	TDS	1 000 max	mg/*l*	Agriculture
Chloride	Cl	250 max	mg/*l*	Agriculture
Pesticide	DDT	0.04 max	mg/*l*	Health
Phenols		0.001 max	mg/*l*	Taste
Suspended solids	SS	100 max	mg/*l*	Colour
Nitrate as N	N	0.9 max	mg/*l*	Eutrophication

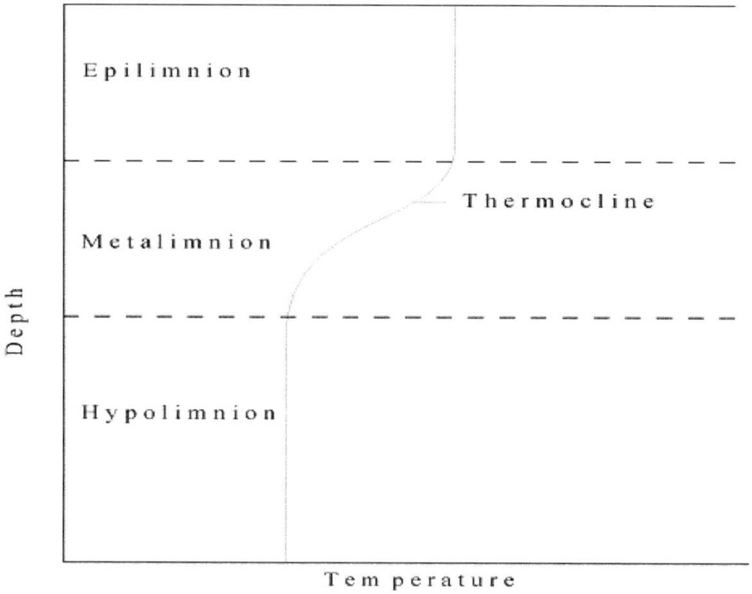

Figure 7.1 Temperature profile in a reservoir

7.4 EUTROPHICATION OF RECEIVING WATERS

Algal growth in water bodies is a nuisance from the health and appearance points of view. Algae may be present as a result of high nutrient loading, i.e. nitrogen and phosphorus. Problems are likely to be experienced if the phosphorus concentration is somewhere above about 0.1 mg/ℓ and nitrogen level above 10 mg/ℓ. Residence time, temperature, carbonaceous matter and cell availability also appear to have a bearing on the formation of algae.

Chlorophyll is frequently used as an indicator of eutrophic level. Thus a chlorophyll level of about 100 mg/ℓ is usually eutrophic while a level less than 10 mg/ℓ indicates am oligotrophic level (underenriched). The intermediate stage is referred to as mesotrophic.

Reservoirs are often stratified as indicated in Figure 7.1. Thermal strata form the epilimnion overlying a hypolimnion. Following a cooling of the upper layers, one may get temperature inversion of the water body resulting in mixing. Wind action can also contribute to the mixing.

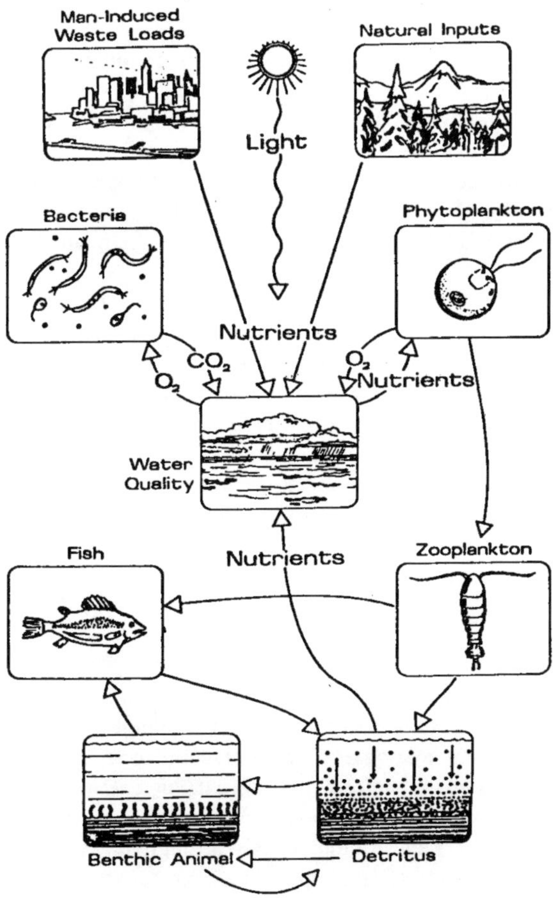

Figure 7.2 Depiction of a healthy ecosystem (after Roesner, 1979)

One would expect oxygen concentration decreasing from the surface to the bed but this may be upset by mixing. In fact, Henderson-Sellers (1979) indicates ways of causing mixing in order to improve water quality. The balance between life and inputs to a water system is delicate and complicated as depicted by Roesner (1979) (Fig. 7.2).

7.5 MASS BALANCE

Catchment water quality can deteriorate over time due to natural and man-made pollution, evaporation and recirculation. It is not easy to predict the rate of build-up of dissolved salts or the equilibrium concentrations in catchments, even with an understanding of the origins and methods of concentration of salts. This is because of the complex nature of water recirculation systems. One way of accounting for all these effects in a real system appears to be by modelling the system on a computer.

The build-up of impurities in a catchment can be simulated mathematically together with the water recirculation cycle. The flows of water in conduits or in vapour form in the air in and out of the system can be calculated. The processes of evaporation, condensation, pollution and make-up can all be modelled.

Once a model is programmed and validated, it may be used to improve the operation of existing service water reticulation systems and for optimizing the design of new systems.

For the purposes of mathematical simulation of water systems, the system must be described in terms of equations. One-stage systems can be described in terms of a mass balance equation which can be solved analytically. In other, more complex, situations it is necessary to express the equations in finite difference form and solve them numerically. Different types of models and the assumptions therein are described below.

Parameters whereby pollution is measured may either be conservative or non-conservative. In a conservative system, input to any part of the system equals outflow. Thus, if the parameter studied is water flow, then evaporation will be neglected in a conservative model. Similarly, if the parameter is a chemical compound, it is assumed there is no reaction, deposition or solution in a conservative model.

The model may be steady-state or time-varying. During the start-up period of a mine as concentrations build up, the system is said to be unsteady. After a while, the system may reach equilibrium. That is, in the case of salts in solution, the increases in mass of dissolved solids in the system due to leaching or evaporation equals the loss by pumping or deposition.

7.5.1 Mixed and plug flow

In a plug-flow system, the water is assumed to travel through conduits or media at a certain rate, conveying impurities at that rate. The salts content at any point can therefore be affected in a series of steps as water with different concentrations arrives at that point. In a completely mixed system, the concentration of salts will be the same at every point. An input is assumed to spread instantaneously through the systems so that the concentration increases by the mass of salt input divided by the total volume of water in the system. This simplified mechanism is often satisfactory to describe systems which exhibit gradual rates of change in concentrations. Real systems will probably be between plug flow and completely mixed, as there will be diffusion and mixing due to turbulence and cross connections. In general salts are conveyed by advection (lateral transport) and dispersion.

Examples:

The simplest illustration of the use of the mass balance equations is for a steady-state system. Q is the flow rate in ℓ/s or Mℓ/d, C is the concentration of mg/ℓ. Inflow of water and of salts per unit time equals outflow rate:

Flow Balance: $$Q_1 + Q_2 = Q_3 \tag{7.1}$$

Mass Balance: $$Q_1C_1 + Q_2C_2 = Q_3 \tag{7.2}$$

$$\therefore C_3 = \frac{Q_1C_1 + Q_2Q_2}{Q_1 + Q_2} \tag{7.3}$$

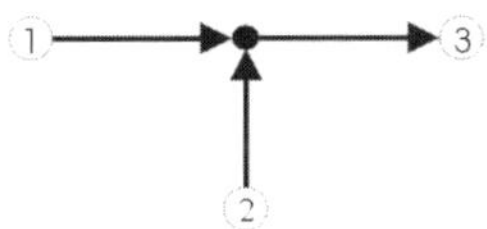

Figure 7.3 Point node

e.g. if $Q_1 = 5$ Mℓ/d, $Q_2 = 10$ Mℓ/d (water flow rate)
$C_1 = 400$ mg/ℓ, $C_2 = 100$ mg/ℓ (salt concentration)
Then $C_3 = 200$ mg/ℓ

And total mass of salt per day $= Q_3C_3 = 15 \times 200 = 3000$ kg/d (7.4)

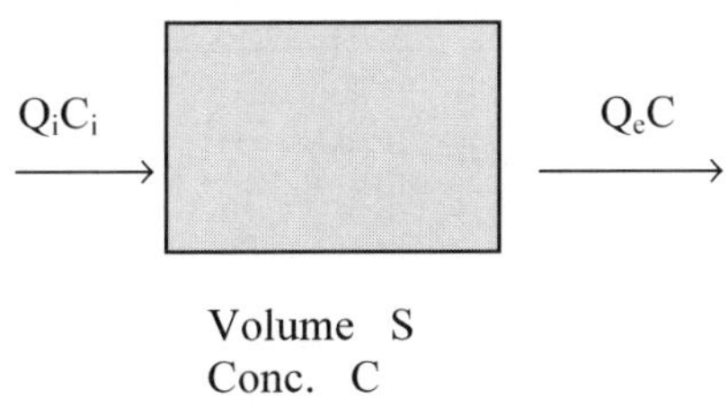

Figure 7.4 Mixed flow node

A completely mixed system can be described by differential equations. Subscript i refers to inflow, e to exit, s to initial conditions:

$$Q_i\, C_i = Q_eC + d(SC)/dt \tag{7.5}$$

$$= Q_eC + S\, dC/dt \text{ for constant } S \tag{7.6}$$

$$\therefore dt = (SdC) / (Q_iC_i - Q_eC) \quad (7.7)$$

If Q is in megalitres a day and T in mg/ℓ, then QT has the units of kilograms of salt per day. Solving for T_p, the salt concentration in the system:
Integrating and evaluating the constant of integration from the fact that concentration C = C_S at t = 0:

$$T = S/Q_e \, \ell n \, [(Q_iC_i - Q_SC_S) / (Q_iC_i - Q_eC)] \quad (7.8)$$

$$\text{Or} \quad C = Q_iC_i/Q_e) - e^{(QiCi/Qe - Cs)\, S/Qet} \quad (7.9)$$

e.g. at $t = 0$, $C = C_S$, and at $t = \infty$, or $Q_e = \infty$ or $S = 0$,

$$C = (Q_i/Q_e)C_i$$

Observe that if Q_i does not equal Q_e, there must be internal gains or losses, e.g. due to evaporation so a non-equilibrium equation should be used.

The above example could be studied numerically. Although this requires specific numbers, it is often the only practical way of solving more complex problems.

Assume $S = 1000 \text{ m}^3$, $Q_i = 1 \text{ m}^3/\text{s} = Q_e$, $C_S = 0$, $C_i = 500$ mg/ℓ.

Choose $\Delta t = 100$ s. The choice of Δt can affect the speed of solution, the accuracy of results and the numerical stability of the computations. It must be determined by trial, from experience or from theoretical considerations.

$$\text{Now} \quad Q_iC_i - Q_eC = S\,((C_2 - C_1) / (\Delta t)) \quad (7.10)$$

$$\therefore C_2 = C_1 + (\Delta t/s)Q_i\,(C_i - C_1) = C_1 + 0.1(500 - C_1) \quad (7.11)$$

The computations can be sent out in tabular form in Table 7.4 as follows:

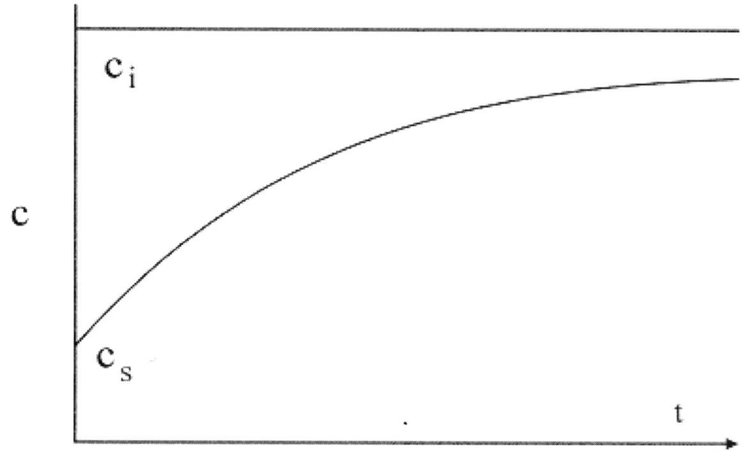

Figure 7.5 Concentration at diffuse node

Table 7.4 Numerical calculation of concentration variations over time

t	C_1	$500 - C_1$	x 0.1	C_2
0	0	500	50	50
100	50	450	45	95
200	95	405	40	135
300	135	365	37	172
o				
o				
1000	326	174	17	343 mg/ℓ

Equation (7.9) would indicate C = 316 mg/ℓ at t = 1000 s, which is comparable with the result indicated by the numerical solution of 343 mg/ℓ.

A plot of C versus t is called a *pollutograph*, and CQ versus t is called a *loadograph*. The pollutograph may peak before the hydrograph (runoff rate versus time) due to the first flush effect. The peak of the loadograph will be more concurrent with the hydrograph. And the hydrograph peaks after the hyetograph (rainfall rate versus time) because of hydraulic lag.

The pollution concentration may also peak after the hydraulic peak due to absorption along the flow path, dispersion or mixing. The concentration may approach the incoming concentration asymptotically if the inflow is continuous. The following diagrams (Fig. 7.6) illustrate the variation in concentration of a conservative (non degradable) salt in a plug flow, a completely mixed system and in a diffuse or slow mixing body of water. The polluted inflow is from the left into a body of initially unpolluted water. The rate of mixing will depend on the rate in inflow compared with the volume of the reservoir. The concentrations will be proportional to the concentration of the incoming stream.

Reservoirs are generally assumed to be completely mixed, whereas rivers are sometimes assumed to be plug flow. In fact, in both there is a degree of mixing due to:

1. Molecular diffusion due to Brownian movement (negligible in most hydraulic systems).
2. Turbulent mixing, due to eddies in the stream.
3. Short circuiting or tracking, e.g. in reservoirs where a track is made across the water body by the flow. The stagnant water in corners is called dead water.
4. Wind mixing.
5. Thermal mixing and inversion (e.g. Henderson-Sellers, 1979).

The degree of mixing can affect concentrations so much that theory cannot be used and monitoring systems are needed to track water quality (Sanders, 1983).

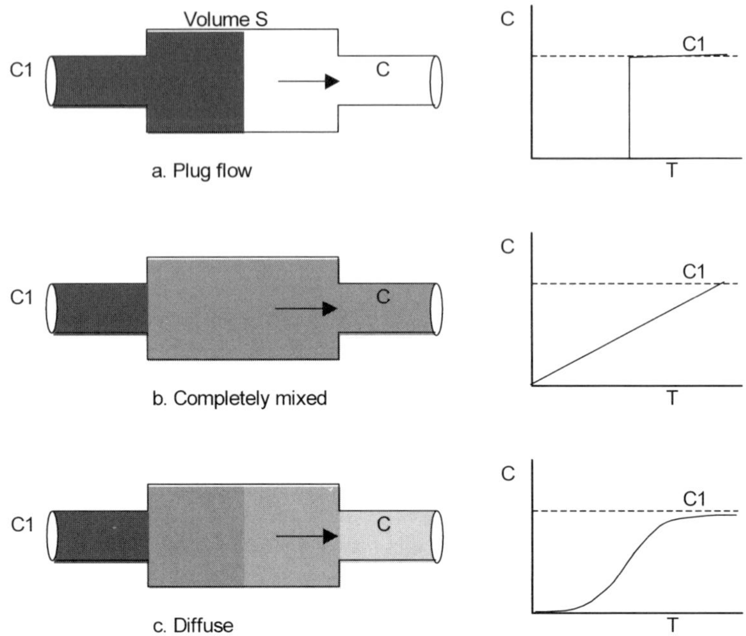

Figure 7.6 Comparison of plug flow and mixed systems

7.5.2 Systems analysis

A more sophisticated approach than the simulation method described above is the use of systems analysis and optimization techniques, with the assistance of computers if necessary. The methods allow an optimum design to be selected from numerous alternatives (Thomann, 1972).

The alternative standard engineering approach is to select the best option from a few selected designs. The latter approach is tedious where there are many alternatives. The design optimization approach involves the creation of a general configuration in which the numerical value of independent variables has not been fixed. An overall economic objective is defined and the system is described in terms of equations of constraints.

7.5.3 Non-conservative parameters

Mass balances are not always possible. Many constituents in still waters change concentration naturally. Some react chemically to result in different salts. If all the salts before and after reaction are soluble, the total concentration of dissolved salts in mg/ℓ in the water remains the same. Sometimes oxygen is taken out of the water to release hydrogen gas, which is more volatile and escapes.

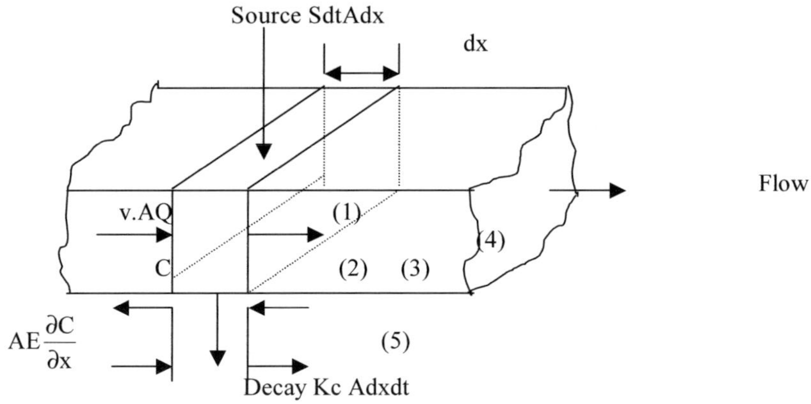

Figure 7.7 Mass balance

Oxygen in water is the cause of many changes. For instance, ammonia is oxidized to nitrites, and these in turn are oxidized to nitrates. The nitrates cannot be eliminated except by chemical replacement absorption or biochemically, as is now done in some wastewater treatment processes.

Absorption of oxygen and other chemicals in water may occur due to biological matter in water. Decay is generally approximated by a first order equation (Velz, 1970):

$$\frac{\partial C}{\partial t} = KC \tag{7.12}$$

7.5.4 Mass balance equation with non-conservatives

The one-dimensional balance equation allowing for dispersion, decay and sources or sinks is derived below. Referring to Fig. 7.7 the numbered parameters are;

(1) $v + \frac{\partial v}{\partial x}dx$ (2) $A + \frac{\partial A}{\partial x}dx$ (3) $C + \frac{\partial A}{\partial x}dx$

(4) $q + \frac{\partial q}{\partial x}dx$ (advection) (5) $E\frac{\partial C}{\partial x} + \left(\frac{dA}{dx}E\frac{\partial C}{\partial x}\right)dt$ (diffusion)

Net increase in mass of C in the element in time dt is

$$dC.Adx = dt\left\{(SAdx) - (KC)(Adx) - C\frac{\partial Q}{\partial x}dx - Q\frac{\partial C}{\partial x}dx + \frac{\partial}{\partial x}\left(AE\frac{\partial c}{\partial x}dx\right)\right\}$$

$$\therefore \frac{\partial C}{\partial t} + KC + \frac{C}{A}\frac{\partial Q}{\partial x} + \frac{Q}{A}\frac{\partial C}{\partial x} - \frac{1}{A}\frac{\partial}{\partial x}(AE\frac{\partial C}{\partial x}) - S = 0 \tag{7.13}$$

For a uniform channel A = constant and t = constant and Q = constant

$$\therefore \quad \frac{\partial C}{\partial t} + kC + v\frac{\partial C}{\partial x} + E\frac{\partial^2 C}{\partial x^2} - S = 0 \tag{7.14}$$

(1) (2) (3) (4) (5)

(1) is rate of increase in concentration of pollutant
(2) is decay rate
(3) is advection
(4) is diffusion
(5) is source

E is the turbulent diffusion coefficient. It is similar to the kinematic viscosity which represents transfer of momentum between layers in a friction model, e.g.

$$\tau = \rho E \frac{du}{dy} \text{ against a wall} \tag{7.15}$$

$$\text{where } E = \frac{\hat{\tau} y}{\rho \frac{du}{dy}} = U_* ky\left(\frac{D-y}{D}\right) \tag{7.16}$$

where: $U_* = \sqrt{(\tau / \rho)}$ = shear velocity, and (7.17)
k is the von Karman constant, 0.4.

But it is not that simple in channels as not only molecular diffusion but macro turbulence, tracking, dead water, stratification, etc., complicate the action, therefore one needs to calibrate models.

Elder (Deiniger, 1973) suggests $E = Ah\sqrt{(ghS)}$ (7.18)

where: h = depth and A = coefficient (averaging 0.07).

Normally, diffusion is negligible in rivers, except estuaries. Thus one gets the Streeter-Phelps equation:

$$\frac{\partial C}{\partial t} = -v\frac{\partial C}{\partial x} - KC \quad \text{or} \quad \frac{\partial C}{\partial t} \cong KC \quad \text{(omitting sources)} \tag{7.19}$$

$$\text{Integrating: } \log_e\left(\frac{C_t}{C_o}\right) = -Kt \tag{7.20}$$

$$C_t = C_{oe}{}^{-Kt} \tag{7.21}$$

K ranges from 0.01 per day in laboratory conditions (Arnold, 1980) to 0.1 per day in rivers.

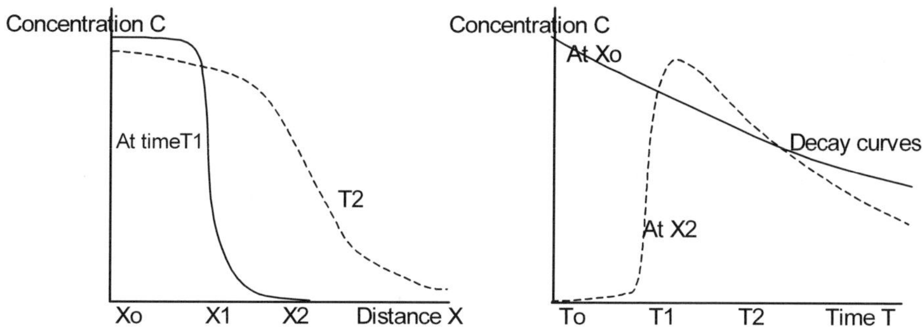

Figure 7.8 Decay curves

7.5.5 Oxygen balance in rivers

Oxygen concentration in a river is measured in terms of DO (dissolved oxygen). Shortage of oxygen is measured as a chemical oxygen demand (COD) or a biochemical oxygen demand (BOD). The long term BOD is about 1.45 x BOD_5. This is the BOD as measured in a laboratory over 5 days, a standard test (AWWA, 1965).

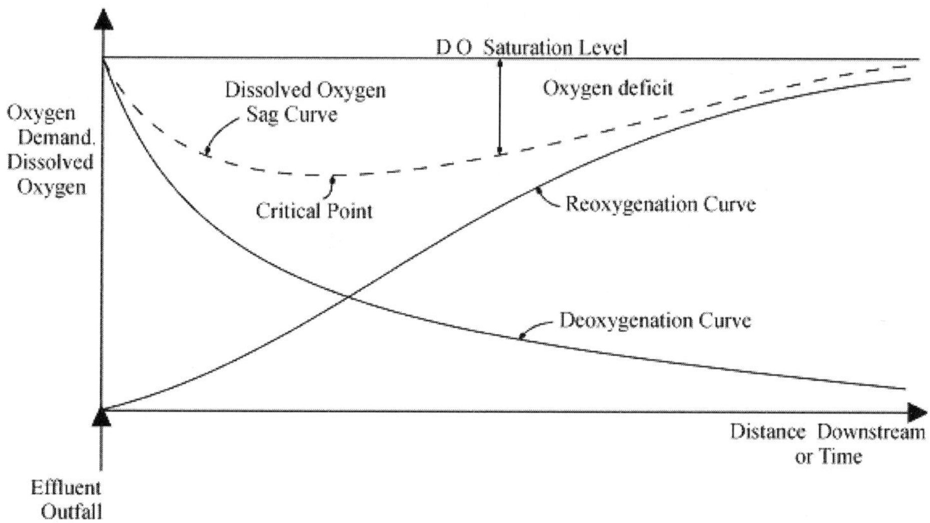

Figure 7.9 The dissolved oxygen sag curve

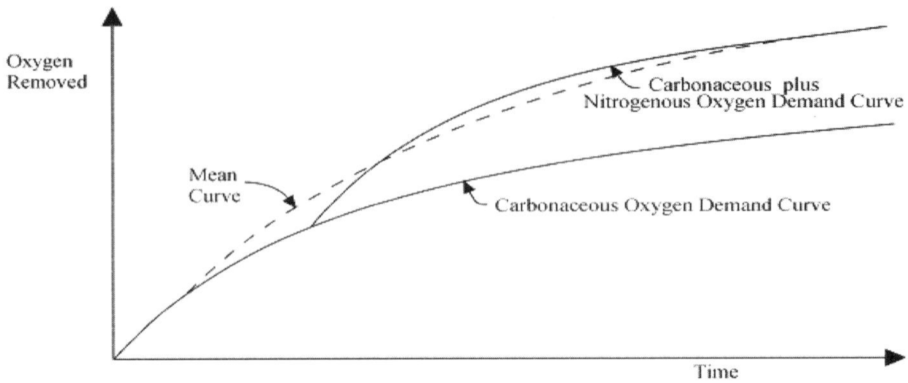

Figure 7.10 Carbonaceous and nitrogenous oxygen demand curves

The equations for BOD and DO represent a pair of coupled equations as follows:
If DO concentration is designated C and BOD is L then

$$\frac{\partial C}{\partial t} = E\frac{\partial^2 C}{\partial x^2} - v\frac{\partial C}{\partial x} - K_1 L + K_2 (C_S - C) \pm S_c \tag{7.22}$$

where C_S = saturation concentration of oxygen, and

$$\frac{\partial L}{\partial t} = E\frac{\partial^2 L}{\partial x^2} - v\frac{\partial L}{\partial x} - K_1 L_2 \pm S_L \tag{7.23}$$

These simultaneous equations can be solved at points along a river and over time increments. $K_2 = K_{20°C}\theta^{(T-20)}$, i.e. it is a function of temperature.

As an example of the solution of these two equations, a two-step explicit method can be employed (Deininger, 1973, p.12). One can get pseudo diffusion where

$\Delta x \neq v\Delta t$ unless a careful numerical procedure is used. (7.24)

Where the river is depleted of oxygen, the BOD equation must be replaced by

$$K_1 L = K_2 (C_S - C) - S_C \tag{7.25}$$

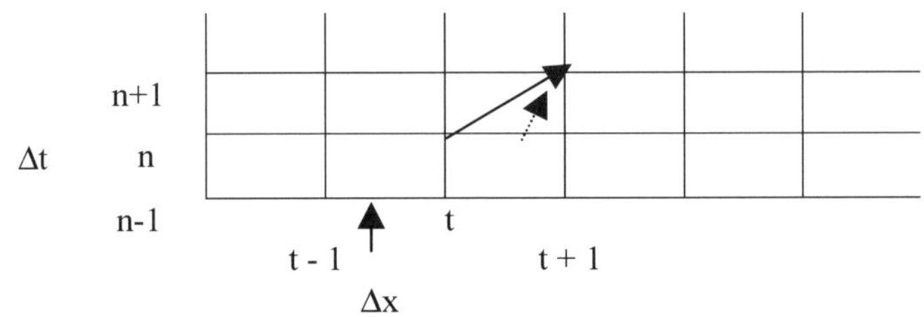

Figure 7.11 Orthogonal solution grid

i.e. the quantity of oxygen consumed is equal to the quantity of oxygen introduced in the same time (Thomann, 1972). An analytical solution is possible, giving an equation for the decay in BOD, or deficit D:

$$\text{If}\quad \frac{dC}{dt} = -K_1 L + K_2(C_S - C) \tag{7.26}$$

$$\text{and oxygen deficit}\quad D = C_S - C \tag{7.27}$$

$$\frac{dD}{dt} = K_1 L - K_2 D \tag{7.28}$$

Integrating gives

$$D = \frac{K_1 L_o}{K_2 - K_1}\left(e^{-K_1 t} - e^{-K_2 t}\right) + D_o e^{-K_2 t} \tag{7.29}$$

One can also evaluate K_1 and K_2 at t_C (Deininger, 1973, p.126).

7.6 NUMERICAL METHODS

Simulation of systems described by differential equations can be done in a number of ways:

Finite elements

Characteristics

Finite difference - Implicit - Four point 

Finite difference - Explicit - Four point 

Leap frog 

Diffusive 

Backward centred 

Lax - Wendroff = diffusive/leap frog

Explicit schemes are simple but not as accurate as implicit schemes. Problems which manifest with explicit schemes include numerical instability and numerical diffusion. Instability can occur if the time step is too great. The accepted stability criterion for diffusive schemes is:

$$\Delta x^2 / \Delta t \geq 2E \tag{7.30}$$

$$\text{or} \quad \Delta t \leq \Delta x^2 / 2E \tag{7.31}$$

There is an additional problem, that of numerical diffusion, i.e. spreading of the pollution gradient due to successive calculations using concentrations at adjacent points. From a second order Taylor expansion, the maximum numerical diffusion is E_P max = $\Delta x^2 / 8\Delta t$. Using the previous expression for Δt, the pseudo diffusion cannot be less than E/4.

7.6.1 Two-step method

The water quality equation including the diffusion term can be solved in two steps to ensure correct advection and diffusion. Thus starting with

$$\frac{\partial C}{\partial t} = E\frac{\partial^2 C}{\partial x^2} - v\frac{\partial C}{\partial x} kC \tag{7.32}$$

$$\text{Use} \quad \frac{\partial C}{\Delta x} = \frac{C_i - C_{i-1}}{\Delta x} \tag{7.33}$$

$$\frac{\partial^2 C}{\partial x^2} = \frac{C_{i+1} - 2C_i + C_{i-1}}{\Delta x^2} \tag{7.34}$$

The last term on the right hand side of the above equation 7.32 for advection, or (2), (3), and (5) in equation 7.14 for advection and decay or source can be used to get the first approximation to $C_{i,\, n+1}$ and then the diffusion term can be applied.

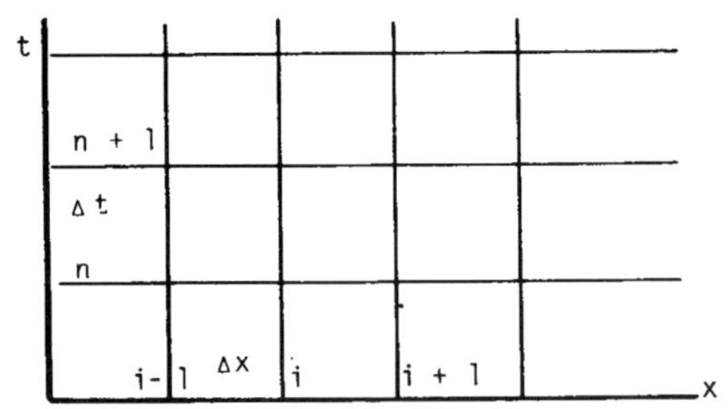

Figure 7.12 Basic rectangular x-t grid

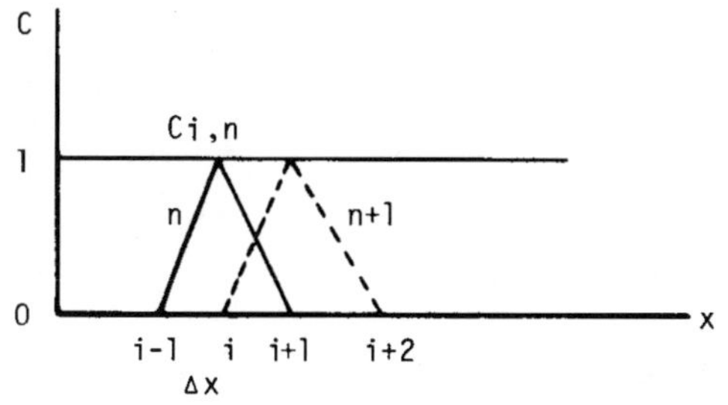

Figure 7.13 Theoretical advection

7.6.2 Demonstration of numerical inaccuracy

The convection term in the water quality equation will be used to illustrate problems and inaccuracies due to an incorrect numerical scheme. Neglecting the diffusion and decay term, we have

$$C_{i,n+1} = C_{i,n} - v\Delta t \frac{C_{i+1,n} - C_{i,n}}{\Delta x} \tag{7.36}$$

We should have a wave of concentration moving downstream at a rate v, unattenuated or changed in concentration.

$$\text{If } \Delta x = v\Delta t \tag{7.37}$$

then using a forward difference explicit method

$$\begin{aligned} C_{i,n+1} &= C_{i,n} - (C_{i+1,n} - C_{i,n}) \\ &= 1 - (0 - 1) \\ &= 2\text{, which is wrong, it should be 0} \end{aligned} \tag{7.36a}$$

i.e. do not use a forward difference $\partial c / \partial x = (C_{i+1} - C_i) / \Delta x$ (7.38a)

Instead us a backward difference $\partial c / \partial x = (C_i - C_{i-1}) / \Delta x$ (7.38b)

$$\begin{aligned} \text{Then } C_{i,n+1} &= C_{i,n} - (C_{i,n} - C_{i-1,n}) \\ &= 1 - (1 - 0) = 0\text{, correct.} \end{aligned} \tag{7.36b}$$

On the other hand, if we use $\Delta x = 2v\Delta t$

$$\begin{aligned} C_{i,n+1} &= C_{i,n} - v\Delta t \frac{(C_{i+1,n} - C_{i-1,n})}{2v\Delta t} \\ &= 1 - \frac{0-0}{2} = 1\text{, is also wrong.} \end{aligned} \tag{7.36c}$$

If we continued with the scheme, the value of C oscillates (see Figure 7.14)

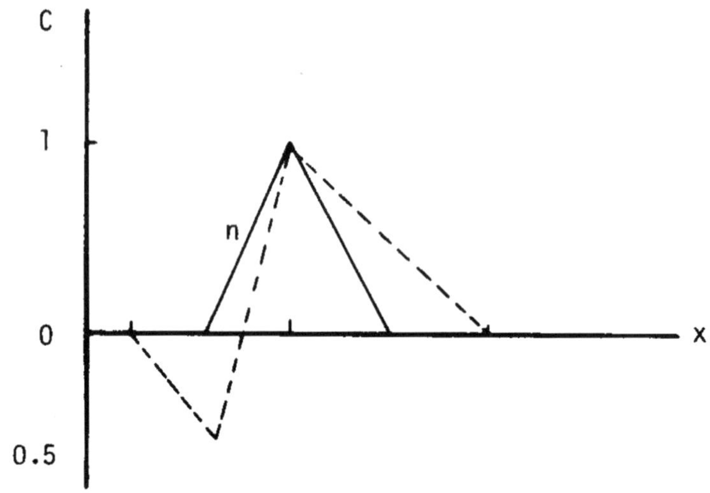

Figure 7.14 Oscillating scheme

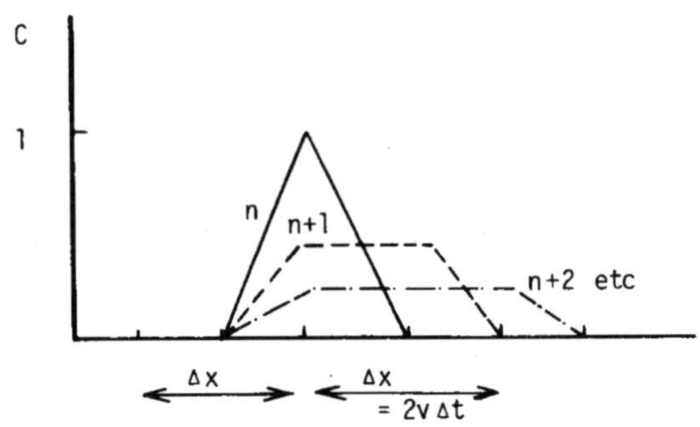

Figure 7.15 Numerical diffusion

On the other hand, if one uses a backward difference with $\Delta x = 2v\Delta t$, numerical diffusion occurs as indicated in Figure 7.15 above.

If $\Delta t > \Delta x / v$, we get numerical instability, e.g. if $\Delta t = 2\Delta x / v$,

$$C_{i,n+1} = C_{i,n} - \frac{v\Delta t}{\Delta x}\left(C_{i,n} - C_{i-1,n}\right) \tag{7.36d}$$

$$1 - 2(1 - 0) = -1$$

Continuing so, an oscillating curve occurs:

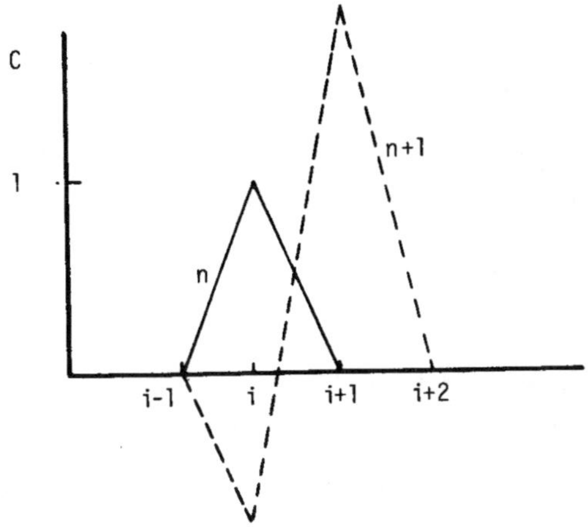

Figure 7.16 Numerical oscillations

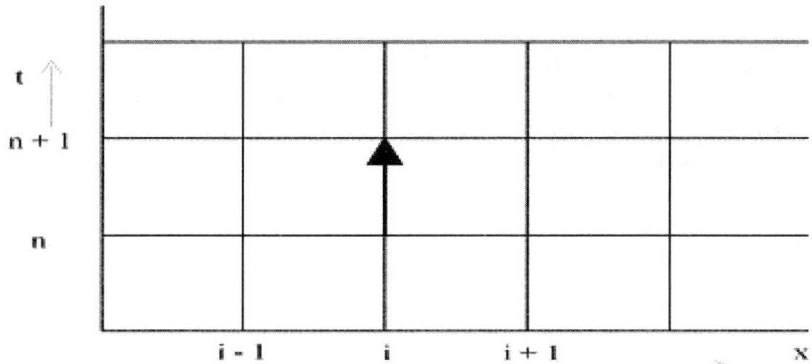

Figure 7.17 Implicit scheme

7.6.3 Implicit finite difference schemes

If the gradient is taken from the unknown future values a more stable solution results.

$$\frac{\partial C}{\partial t} = -v\frac{\partial C}{\partial x} \tag{7.39}$$

becomes $C_{i,n+1} = C_{i,n} - \frac{v\Delta t}{\Delta x}\left(C_{i,n+1} - C_{i-1,n+1}\right)$. All values at n+1 are unknown and a set of i equations is established for i unknowns. The method is unconditionally stable but solution of the i simultaneous equations can be laborious, especially for non-linear systems, e.g. if we use the hydrodynamic equation with the term $v\frac{\partial v}{\partial x}$, it is non-linear since $v_{i,n+1} (v_{i,n+1} - v_{i-1,n+1}) / \Delta x$ is parabolic. (7.40)

So rather use $v_{i,1} (v_{i,n+1} - v_{i-1,n+1}) / \Delta x$, which is linear. (7.41)

Methods of solution of i equations include (Fried, 1975):
1. Direct methods, e.g. matrix methods and Gauss elimination.
2. Iterative method – i.e. assume reasonable values for all C's and iterate the equations substituting assumed values on the right hand side until the left hand side agrees with assumed values. This only converges if $\Delta t < \Delta x/v$.
3. Relaxation methods.
4. Alternating direction implicit procedure (Fried, 1975), i.e. compute derivative with respect to x implicitly and y explicitly and then vice versa (stable).

One also gets combined explicit/implicit methods for more accuracy (e.g. McDonnell and O'Connor, 1977). The following should be noted:

Explicit method:
1. This must be designed to be stable, i.e. any errors due to second order terms in the Taylor expansion (we took just the first order) must decay during computation.
   The time interval must therefore be smaller than for the implicit method.
   For explicit hydrodynamic equation, using Fourier series it may be shown to be stable if $(\Delta x) / (\Delta t) \geq \sqrt{gy}$ = wave celerity, i.e. speed of computation greater than speed of a disturbance in the system.
2. It must be accurate. Check with a few space and time intervals and against an analytical solution if there is one.
3. It should minimize numerical diffusion.
4. One can use varying grids where greater accuracy is required:

The availability of commercial models for studying pollutant transport and reactions is such that only simple models or newly discovered processes will be compiled individually, and standard problems will be modeled on commercially available software.

7.7 SOIL EROSION

The erosion of soil in catchments, in the form of sheet erosion or rill and gully erosion, is a complex phenomenon. Research into the process is manifesting itself in the form of mathematical models. Some of the relationships thus produced are outlined below.

The amount of sediment detached by rainfall is based on kinetic energy of the rain and may be approximated by an equation of the form

$$D_r = S_{dr} A_S I^2 \tag{7.42}$$

where: A_S is the land surface area,
I is the rainfall intensity, and

D_{dr} is a constant dependent on soil type and land surface conditions

Sediment detachment by overland flow is assumed proportional to the square of the flow velocity which in turn is proportional to the cube root of the slope and flow rate:

$$D_f = S_{df} A_S S^{2/3} Q^{2/3} \tag{7.43}$$

Where: S is the land surface slope,
Q is the flow rate, and
S_{df} is a constant.

To account for imperviousness (or ground cover), the above equations may be multiplied by (1 – IMP) where IMP is the proportion of impervious area.

The sediment transport capacity due to rainfall is:

$$T_r = S_{tr} SI \tag{7.44}$$

The sediment transport capacity due to overland flow is

$$T_f = S_{tf} S^{5/3} Q^{5/3} \tag{7.45}$$

Thus provided all the constants could be evaluated, the computational procedure is to estimate the erosion rate due to rainfall and overland flow. The transport rate will increase as indicated by equation. (7.42) plus equation (7.43) until the limit indicated by either equation (7.44) or equation (7.45) is reached (Meyer and Wischmeir, 1969).

An empirical equation for the prediction of soil erosion rates was developed at Purdue University in the 1950s (Wischmeir and Smith, 1965). Although it was originally developed for croplands, it has been adapted to other erosion loss problems. The equation, termed the Universal Soil Loss Equation (USLE) is:

$$A = R\,K\,L\,S\,C\,P \tag{7.46}$$

Where: A = soil loss in tons per acre per time period
R = rainfall factor per unit time period
K = soil erodibility factor
L = slope length factor
S = slope gradient factor
C = cropping management factor
P = erosion control practice factor

The rainfall factor R is a function of the kinetic energy of a storm times its maximum 30-minute intensity, summed over the time period, which is usually a year. Values of R were given by Wischmeir and Smith (1965) for the United States. These values range from 20 per year on the west coast and in the arid north-west, to 350 in the wetter south-east.

The soil erodibility factor K depends on soil size distribution, structure and organic content. It varies from 0.02 for sands to 0.2 for clay and it may be as high as 0.6 for silt. It decreases slightly for high organic content.

The slope length factor L and slope gradient factor S are usually combined to give a topographic factor, which may be estimated from the formula

$$LS = \left(\frac{X}{72.6}\right)^m \frac{\left(430Z^2 + 30Z + 0.43\right)}{6.57} \tag{7.47}$$

Where: X = field slope length in feet
Z = slope in feet per feet
m = 0.5 for slope equal to or greater than 5%
0.4 for slope of 4%
0.3 for slope less than or equal to 3%

It may be necessary to break the catchment into a number of planes to account for varying topographic factors.

The cropping management factor C is also called the cover factor. It varies widely, from 0.01 for urban land or pasture to 0.1 for cropland. It may be as high as 1.0 for fallow land. The erosion control practice factor P allows for reduction in erosion due to practices such as contouring, terracing and strip cropping. For land slopes less than 2% it may be as low as 0.3, increasing to 0.5 for slopes over 20%. The factor is halved by terracing but increases for contour strip cropping. Considerable experience is obviously necessary in applying the USLE. It does however hold promise for urban systems subject to further research into the various factors.

7.7.1 Desertification

Desertification is a world–wide phenomenon which causes the earth's ecosystems to deteriorate. It affects about two-thirds of the countries of the world, and one-third of the earth's surface, on which one billion people live, namely, one-fifth of the world population.

The vulnerability of land to desertification is mainly due to the climate, the relief, the state of the soil and the natural vegetation, and the ways in which these two resources are used. Climate affects soil erosion and the chemical and biological deterioration of the soil. The state of the soil (texture, structure and chemical and biological properties) is a major factor, particularly in the sub-humid zones where the influence of climatic factors is less marked. It plays an essential role in causing vulnerability to desertification caused by human activities.

Climatic changes are both a consequence and a cause of desertification. The destruction of the natural grass and woody vegetation cover in dry areas affects the topsoil temperature and the air humidity and consequently influences the movements of atmospheric masses and rainfall. Furthermore, the drying of the soils and the destruction of soil cover encourage air erosion.

Even though the cycles of drought years and climatic changes can contribute to the advance of desertification, it is mainly caused by changes in the ways man uses the natural resources, mainly by over-grazing, land clearance, over-cropping cultivated land and wood formations and more generally using land in a way that is inappropriate for the local conditions. Human activities connected with agriculture, livestock and forestry production vary widely from one country and from one type of society to another, as do the strategies for land-use and the technologies employed.

Desertification directly reduces the world's fresh water reserves. It has a direct impact on river flow rates and the level of groundwater tables. The reduction of river flow rates and the lowering of groundwater levels leads to the silting up of estuaries, the encroachment of salt water into water tables, the pollution of water by suspended particles and salination, which in turn reduces the biodiversity of fresh and brackish water and fishing catches, interfering with the operation of reservoirs and irrigation channels, increasing coastal erosion and adversely affecting human and animal health. Lastly, desertification leads to an accelerated and often uncontrolled exploitation of underground fossil water reserves, and their gradual depletion.

Catchments sediment yields due to soil erosion can vary from 10 to 1000 tons per square kilometer per year. Typical figures in high runoff regions are over 100 and in poorly managed areas 500 t/km^2/annum are common (Petts, 1984).

It is difficult to measure sediment load in a river because the concentration varies with depth and flow rate. Assessment methods preferred are sediment volume measurement in existing reservoirs using depth sounders and GPS, or assessment of landform loss at intervals.

Two types of erosion are common: Sheet erosion, which is a function of rainfall, geomorphology and agricultural practices, and Rill erosion, which is affected by seepage, ploughing and soil type. The rate of erosion is highly dependent on the rainfall and runoff rate. Soil is transported to rivers and may settle out in low velocity zones such as reservoirs. But there is a sorting process and it is the coarser particles which settle first (see Fig. 7.20).

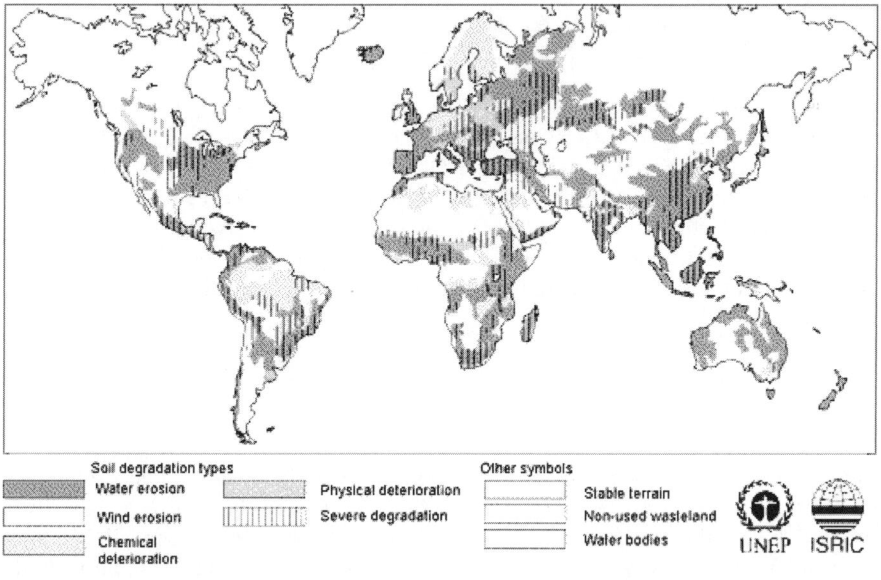

Figure 7.18 Soil degradation map

CLASS	SEDIMENT YIELD ($t.km^{-2}yr^{-1}$)	RUNOFF (mm)
1	ARID	< 50
2	0 - 50	50 - 500
3	0 - 50	500 +
4	50 - 100	50 - 500
5	50 - 100	500 +
6	100 +	50 - 500
7	100 +	500

Figure 7.19 A general classification of world rivers. Sediment yield is indicated in tonnes per square kilometre per year (Petts, 1984)

The larger particles settle at the head of a reservoir and are stable against erosion . The finer particles remain in suspension or may settle at the dam wall. These may be cohesive and also difficult to re-suspend. On the other hand the larger particles can armour the upper layers against subsequent resuspension. Vegetation can become established and further consolidate the fill. Hence unless an active sediment discharge programme is maintained the reservoir could one day become useless. The cost of sediment control must be compared with new reservoir construction, unless a limited economic life is all that is required.

7.7.2 Reservoir sedimentation

Sediment deposits in reservoirs are a problem particularly in high yield catchments. The sediment load is independent of reservoir size. Where the reservoir capacity is considerably less than the mean annual river flow the reservoir can practically fill up in a few years unless an adequate transport velocity can be maintained through the reservoir. The following problems arise:

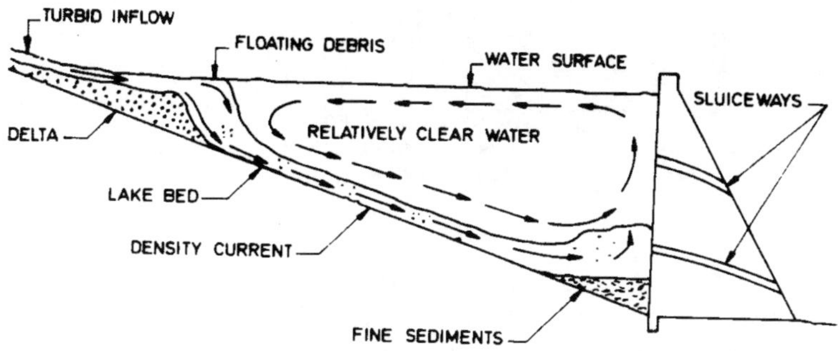

Figure 7.20 Sediment accumulation in a typical reservoir

- Loss of storage capacity
- or alternatively larger storage requirements
- Change in surface area-storage relation and more evaporation loss
- Increased backwater effects because sediment is deposited in the upper reaches of a reservoir
- Loss of land upstream because of bad soil deposits during flood
- Less soil deposits on flood plains downstream
- Blocking of outlets
- Bad water quality
- Abrasion of turbines and equipment
- Erosion downstream because the load of the river is below equilibrium
- Local scour downstream of dam due to energy dissipation over spillway
- Additional structural load on dam wall

Regular sediment surveys of reservoirs are useful. Permanent ranges or cross sections at beacons can be established at likely deposit or erosion localities. Then it is a simple matter to sound the cross section from a boat at intervals. Nowadays with rapid electronic depth sounder-GPS instruments, large areas can be surveyed in a day and the data can readily be converted to a contour plan or cross section drawings for assessing volumes of deposit. Topographical programmes are available for plotting sections, drawing depth-capacity curves and depicting in wire frame diagrams the location of deposits.

Reservoir surveys should go hand in hand with catchment surveys to detect the source of sediment and to attempt to reduce erosion rate. An agricultural education programme is needed to reduce erosion, or land use should be controlled, e.g. by changing ownership or cropping.

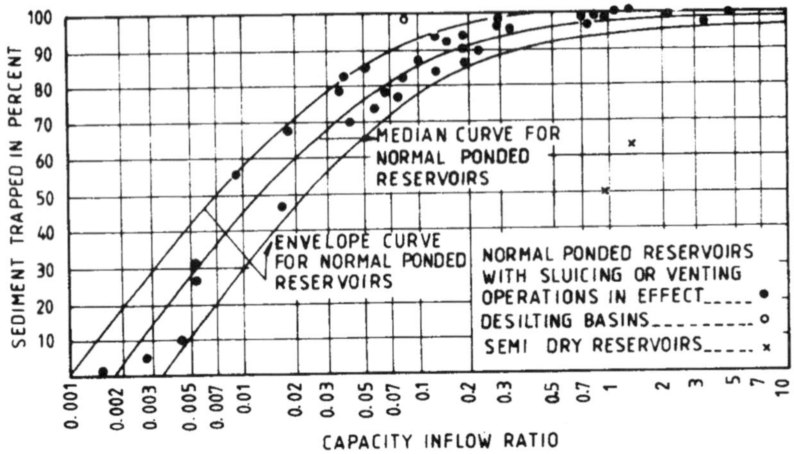

Figure 7.21 Reservoir trap efficiency (after Brune, 1953)

It is not easy to get rid of sediment once it has deposited in a reservoir. It deposits because the water slows down and will not be re-suspended unless the velocity is high enough (even high than the transport velocity). Small gates in dam walls have rarely much effect. Large gates counter flood control effects, and could lose water which should be stored. Measures to control sedimentation or its effects include:

- Catchment management to reduce erosion (the first priority). Farmer education and inducements may be required.
- Sediment harvesting bunds in fields
- Catch weirs in gulleys and streams
- Monitoring and modeling to predict and take corrective action
- Bypass sediment laden waters by diversion channel, loops or off-channel storage
- Provide additional storage in reservoirs
- Build new dams at intervals
- Dredge sediment mechanically from barges or siphon pipes over the dam wall
- Provide large desilting gates on the dam wall
- Design the dam as a barrage to be lowered before high inflow occurs. Radial gates and inflatable cylinders can be used.
- Lower the water level behind the dam wall prior to a sediment laden inflow.

7.8 ENVIRONMENTAL AND SOCIAL IMPACT ASSESSMENT

The effects of construction of water resources projects on a basin and its environment need careful assessment. It is standard practice to evaluate the impact of dam construction on river systems, and guidelines for this have been issued by U.S. agencies, e.g. AID, the World Bank, and a sub-committee of the International Commission on Large Dams (ICOLD, 1982).

The U.S. Water Resources Council attempted to achieve a balanced assessment by requiring account of effects of four factors (Petersen, 1984):

1. National economic development
2. Environmental quality
3. Regional economic development
4. Social effects

While not deprecating environmental effects, project construction in developing areas is likely to have more social and economic effects, and to be desperately required in many cases. The local population thus tends to attach less importance to environmental factors than in developed countries. Short-term aesthetic impacts may be of almost no local concern, but guidance is needed where long-term impacts result, e.g. soil erosion or deforestation. On the other hand, social impacts can be enormous and warrant careful evaluation. In fact, it is quite likely environmental effects usually will be positive as economic and social improvement will facilitate better farming practices and a more conscientious population. The most glaring environmental problem in many developing basins is soil degradation and erosion which must be tackled with positive agricultural and reclamation policies. Without such policies, further deterioration can render more tracts of land useless and result in sedimentation of rivers and reservoirs on an increasing scale (Goodman, 1984).

Environment quality is generally associated with cultural, ecological and aesthetic properties. Culture can include lifestyle, history, sociological and economic aspects. Ecology can include flora, fauna, water, geophysics and climate. Aesthetics embraces the senses, including visual and tourism potential.

A matrix type analysis is often used, done over four stages, namely:

1. Preliminary identification and data identification
2. Preliminary appraisal to identify likely environmental effects
3. Identification for and inventory of data
4. Assessment of effects

The effect of both construction activities and the presence of a reservoir need assessing. The effects of the following actions were suggested by ICOLD for study:

- Construction
- Water use
- Presence of reservoir
- Operation
- Socio-economic activity

The following effects have to be considered:

- Socio-economic impact
- Geophysical impact
- Impact on water
- Impact on flora
- Impact on fauna

The standard environmental impact assessment format of first world countries is not necessarily the way to assess impacts in developing areas because priorities may differ in a developing country. The desire for development may be very important at this stage, giving environmental issues a lower priority. Social and economic impacts may be so important that aesthetic conservation and pollution may receive less attention and expenditure.

On the other hand, some environmental aspects, particularly those with long lasting and economic consequences, may receive a high priority. Soil erosion is such an example. Another is the effect of river control on flood plains used for agriculture. Others are the relocation of people, and the intrusion of development into a relatively stable but undeveloped society. Whether development is a desirable feature may be debatable by some first world conservationists, but to the people who have no opportunity now, development may be all important because it may be accompanied by education, jobs and money.

Cada and Zadraga (1981) presented much of the following considerations in their report 'Environmental Issues and Site Selection Criteria for Small Hydro Projects in Developing Countries.'

7.8.1 Impacts on water and related resources

There are topographic, hydrological, ecological and socio-economic consequences of dam building. Following are some effects on the river and water quality:

Physical impacts:

(a) Streamflow patterns will change with the construction of dams. Water will be released through turbines over certain hours of the day, and the river will therefore b dry at other times for a number of kilometres downstream to where other inflow occurs, except for pools. Minimum low flow releases may be required during non-generating periods, reducing flows available for power production.

(b) Flood flows will be less frequent and only those overtopping the spillway will be passed downstream. Therefore, inundation of flood plains downstream will be less frequent and river bank agriculture may be reduced.

(c) During filling of a large reservoir (which may require a year or more), downstream flows would be reduced to minimum low-flow requirements except for releases through turbines.

(d) River diversion during construction will only affect a limited length of river bed but flow rates will be essentially unchanged.

(e) Evaporation from reservoirs could decrease the mean annual river yield, especially in arid and semi-arid areas.

(f) Groundwater recharge is unlikely to be affected at dam sites as solid rock exists beneath reservoir walls and basins in many cases.

Chemical impacts:

(a) The quality of water from upstream could be important as it may contain fertilizers and insecticides that could affect the water quality in reservoirs.

(b) Reservoir stratification by chemicals or temperature should be monitored but at small hydro power developments is not normally expected to be significant as the water level will fluctuate a lot. Multiple level outlets can be used to average outflow quality so reservoir quality will not deteriorate appreciably over time and so that any downstream water temperature criteria can be met.

(c) Contaminants are expected to be low in concentration causing little adverse conditions to whatever fish or biological life there is.

Biological impact:

(a) Changes in flow or load could change river bed conditions and adversely affect fish breeding or aquatic vegetation.
(b) Migratory fish could be blocked by construction of a dam wall. The reservoir may be stocked with fish and these could escape down the penstocks unless the screens are adequate.
(c) The type of fish in the reservoir may be different from that in the river owing to the more steady water level and quality, and different water temperature. Stocking of the reservoir should be considered, particularly if the local people could fish.
(d) Lagoons and tidal mouths sometimes exist where life could be affected.
(e) Dredging and in-stream excavation during construction are likely to affect water turbidity and consequently fish life downstream.
(f) The construction of a reservoir could result in the deposit of vegetation and trees in the basin but these do not form a noticeable load.

Sedimentation impact:

The biggest impact of a large reservoir could be on the suspended load in the river downstream of the reservoir. Over 90% of silt (primarily the coarser particles) could be deposited in the reservoir so that the water downstream will be deficient in sediment. There may therefore be potential for degradation of the river bed and erosion of the banks. The frequency and amount of deposition of sediment on the river banks downstream will decrease. High suspended sediment loads can adversely affect operation and maintenance of the turbines.

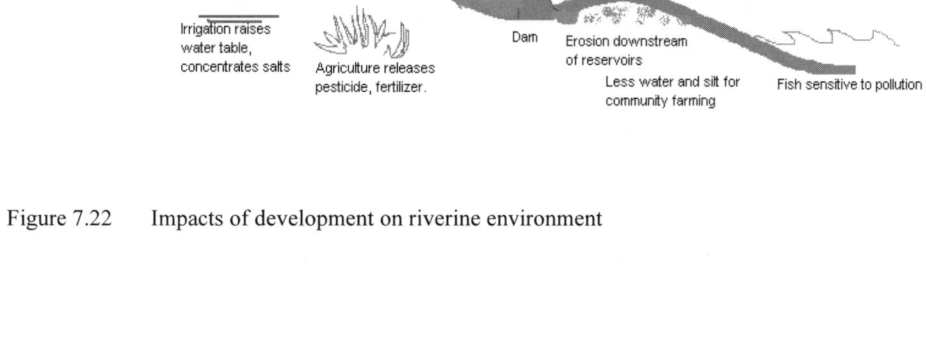

Figure 7.22 Impacts of development on riverine environment

7.8.2 Impacts on land and related resources

Physical impacts:

(a) The reduction of silt deposits was discussed above.

(b) Upstream of a dam there will be no change unless agricultural patterns change. It is only likely to be dry-land farming or pumped irrigation which could affect the catchment.

(c) If irrigation or some form of intensive farming does occur in the catchment, soil erosion could increase unless soil conservation measures are used.

(d) Arable land upstream of a dam could be inundated. This land may be farmed for subsistence purposes or used for grazing, and the farmers should be compensated, for example, by offering pumped water supplied to higher lying land.

(e) Settlements could be inundated and a survey of the water-line should be made to establish if isolated buildings will be affected.

(f) Roads across the river may exist and the dam wall may have to be built to accommodate a realigned road. Service roads will be required to reach settlements cut off.

(g) Archaeological resources may be inundated by the reservoir.

Biological impacts:

(a) Site clearing will destroy vegetation. Grazing is included in arable land above.

(b) Wild birds or fauna could be affected by construction.

(c) Existing migration routes could be affected.

(d) The reservoir shore line will often be more accessible than the prior river bed for livestock or game to water at.

(e) Wetlands or nature areas downstream could be affected.

(f) Sediment deposits in the reservoir headwaters during floods may be arable during non-flood periods.

7.8.3 Impacts on public health

(a) The reservoir may be a breeding ground for mosquitoes and other insects.

(b) Interbasin transfer of water could introduce new diseases, e.g. bilharzia.

(c) Influx of construction works may result in spread of disease, e.g. AIDS.

(d) The construction camp and completed reservoir will result in net immigration as well as tourist points. Sanitary facilities and water supplies will be required for these people.

7.8.4 Socio-economic impacts

(a) The injection of money by way of construction and provision of a water supply and other facilities at the project sites will serve as a node for growth. Migration from outlying areas can be expected, and concentrations of people will make it easier to provide water supply and sanitation, transport and health and educational facilities.

(b) Being a convenient artery, the reservoir may attract tourism, fishing and possibly commercial and recreational boating.

7.8.5 Impacts of dams

ICOLD (International Committee on Large Dams, 1982, 1985) has initiated considerable research into the impact of dams and the assessment of such impacts. It has published a number of booklets assisting planners of dams and other major water works. Apart from technical references such as the matrix method description (ICOLD, 1982), there are guides on what to look for in making assessment. ICOLD (1985) has through careful research and documentation provided a wealth of information from dams in temperate, tropical, subtropical, and severe winter regions. The chapters on tropical, semi-tropical, and regions are probably of most relevance to developing countries.

Although the committee's report concentrates on natural aspects (e.g. water, land, fauna, flora, climate) there are sections on economic and social aspects, and on man. However, theses are primarily from a first world point of view. The fact that many dams are designed by first world engineers for third world countries is a problem very few recognize unless they understand the way of life of the third world population. It is not necessary to preserve the rural way of life, but to see how those people can best adopt to the new lifestyles and facilities thrust upon them. In Chapter 10, the impact of dams is considered in greater detail.

REFERENCES

American Water Works Association (1965). Standard Methods for the Examination of Water and Wastewater.

Arnold, R.W. (1980). Modelling water quality in the upper Klip river. MSc(Eng) Dissertation, University of the Witwatersrand, Johannesburg.

Brune, G. M. (1953). Trap efficiency of reservoirs. *Trans Am Soc Geophys Union*, 34 (3)

Cada, A. and Zadraga, Z. (1981). Environmental issues and site selection criteria for small hydro projects in developing countries. Oak Ridge Natl. Lab., Tennessee.

Deininger, R.A. (1973). Models for Environmental Pollution Control. Ann Arbor Science.

Fried, J.J. (1975). *Groundwater Pollution*. Elsevier, Amsterdam, 330 pp.

Goodman, A.S. (1984). *Principles of Water Resources Planing*, Prentice Hall, H.J., 563p.

Henderson-Sellers, B. (1979). *Reservoirs*. MacMillan, London, 128 pp.

ICOLD (International Committee on Large Dams) (1982). Dams and the Environment, Bull. 35, Paris.

ICOLD (International Committee on Large Dams) (1985). Dams and the Environment, notes on regional influences, Bull. 50, Paris.

McDonnell, D.M. and O'Connor, B.A. (1977). *Hydraulic Behaviour of Estuaries*. MacMillan.

McPherson, D.R and Sharland, P.J. (1979). River quality tests. Undergraduate project, University of the Witwatersrand, Johannesburg.

Meyer, L.S. and Wischmeir, W.H. (1969). Mathematical simulation process of soil erosion by water. Trans. American Society of Agricultural Engineers, Dec. 12 (6).

Munn, R.E. (Ed) (1975). *Environmental Impact Assessment, Principles and Procedures*, Scope 5, 2nd Edn., John Wiley & Sons.

Petersen, M.S. (1984). *Water Resources Planning and Development*, Prentice Hall.

Petts, G.E. (1984). *Impounded Rivers*. John Wiley & Sons NY

Roesner, L.A. (1979). Response, Water problems of urbanizing areas. *Proc. Res. Conf. ASCE*.

Sanders, T.G. (Ed.) (1983). Design of Networks for Monitoring Water Quality. Water Resources Publics., 328 pp.

Thomann, R.V. (1972). *Systems Analysis and Water Quality Management*, McGraw Hill, N.Y.

Velz, C.J. (1970). *Applied Stream Sanitation*, Wiley Interscience, N.Y.

Wischmeir, W.H. and Smith, D.D. (1965). *Predicting rainfall erosion losses from cropland east of the Rocky Mountains*. Agricultural Handbook 282, ARS-USDA.

WHO (1984). *Guidelines for Drinking Water Quality*. World Health Organization, Geneva.

Yalin, T.S. (1963). An expression for bed-load transportation. *ASCE J. Hydraul. Civ*., 89 (HY3), 221-250.

CHAPTER 8

Water Use

8.1 DOMESTIC AND URBAN USE

Water used for domestic, commercial and industrial purposes may not be large per capita compared with irrigation, but it is inflexible and the value is high. A major problem of urban use is it causes heavy pollution. Sewage, industrial wastes, urban washoff and air pollution account for the majority of pollutants in surface and groundwater. Most countries now treat wastewater, and remove settleable and biological pollutants. Dissolved salts remain a problem as demineralization is expensive and these build up in recycling systems or discharge to the sea.

The temporal pattern of urban water use means balancing storage is usually required. Peak consumption occurs in spring, and in most countries this is the end of the dry season so river storage is required as well.

The following factors need consideration for urban supplies:

- Quantity
- Peak rate
- Storage
- Emergency
- Fire fighting
- Pressure
- Quality
- Reliability

The total volume of water required for domestic use in the world is some 200 km^3/a, which is only 0.5% of the average total river runoff of 40 000 km^3/a. The total urban water consumption including industry is some 2 000 km^3/a. However, the largest user of water is agriculture, accounting for 4 000 km^3/a.

Theoretically there is sufficient water to go around, but the problems of distribution in space, time and affordability, result in water shortages. However, with growing world populations (Fig. 8.1) and increasing per capita consumption, the total consumption could rapidly reach the limits of availability. The problem is likely to get worse in developing communities in particular.

The majority (4 billion) of the world population has inadequate water for healthy living. Even in developed economies, there are often water shortages, due to droughts, inadequate local resources, pollution, finance, or simply poor planning.

Many other factors come into the equation, such as water resources available, sustainability, politics and ownership. Figure 8.2 shows the populations percentages with inadequate access to safe drinking water in developing countries. An (optimistic) projection is made for a decrease in the percentage in the future.

Concern has been expressed as to the carrying capacity of the earth primarily from the point of view of food and water (Meadows et al., 1972). Populations continue to grow and the world population in 1990 was estimated to be 6 000 million. The biggest concentrations are in China (1 000 million) and India (700 million) followed by the USSR and USA.

Fortunately the per capita water requirement for domestic and industrial use are relatively low in the developing countries but those are the demands which need watching over the next century.

The availability of resources for larger populations is only a problem if the countries are to develop to the present consumption levels of the first world. The populations of the USA (220 million), Europe (400 million) and Japan (120 million) total only 25% of the world population, so one could expect usage of metals, timber and oil and water to increase many fold if the entire world reached the consumption levels of these countries.

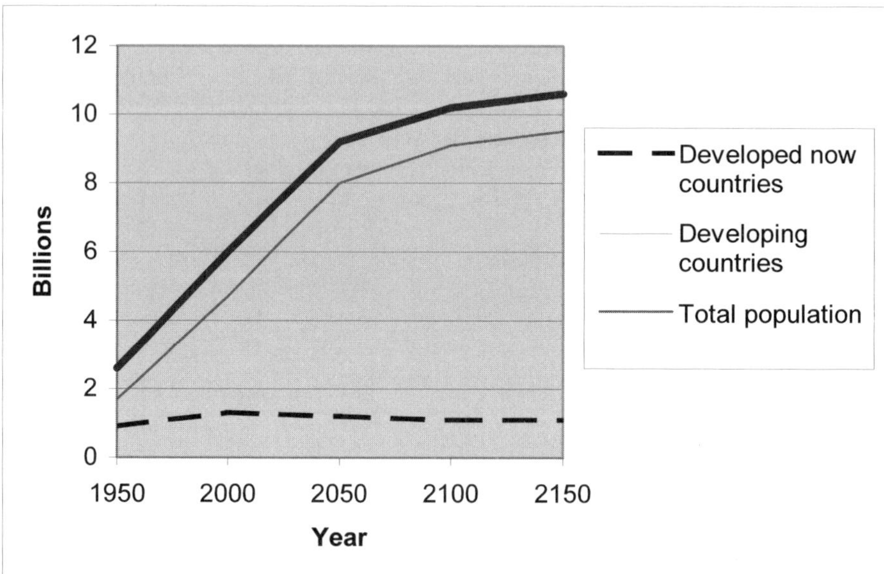

Figure 8.1 World population projections (United Nations, 2002)

8.1.1 Volumes required

Water requirements vary with type of user and various other factors. The following urban users need to be considered.

- Domestic – drinking, washing, gardening, pools.
- Municipal – institutions, education, washing, fire fighting, power, transport.
- Commercial – shops, offices.
- Industrial – mining, heavy users, washing.
- Losses – leaks, misuse, testing, overflow.
- Irrigation – parks, recreational, agriculture, stock.

Very little of the water used in the urban environment (10–20%) is lost, but the balance which is returned to the system is generally polluted by the user.

The amount of water used by each type of consumer can be influenced by:

- Income
- Standard and type of living, e.g. bungalows versus apartments
- Size of dwelling unit and stand
- Climate
- Type of supply, e.g. communal or multiple household connections
- Price
- Whether it is metered
- Who pays, e.g. public water is not paid for by the user
- Pressure
- Type of sanitation, e.g. waterborne
- Social customers
- Availability from the supplier, e.g. intermittent

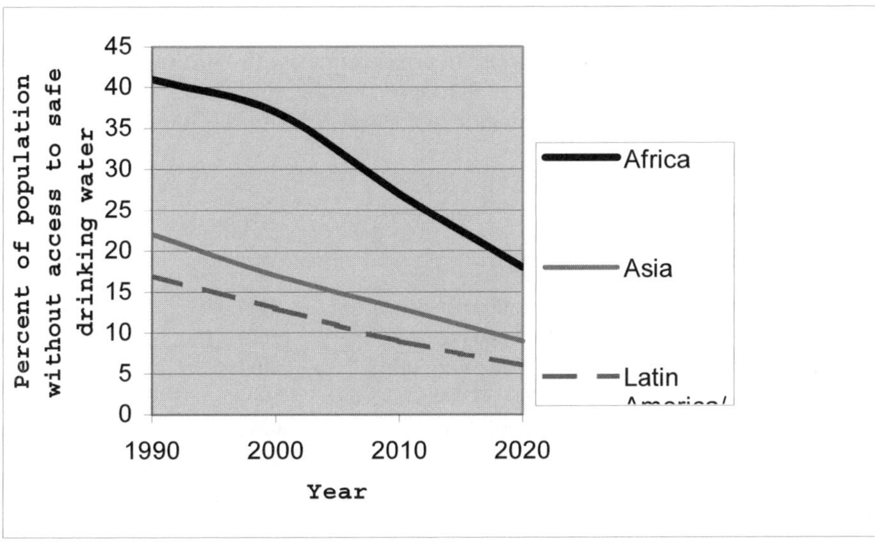

Figure 8.2 Poor access to safe drinking water in developing countries

Table 8.1 Domestic water usage in warm climates – according to standard of housing

(i)	Areas of high quality housing	200 – 250 ℓ/c/d
(ii)	City residential areas including high standard flats	160 – 200 ℓ/c/d
(iii)	Suburban; tenement dwellings; low cost housing	90 – 100 ℓ/c/d
(iv)	Urban areas served by standpipes	50 – 70 ℓ/c/d
(v)	Rural: standpipes	30 – 50 ℓ/c/d
(vi)	Rural: distance to source > 1 km	10 – 30 ℓ/c/d

The possibility of variations in demands due to changes in population, etc., should be considered and risk may affect the time horizon or type of project installed (Saparta and Munoz, 1995). There may also be a technical upper limit to which a project can be installed, for example limited availability of water, or size of pipes available. In planning to meet demand, it is necessary to add losses which can add 10–30% to raw water requirements.

The demand could be reduced by pricing or pressure control, but this is more an economic supply and demand situation than supply to poorer communities which may be limited purely from the affordability point of view.

An approximate breakdown of the use per capita in the U.K.(Twort et al., 1994) is:

Drinking	1 ℓ/c/d
Toilet	45
Bathing	23
Laundry	17
Kitchen	50
Car washing	4
Garden	40
TOTAL	180 ℓ/c/d

Water demands given in Table 8.2 apply to urban centres with flush sanitation. Where upper and lower limits are given, it is envisaged that the upper limit would generally apply to the high income level township, and lower limit to the lower income level township.

Table 8.2 Municipal water requirements – Annual averages

Type		Per coverage ℓ/d	Per person ℓ/d
Houses		600–3 000 per stand	200–300
Multiple units		600–1 000 per unit	160–200
High rise		450–700 per dwelling	90–100
Low cost		10 000–50 000 per hectare	50–70
Offices, shops		400 per 100 m^2	60–100 per employee
Hotels			300–500 per bed
Hospitals		500 per 100 m^2	200–300 per bed
Schools (day)			25–75 per pupil
Parks		100 per 100 m^2	
Factories	- light		200–500 per worker
	- heavy	10–50 m^3/ton of product	

The actual demand can vary widely according to circumstances. In affluent societies, people may demand water for drinking, ablutions, cooking, washing in kitchen, laundry or garage, gardening or recreation. They may be relatively unconcerned about wastage. The drawoff could be over a few hours a day in total, so that the distribution system has to be sized for peak rates.

8.1.2 Consumption pattern

The peak rate of drawoff can be many times the average. This is because of seasonal variations, temporal usage patterns and the fact that for a large city the peaks are unlikely to occur simultaneously. An individual household can draw at 40 times the average whereas a town's peak will be something like 4 times the annual average (Stephenson and Turner, 1997).

The occurrence of domestic peaks is generally in the early morning and evening. Industry draws a peak in the late morning and commerce in the afternoon. The juxtaposition of different types of consumer can therefore alleviate peak drawoffs from pipes and reservoirs and reduce costs. However other services such as transport and property prices will influence the desire of users to locate.

8.2 WATER DEMAND PROJECTIONS

The cost of water is highly dependent on water consumption, and indeed, correct estimates of water consumption. In planning and design of water supplies the sizing of works, the locality of sources and discharge of wastewater are dependent on the planned demand. The procedure is to plan for a number of years into the future, e.g. 10 to 20 years for unlimited sources and demand. The actual planning horizon can be selected by economic procedures.

The classical method of demand projection was to extrapolate historic demands using a logarithmic plot. A constant growth rate will project as a straight line. Deviations due to economic, hydrological or technical problems will be ironed out in the projection. Global conditions are changing and a number of new factors appeared.

Reduced growth may be upon us due to:

- Economic slowdown (international and regional)
- Lower fertility rates, particularly in developed countries
- Lower standards of living
- Limitations of services in built up areas
- Increased water costs due to greater distances, poorer quality and higher standards
- Improved loss control
- HIV/Aids

This is offset by:

- Migration to cities
- Increased standards for developing communities
- Increased political freedom and hopefully less oppression.

A net population growth rate of 2% p.a. was commonly accepted. This was compounded by increased water consumptions, industrialization and greater environmental use. Many counties are now experiencing negative growth due to smaller families, less marriages and increasing costs of living. On the other hand the potential increase in living standards in developing counties could eclipse these.

Allowing for uncertainty in projection a number of forecasting methodologies are used:

1. Extrapolation of historic trends
2. Using questionnaires completed by consumers
3. Demographic modelling which uses population modelling and scenario planning.
4. Stochastic projections with confidence bands

In general there appears to be a decline in growth rate and greater uncertainty in projections. To cope with these new conditions the planner should change his approach. Shorter time horizons may be required in designs, Preference should be made for operating intensive rather than capital intensive projects. To assist planning a higher interest rate could be adopted favouring lower capital expenditure. Above all, a stochastic planning procedure will ensure least cost planning. New ideas are likely to emanate. Greater recycling rates will solve a number of problems. Pollution control, resource conservation and a switch to operating cost oriented works will result.

To make projections, the economy should be divided into sectors and probable water demands estimated for each for short, medium and long terms. Both population growth and per capita growth should be estimated separately for each sector and compounded. The overall demand is a function (sum) of the means of individual sector demands but the variance of the total is less than the sum of the sector variances.

8.2.1 Statistical analysis

Continuous random variables that tend to cluster close to a mean value, μ, with few extreme values, both small and large are often distributed according to the normal or Gaussian distribution. It is symmetrical and described by the following equation:

$$f(x) = \frac{1}{\sqrt{2\pi}} e^{\frac{-x}{2}} \tag{8.1}$$

or

$$z = \frac{x - \mu}{\sigma} \tag{8.2}$$

Where $X \sim N(\mu, \sigma^2)$

z = Probability

μ = Arithmetic mean

σ = Standard deviation

σ^2 = Variance

Approximately 68% of x-values lie within the $\mu \pm \sigma$, 95% of x-values lie within $\mu \pm 2\sigma$ and 99.7% of x-values lie within $\mu \pm 3\sigma$. Tables for a $X \sim N(1, 0)$ are readily available and give z values which are used to obtain the confidence limits, using the following equation:

$$X = \sigma (\pm z) + \mu \tag{8.3}$$

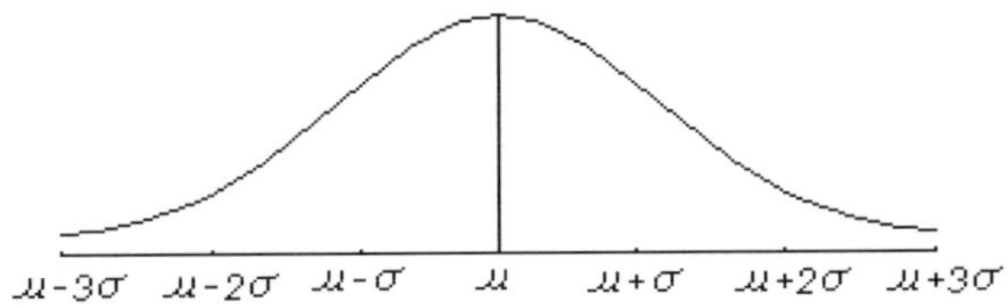

Figure 8.3 Normal distribution

The arithmetic mean, which is often referred to as the average is calculated from:

$$\bar{x} = \frac{\sum_{i=1}^{n} x_i}{n} \qquad (8.4)$$

The standard deviation is the measure of the variability of the data, and is calculated using the following equation:

$$s = \sqrt{\frac{1}{n-1}\sum_{i=1}^{n}\left(x_i - \bar{x}\right)^2} \qquad (8.5)$$

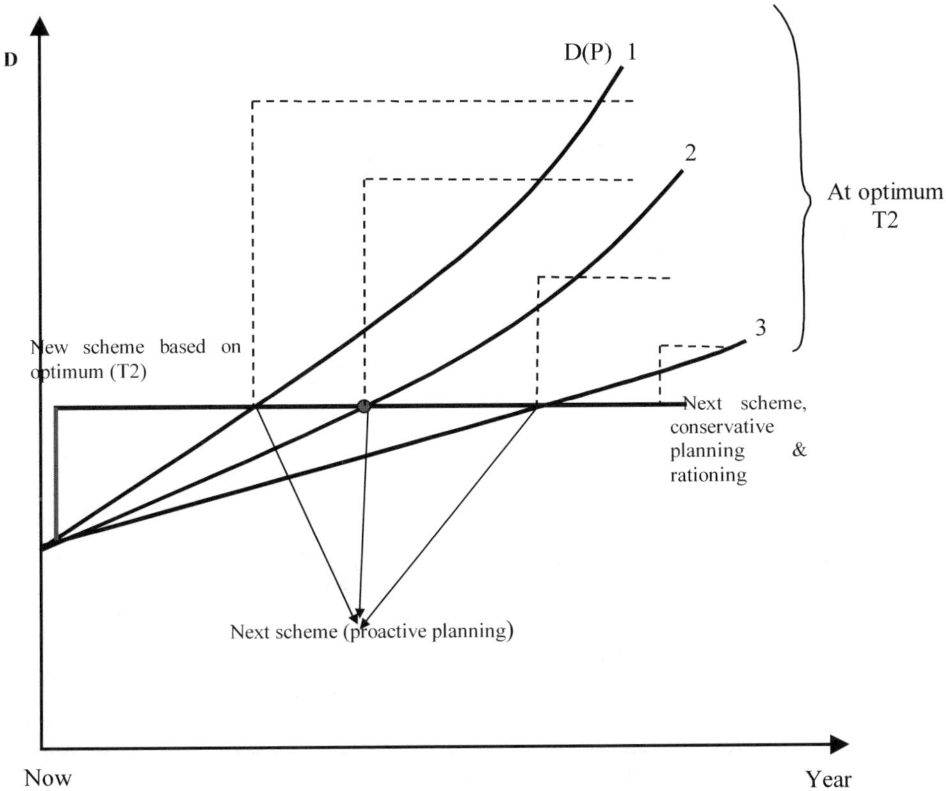

Figure 8.4 Alternative dates for new stages in supply

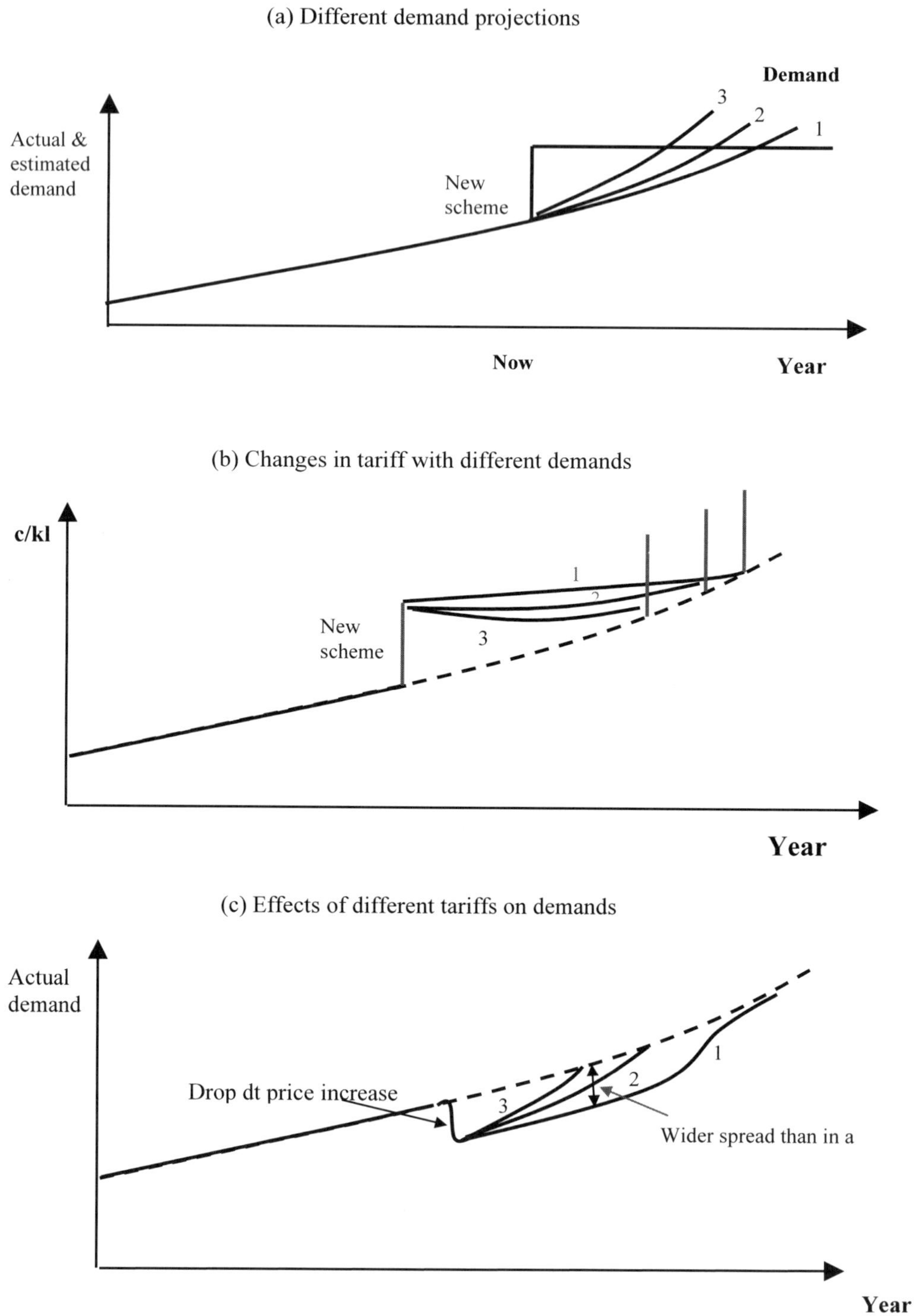

Figure 8.5 Effect of demand variation on cost of water and size of new scheme

Thus in estimating future water requirements the demand should be broken down by sector, e.g. industrial, commercial, agriculture and domestic. Each sector should be projected individually, using probability theory, i.e. a mean and standard deviation estimated, and the confidence bands of each sector demand estimated. Then the combined demand, with a mean and new standard deviation, and corresponding composite confidence band can be evaluated.

8.2.2 Planning alternatives

In Figure 8.4, the demand is projected ahead for three different possible growth rated at the optimum tariff. The year at which subsequent new water supply schemes have to be brought in varies with the growth rate and the approach to demand management. If the demand could be managed by rationing, the next scheme could be delayed, saving interest.

On the other hand (Fig. 8.5), if the expected demand does not materialize there could be a compound effect. In Figure 8.5a, three different future demands are plotted. The corresponding effect on water price is illustrated in Figure 8.5b. If the cost of the new scheme is to be recovered in its entirety, the lower the demand, the higher will have to be the water price. This affects the water consumption even more. A large price increase, in the case of the low demand increase, will compound the effect on demand and reduce demand even more. This feedback effect should be incorporated into demand models.

8.2.3 Disruptions

Projections assume a continuous growth in population and demand. The effect of catastrophies could be a drop in demand, and they never increase demand. Events not accounted for include:

- World economic depression
- War – regional or local
- Drought
- Sabotage of supply
- Political radicalism
- Regional deterioration

Any of these would reduce consumption and therefore reinforce the conservative approach. In some case the demand reverts to the original trend, e.g. after financial change such as interest rates. In other situations there may be a permanent part savings, e.g. after drought where the public were made aware of water scarcity and cost.

Whether the demand will revert to the original line and how long it will take depend on the resilience of the system and the consequences of the cutback. Some consumers may relocate if conditions become too severe. Others may change their method of operation, because cutback often means loss of income.

8.3 HYDRO ELECTRIC POWER

Hydro electric power generation does not consume water. The water is discharged downstream after being passed through the turbines. However, large storage dams are required to generate head, and to store water from times of high river flow to times when the power consumption is high. This means typically storing from the wet season (often summer) to the cold season when power demand is highest.

As a result of the storage there are changes in the river regime, most particularly the energy which would otherwise have been used in transporting sediment or eroding the bed of the river is used for generating hydro electricity. In addition, storage is often needed in order to enable the water to be used when power is required. This changes the flow pattern of the river and in many cases downstream storage may also be provided to re-establish an unacceptable regime of the river. The construction of storage often aggravates evaporation and there is a net reduction in the downstream river flow owing to this.

There is considerable operating agility needed to optimize the use of water available and storage. Coupled with the seasonal storage (months runoff) and drought storage (years of storage), there is also smaller overnight storage required. This is because the hydro power available operates to meet peaks in demand.

The advantages of hydro electric power are the following:

- The cost of hydro electric power stations is generally much lower than the same capacity in fossil fuel power stations. This is because fossil fuel power stations may require furnaces, boilers, steam conduits, cooling towers and chimneys. The turbines are larger in the case of steam generation than water turbines because the density of steam is much less than water.
- Hydro power is sustainable, i.e. as it is a naturally replenished source there is no mining of fuel.
- Hydro power is generally environmentally friendly. There are no polluted emissions to air or water and no unsightly structures. The reservoir or lake is often scenic and can be stocked with fish.
- Hydropower can be switched on and off rapidly whereas fossil fuel burners require a period of time to heat up and die down and the generation cannot follow the peak demand entirely.
- Since the water is not consumed, a hydropower dam can become the center of a multi-purpose development as water can be re-used downstream. Multi-purpose development saves costs such spillways and infrastructure, and the cost of the dam to other consumers is reduced.

The main environmental objections to hydroelectric power are related to the construction of dams and the inundation of large areas of land. This leads to the problems of relocation of residents from the area and sometimes the destruction of habitats for flora and fauna. However, proper environmental studies should be made and it may be that alternatives or even more preferable habitats result with the construction of dams and lakes.

The main disadvantage is the necessity for a dam, in many cases. This can be the most expensive component, and the associated environmental and social problems are often debated. Dams are biggest in many semi-arid areas where rivers dry up during winter (when power demand is highest). There may also be labour disruption as labour is only required during construction. Thereafter, the labour content is low compared with coal mine sources.

The cost of hydro power stations is so low compared with thermal power stations that it pays to operate hydro power on peak loads, i.e. at low load factors, where there is limited water available. In some cases, hydro power is used with pumped storage. Water is pumped into a high level reservoir using thermal energy during off peak periods, and the flow is reversed to generate hydro power during peak demand periods. Even though there is a net loss of energy due to inefficiency in pumping and generating, the cost saving in power station machinery makes it economic.

The *load factor* is defined as the ratio of average power generation to peak over a specified time. The volume of night water storage required increases the lower the load factor, for any average flow rate. The load factor can have a considerable effect on the viability of a project because if hydro energy is used only on peaks it saves the additional cost of thermal plants.

Hydropower can be considered from a number of points of view. Large-scale hydro power plants of the order or hundreds or even thousands of megawatts can be installed on major rivers where there is a large flow. Large reservoirs are constructed with dams in order to store water from wet season to when the hydroelectric power has to be generated. The dams also provide head and enable higher velocities to be obtained through the pipework and turbines for maximizing the power generation.

8.3.1 Pumped storage

Owing to the economy of hydro electric plant compared with alternatives, it is often worthwhile to pump water uphill during off peak periods, to be stored for generating during peak periods. The savings in thermal power plant usually more than offsets the loss of efficiency in pumping and generating. There is no net consumption of water as it is continuously recycled, being pumped during off peak periods by using energy from the thermal power plant and generating during peak periods to reduce the cost of the thermal power plant.

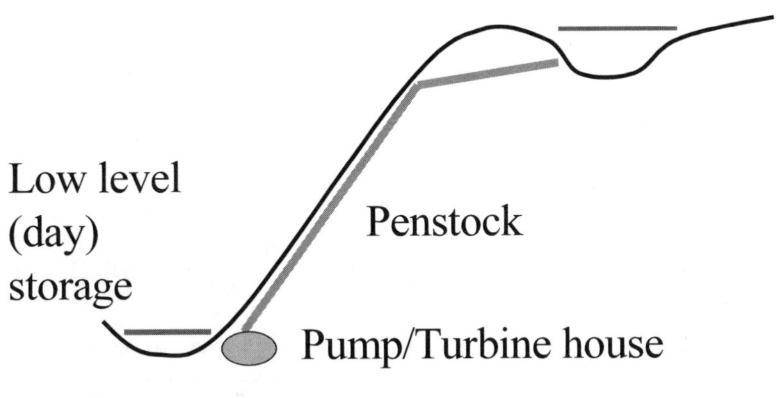

Figure 8.6 Pumped storage system

In some cases, it is possible to use the same machinery for pumping and generation. Centrifugal type machines, i.e. centrifugal pumps, are very similar to Francis turbines and, with some loss of efficiency, can be used for both purposes. However, for high heads it is necessary to use Pelton wheel turbines. In this case, it is not possible to pump with the same machine and either a separate pumping plant is needed or else a pump can be connected with a coupling to the same electrical machine as the turbines. This also reduces the starting up time of the turbine as the momentum of the machine is reduced when the clutch is decoupled.

8.4 ENERGY CALCULATIONS

The power which can be generated from a flow of water Q over a head H is $\Upsilon HQ\eta$, where Υ is the unit weight of water 9.8 (kilonewtons per cubic metre) and η is the fractional efficiency of the system.

Owing to the low cost of hydro electric power stations compared with alternatives, it is preferable to put the hydro electric power on peak load to maximize the value of the water energy, and to generate from alternative sources on base load. This will require certain storage, although overnight storage is very small compared with reservoir storage requirements for seasonal and drought storage.

8.4.1 Example

Consider a town which is to be supplied with power from a small river to whatever degree it can, plus the balance from a coal fired power station. The average flow of the river is 5 cubic metres a second and the head available is 30 metres, while the turbine set efficiency is 85%. The power demand curve of the town is given in Figure 8.7. This graph is replotted as a load duration curve on the right of Figure 8.7.

The energy available from the river per day is $\Upsilon HQ\eta t = 9.8*30*5*0.85*24 = 30000$ kWh/d

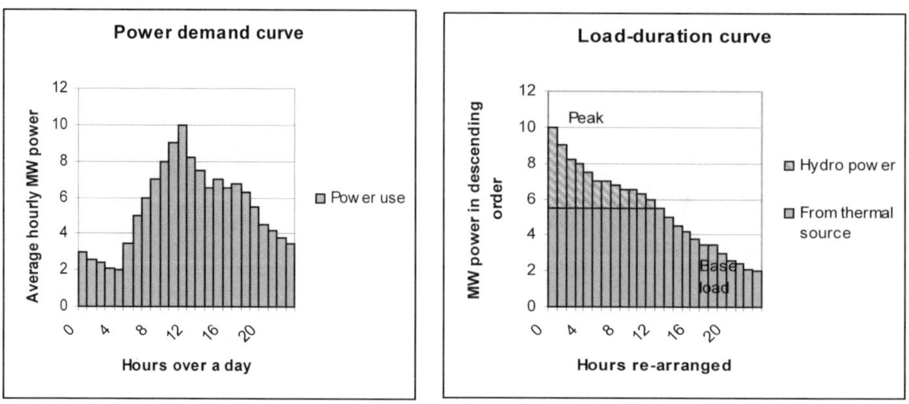

Figure 8.7 Power demand curve of sample town

It would be most economic to operate the hydro power plant on peak loads, thereby saving thermal power plant costs. To do this, it is necessary to store the water for a few hours a day.

The energy available from the water is put onto the peak under the load duration curve and is represented by the hatched area. This indicates that 4.5 Megawatts of power plant can be installed at the hydro power station.

The load factor of the town as a whole is the area under the load duration graph divided by the peak x 24 hours = 60%. The load factor of the hydro power station is much less, i.e. 22%

The volume of river flow per day is Qt =5*86400 = 432 000 m^3. The volume of water which has to be stored between generation periods at the hydro power station is 188 000 m^3, which is the average time to the right of the hatched area in seconds times 5 m^3/s.

8.5 DEVELOPMENT FACTORS

There are a number of factors supporting the development of hydroelectric power in developing areas which come about from policies established for or independently of the necessity for development. There are, however, a number of factors which disfavour hydropower development in these areas.

One of the factors which strangely encourages developing countries to develop hydro power is first world countries' tariff structures imposed on developing countries. On international grids, charges to neighbouring countries are equivalent to those of bulk consumers, but the latter includes certain amounts for distribution, transmission and administration as well as financing. On the other hand, when a country does an economic analysis, it will compare its own costs with the prices which would otherwise be charged to that country, and this tends to favour local development rather than importing electricity.

It is also possible to use tariffs in such a way as to minimize the payments by keeping purchased power on base load so that the energy is at a high load factor. Thus hydropower is generally more suitable on peak load where the power tariff includes an energy component and a monthly power peak cost.

Hydropower needs fairly large scale development, generally, to be competitive, however. That is, it is often difficult to justify micropower unless transmission costs are added in for that particular area only. This is generally not done, and organizations in particular charge a total tariff (including a large proportion of the transmission cost) wherever power is required.

Hydroelectric power is in general capital intensive and requires large investments in the way of dams, tunnels and power stations before the project can be productive. It is, however, difficult for developing countries to obtain this capital as their revenue is insufficient to generate it. On the other hand, developing countries also are regarded as a poor risk investment in many cases so that they have to pay premium interest rates. All this tends to discourage hydroelectric development. Economic justification is also high susceptible to initial estimates, and often final costs exceed the initial estimate on which a scheme was justified. After that, difficult ground conditions encountered may increase engineering costs, floods may cause damage or bring silt, damages may render the scheme unpayable, etc. Potential construction problems in developing countries also result in unknowns. Spares are difficult to obtain, and labour problems are more complicated. All these factors disfavour hydropower development when considered on a financial basis only.

A secondary argument in favour of hydroelectric power and, in particular, dams and water development in these countries is that project construction will bring jobs to many people and, therefore, money to the country. Unless a deliberate attempt is made to do this, however, generally the most competitive bids, when the job is put out to tender, are from mechanised contractors. Even if labour intensive methods are specified, the amount of money put into circulation is often short-lived, and after a few years the labourers are without employment again. Thus, it is probably not as effective as if a national training programme and development facilities for local contractors had been established.

Apart from the fact that developing countries often have unstable economies that discourage investment (Cabora Bassa is an example of a big development which to some extent is wasted owing to the political scenario in Moçambique), there are a number of other risks to be considered in development of hydropower. The hydrology of many rivers in remote areas is highly variable and there is always a chance that an extreme drought will occur, leaving the country at the mercy of neighbouring countries for power requirements or else bringing the country to a standstill. On the other hand, this may be a reason for linking into a larger grid, to prevent this happening.

The larger and more sophisticated a power station, the more likely it is to be designed by a country with a sophisticated infrastructure and technology, with the design and manufacture fees retained outside the developing country's borders. The same often applies to contractors who (if highly mechanised as required for many large projects) are often based at the larger centres.

The problem then arises as to how to involve the local people in order that expenditure for the project is circulated within the country. It is perhaps just as inefficient when expenditures are paid to designers and constructors outside the country as when electricity is purchased from an outside electricity corporation.

Local expertise should be employed at all levels of development of such projects, in particular at the planning stage. Local people should have a responsible role in planning, economic investment and technical factors, in order to become familiar with the responsibilities and operation of the system in subsequent years. It is considered that construction time could be sacrificed, if necessary, in order to preferentially use local people even if they have to be trained prior to use. It will also be necessary to train operators, managers and maintenance crew, and the availability of such staff needs to be considered at the planning stage.

8.5.1 Economics of hydro power development

Interest rates applicable to the development of a dam or other major capital investment have an important bearing on the economic viability of the project. Current interest rates in developing countries are relatively high, i.e. over 12%, but on the other hand inflation rates are often higher. In fact, with present inflation rates, the real interest rate could be negative although the development banks and other development sources have indicated rates between 6% and 3% stupid, they should be considered. These are real interest rates, i.e. the time preference rate for money which otherwise is static in value. It is the same as borrowing money from a country with a very low inflation rate at this interest rate.

It may be that developing countries should choose a lower real discount rate which would justify larger capital intensive projects in preference to operating intensive projects. This, however, could only be the case if the capital is spent within the country. Shadow values could only used however to distinguish between preferences. The rate of interest and the rate of inflation in developing countries should probably be assumed higher than for developed countries in this type of analysis, because development can often proceed at the expense of inflation, i.e. savings are used. Whether real interest rates, or financing rate plus inflation, are used may not affect the results of an economic analysis, but it will affect a cash flow study. It is a shortage of cash which can affect the viability of large projects to a small power company.

It is important that limited water resources be used to their best advantage on a regional basis. Planning of developments such as hydroelectric schemes should, therefore, be done to realize the maximum potential of this limited resource. Planning should begin with a thorough basic study of hydrology and water resources. Many schemes have been and are being proposed with inadequate data. An integrated data collection system to develop water resources must be developed. Continuous records are required of the following:

1. Precipitation in the various catchments with both totals and rates.
2. River flow data including variations from season to season, year to year and peak flood flows.
3. Sediment loads in rivers. This again will have to be on a continuous basis as concentration of silt depends very largely on the river flow rate.

In parallel with the hydrological investigations, an in-depth analysis of local and regional needs is required. This includes:

4. Local needs, viz.:
   - (a) Local power and energy requirements.
   - (b) Local water requirements to meet existing and proposed agricultural, industrial, commercial and domestic developments.
   - (c) Environmental requirements, and
5. Regional needs analysis.
   Sensitive consideration of neighbouring countries' electricity and water supply needs and economics is needed.

Once resources are identified and needs covered, both local and regional, then potential developments should be classified within the regional or local context. Detailed objectives for the evaluation of proposed developments need to be stated and guidelines established. Expansion of hydroelectric facilities should then take place within this framework. In general ad hoc proposals could spoil the ultimate plan and even not meet immediate demands in the most efficient manner.

8.6 MACHINE SELECTION

The head and flow rate will determine the most efficient type of hydraulic turbine. Various types of machine are listed below, but more details information must be sought from books or suppliers (see Fig. 8.8).

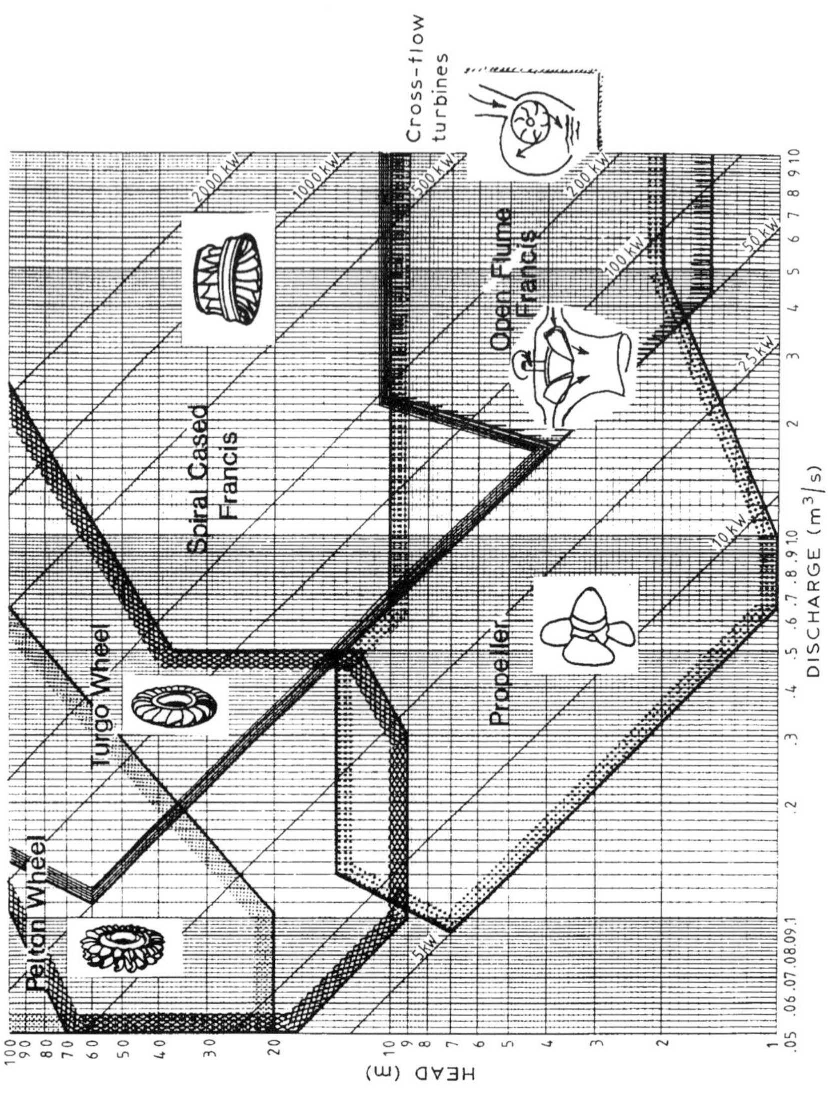

Figure 8.8 Range of application of various turbine types (University of Salford, 1983)

Impulse turbines: water head is connected to velocity through a nozzle, and drives a *Pelton wheel* with buckets (heads up to 200m) or a *Turgo wheel* with blades (heads up to 20m), or a *cross flow* turbine with curved blades. The latter are particularly viable and run with flows down to 16% of design flow rate. Heads from 2m to 100m can be used.

Reaction turbines: have a pressurized casing, and include the *Francis turbine* with radial flow (operating at heads from 30m to 500m) and the propeller type turbine with axial flow (operating at heads from 3m to 60m). Centrifugal pumps in reverse operate similarly to Francis turbines.

8.7 SMALL HYDRO

Micro-hydro power, at the other extreme of hydropower generation, is often considered for off-grid power demands. The economics is such that the relative costs of the micro turbines are higher than for large scale hydro power. The result is that savings have to be made elsewhere. The biggest saving is generally in the elimination of long transmission lines to rural areas from the grid or from larger fossil fuel type power stations. The biggest cost of micro-hydro power is often the dam to store water. For this reason, micro-hydro power is often constructed using run-of-river and therefore has to be built in mountainous areas where there is catchment storage in the ground or a continual flow of water. Alternatively, hybrid systems have to be considered. Hydro power could then be used for a proportion of the time and in fact the river flow is likely to be in excess of that required for the power demand. Efficiency of the system is therefore not of great concern and the capital cost is the major component. During periods of low river flow, it may be necessary to resort to alternative power sources such as diesel or biomass energy. Wind and solar energy have the same disadvantages as hydro power in that they are not continuous.

There has been considerable interest in small hydro electric plants. The cost of such plants is often justified by the savings in transmission costs. Thus mini plants (< 2kW) may serve rural communities or industries and micro plants (< 500W) could serve individual houses away from other sources of electricity. The civil and mechanical costs per unit of such plants are, however, generally higher than for large-scale plants, and this prohibits their use for rural and poorer communities in particular, the very people they are often intended for.

Use of pumps in reverse has been advocated by Dutkiewicz (1986), but again such applications are confined to use on suitable streams where there is adequate water, as mechanical efficiency is then not important. Efficiencies of 50%-80% can be expected, compared with over 90%, i.e. for large plants.

Small-scale hydropower development is not new. It has been used for centuries for driving mills and more recently for electricity generation for specific purposes, e.g. a factory. With the advent of subsidized rural transmission systems from larger plants to meet more general electrical demands, such small-scale plants become redundant.

The energy crisis of the 1970's revived the interest in small scale hydro power (defined as less and 25kW). The linking of small plants to national grids has become relatively easy. However, this requires constant speed of rotation of the generator at a synchronous speed. The buy back price paid by electricity authorities is generally less than their selling price as the energy is not regarded as reliable.

Generally small plants do not have storage, i.e. they are run of river, because the flow requirement is well below the flow of the river. If the small plant is only for a local use, low voltage distribution is often economical, compared with the necessity for very high voltage over long distances to minimize volt drops.

Not only rivers provide hydroelectricity. Frequently canals with falls, or outlets to irrigation areas, can generate small heads. Such may be suitable for reversed flow pumps acting as turbines. Alternatively, pumps may be installed to provide water supply to higher lying villages, or to provide water for a mini pumped storage arrangement with higher head and ability to generate as called on.

Intake screens are an important aspect of small as well as large hydro plants. Debris can damage turbines and blockage of a turbine could cut off power supply, with the loss of

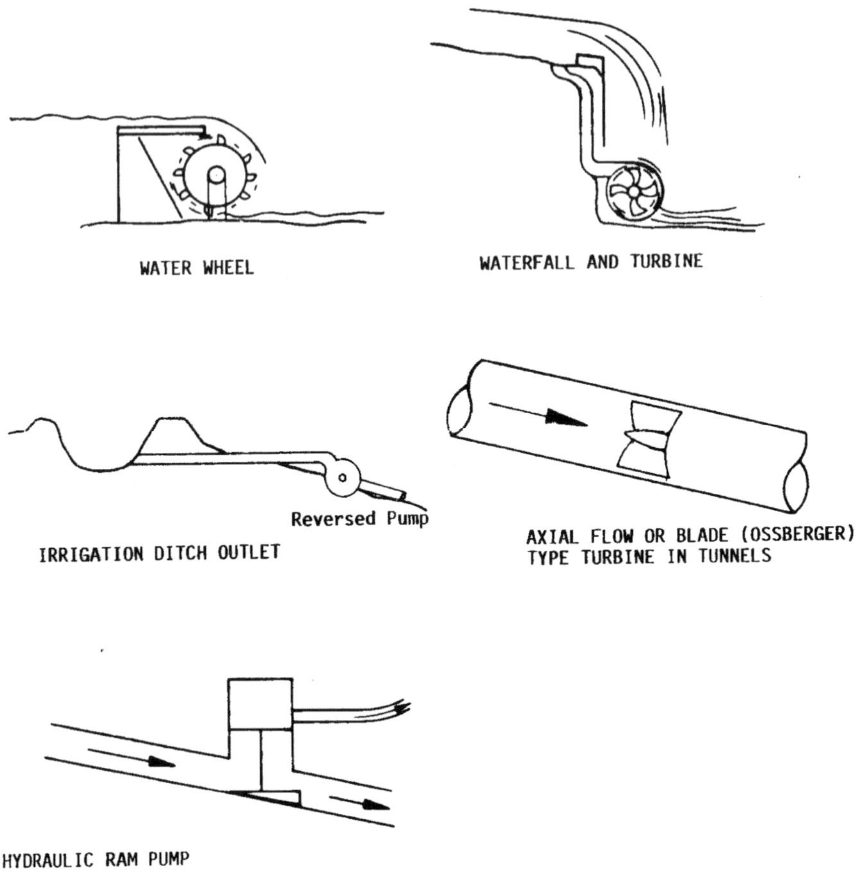

Figure 8.9 Micro hydro options

reliability in the system. Selection of pipe sizes is another area requiring careful engineering as pressures can be high, and friction losses may reduce initially envisaged outputs.

Hydropower is considered as one of the options for generating power for rural communities, as it is less damaging environmentally. Hydro power is sustainable in that water is a renewable resource, and there are no emissions or effluents associated with hydro power generation. For micro-hydro power, diversion and damming of rivers may not be necessary; therefore there are few, if any, of the environmental impacts associated with large dams. Damage to fish in turbines is thus reduced, and barriers to migration of aquatic organisms are less significant than those caused by macro-hydro power generation (Kubečka et al., 1997). Provision of electricity through installation of micro-hydro power generators to communities currently dependent on wood for fuel will reduce the collection pressure on woody biomass.

There is no clear consensus regarding the classification of scales of hydro electro power. However, a small hydro plant could be less than 10 MW, a mini plant is generally less than 1 MW, and a micro plant is generally less than 100 kW. (The latter figure would be sufficient typically for 100 rural huts and a community centre, based on 1 kW per hut and a typical diversity factor.)

The costs of civil works (dams, penstocks, power stations) can be relatively high in a macro-hydroelectric installation and these costs cannot always be scaled down proportionally to smaller stations. Therefore the total project cost per kW installed may be higher for mini-hydro and micro-hydro stations. (Turbine costs are typically in excess of \$1000 per kW of output.) The advantages of micro plants manifest particularly in the saving in transmission network for remote users, as the typical micro-hydro plant is designed just to serve a local community; thus long, high voltage lines and associated energy losses are avoided.

Other infrastructure development costs can sometimes be ameliorated on micro schemes. Seasonal dam storage costs can be avoided if the required flow is less than the low river flow. If the water source does dry up, an alternative power source would be necessary. Flood spillway costs can be avoided if the power station flow is diverted off the main channel to a gorge or ravine. Sealed turbine units may obviate the necessity of power stations. Low set efficiencies may be tolerable if there is surplus river flow. Transformer costs can be reduced by reticulating at low voltage, which also enhances safety.

Micro-hydro power therefore has some distinct advantages over macro-hydro power (Garrad Hassan, 2001). Some of the hydraulic problems associated with small hydro developments were discussed by Stephenson (1999). In a paper, Balance et al. (2000) describe a search for both macro- and micro-hydro power sites in South Africa using a Geographic Information System (GIS) and some statistics:

The micro hydro potential is a function of the mean annual runoff and river gradients:

$$\text{AEPL (kWh yr}^{-1}\text{ km}^{-2}) = 9.8\text{ (m s}^{-2}) \times \tan(\Theta) \times \text{Reach} \times \text{MAFV (m}^3)/3600\text{ (s)} \quad (8.6)$$

Where: AEPL = Local Annual Energy Potential, in kilowatt hours per year per km^2
9.8 = acceleration due to gravity, in metres per second squared
Θ = river gradient
Reach = river reach length (1 000m)
MAFV = mean annual flow volume in m^3 per km^2 of catchment
The mass density of water is assumed 1000 kg/m^3 in the above.

The calculation of micro-hydro energy potential uses MAFV per unit area, rather than cumulative flow. With the aim of supplying rural communities with reliable power from micro-hydro, the total volume of river flow was thought not to be as important as the convenience of a local power source. The proximity of fast flowing water, even if not at a great flow rate, could enable sufficient local micro-scale generation to be exploited. For example, at the source of a river (high in the mountains), there is relatively less volume of water, but the steep gradient provided by the mountains results in a higher potential for power generation, because of the energy of the falling water.

Micro-hydro power, which is dependent on instantaneous stream flow, is susceptible to variation in flow. Small rivers are susceptible to highly variable flow.

The generating capacity of a run-of-river micro-hydro operation would be constrained by low flow conditions. A measure of flow variation and low flow values were therefore included in this assessment of micro-hydro suitability. In addition to plots of AEPL the river flow COV (coefficient of variation) of annual totals and LFI (low flow index) of seasonal variation over a year were mapped. In the case of micro hydro only the LFI would be of relevance as storage would be small. On the other hand the COV would be more relevant to macro hydro plants.

For macro hydro potential study, a cumulative mean annual flow volume (CMAFV) was calculated by aggregating the mean annual flow volumes of all cells upstream to any given cell along simulated river channels. These channels were developed by overlaying altitude, gradient and slope direction, and synthesizing a flow direction for each cell (i.e. determining whether a cell has a net inflow or outflow of water). Sequences of cells with high cumulative flow volumes were considered analogous to rivers.

The macro-hydro power potential was then estimated using equation 8.8:

$$\text{AEP (kWh yr}^{-1}) = 9.8\ (\text{m s}^{-2}) \times \tan(\Theta) \times \text{Reach} \times \text{CMAFV}\ (\text{m}^3\ \text{yr}^{-1}) / 3600\ (\text{s}) \quad (8.7)$$

Where: AEP = Annual Energy Potential, in kilowatt hours per year
9.8 = acceleration due to gravity, in metres per second squared
Θ = river gradient
Reach = river reach length (1 000m)
CMAFV = cumulative mean annual flow volume, in m^3 per year

The areas most suitable for micro-hydro power generation were found to be in the foothills of mountains, and in source areas of the catchments However, site-specific studies are required for final selection, as proximity to settlements requiring power supply and construction possibilities are important influential factors. Macro hydro potential was biggest in large and steep rivers.

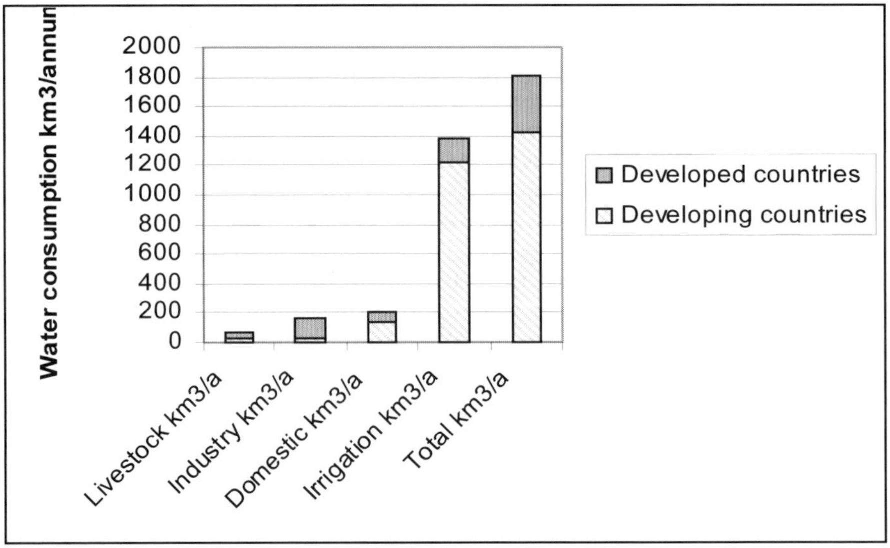

Figure 8.10 World water consumption by sector.

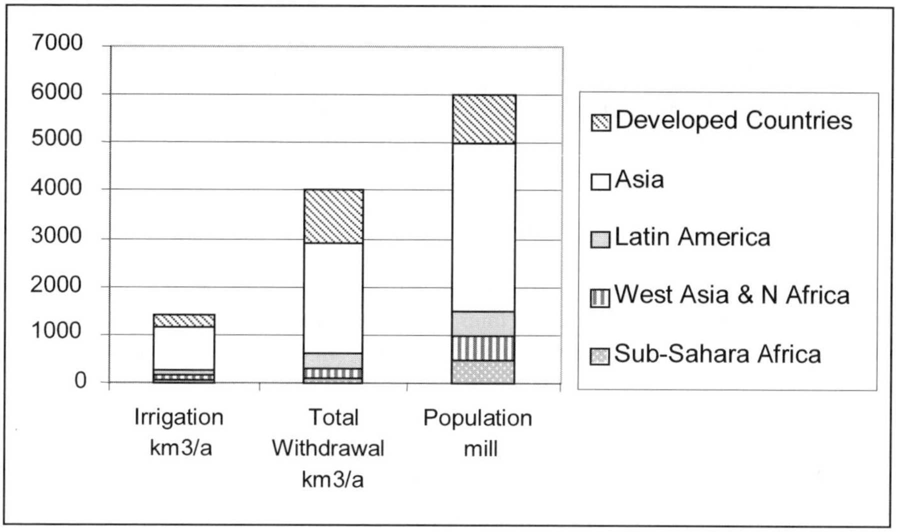

Figure 8.11 Irrigation water consumed by region and total world withdrawal, 2000.

8.8 IRRIGATION

Irrigation traditionally uses up more water than any other user. The return flow is low, less than 25% and most water is evaporated. So not only does it use much water (see Fig. 8.10), it actually loses it from the catchment. Figure 8.11 shows water abstracted and water consumed.

Agriculture is the dominant productive sector in Africa and Asia. It provides 75% employment, 30% exports and accounts for 87% of total water use in Africa. However, agricultural production is limited by:

- Lack of water in much of Africa – short rainy season
- High potential evapotranspiration.
  - e.g. potential evapotranspiration up to 2m/a exceeds rainfall in many areas
- Short growing season (when water is available and temperature >5°C).

Irrigation can result in water pollution. Fertilizers contribute large volumes of nitrates and phosphates, nutrients which cause eutrophication in tropical climates, and are dangerous to life in large doses.

8.8.1 Impact of irrigation

Experience with some projects, even those funded by international agencies, illustrates typical problems associated with inadequate regard for existing social and cultural patterns, for example:

1. A south Asian irrigation project to promote cultivation of onions and chillies expected those crops to be fitted into an existing labour-intensive rice-growing system with peak labour requirements at transplanting and harvest times. Those

labour peaks competed with the time required for chilli and onion production, and the farmers gave priority to their subsistence crop of rice. Because the cash crops were new to the local culture and conflicted with existing crop priorities and interests of the farmers, they were not adopted.

2. Another Asian irrigation project ignored known social obstacles to forming water users' organizations and relied on the force of ministerial decrees that the farmers refused to follow.

The World Bank completed six detailed evaluations of agricultural programs based on farm surveys as part of a 1985 revue. Surveys of several of those projects that are closely tied to water resource development are summarized on the following pages (The World Bank, 1987) (and tables 'Reasons for satisfactory/unsatisfactory performance of agricultural projects).

Social impact of three irrigation projects in Korea, Turkey and Sri Lanka

In the Korea Pyongtaek-Kumgang Irrigation Project, it was found that project farms were about 17.8% larger than the average national holdings, but farmers only had average incomes 6% above the national level. This disappointing result can be explained by Korea's rapid industrialization and the proximity of Seoul to the project area, which caused a drain on farm manpower and large increases in farm labour wages. Though spreading, farm mechanization has not kept pace with the reduction in farm labour supply. Because of migration to urban areas, fewer young people are now employed in agriculture, while older males increasingly do the work with the help of farm machines and female labourers. Revision of farm size ceilings (now 3 ha) will become essential to prevent further migration. A farm large enough to allow a potential farmer both to have a suitable income and be a part of Korea's modern culture may be necessary to attract younger people back to agriculture.

The Turkey Seyhan Irrigation Project has contributed to significant increases in farm incomes. Living standards greatly improved, demonstrated by better health, education and, to a certain extent, lower birth rates. Although land tenure in the project area is highly skewed (23% of the families hold 80% of the land), all smallholders have benefitted from the project. It also helped improve permanent farm labour incomes and provided about 30,000 man-months of employment for seasonal labour.

People in the area are convinced that the project is the source of their great fortune, characterized by a levelling up: poor before, they now are free from debt and risk, they have access to innovation and technology, high living standards, and a stake in the system that only the richest farmers had before.

However, problems loom. There is room for only some children to take over the land, and the fact that good technical education can be rewarded by high farm incomes, provides incentives for all children to assert their claim to farm shares. Some technical problems remain unsolved. Complaints about insufficient credit are common. Policies greatly favour farmers, but subsidy policies could change, as the Bank has sought over the years.

The Sri Lanka Lift Irrigation Project failed to provide irrigation water adequately, dependably or equitably. Systems were underdesigned; water supply was inadequate or irregular; canal deliveries were unsynchronized with requirements of lift systems; broken down pumps took too long to repair; and water distribution was poor. These problems were due to design and implementation deficiencies, some of which were overcome in time, but not until many farmers had justifiably lost confidence in the schemes.

But the technology of chilli and onion cultivation has begun to spread within the project areas and limited export markets for green chillies have developed. Thus, although the market prospects are narrowing, lift irrigation continues to be used by farmers for the cultivation of chillies and other high-value crops. In such cases, irrigation water is likely to supplement rainfall or gravity irrigation and lift mechanisms may continue to be individually controlled.

Project farmers enjoy incomes distinctly higher than non-participating smallholders. However, with no funds to replace pumps/engineers in the next two years when their economic life comes to an end, the project may become unsustainable unless farmers themselves, or government, take action, to replace the pumps/engines in time. The lined channels, pump houses, pipes, etc., which have another fifteen years of useful life, may become redundant without operating pumps.

Case study, Khashm el Girba irrigation scheme, Sudan

The Khashm el Girba project in eastern Sudan, west of the Atbara River, is part of the resettlement program for the High Aswan Dam project. It involves irrigation of about 200,000 ha of land by gravity flow from a storage reservoir on the Atbara River. The project began operation in 1964 (Abu Sin, 1985).

Because provision of irrigation water is frequently one of the major purposes of water resource development projects in developing areas, and resettlement is one of the major problems associated with construction of storage reservoirs, study of the formulation of this project and the ensuing problems can be instructive for all water resource planners.

Abu Sin (1985) postulates that the root of the problems is the different between how planners and settlers define "development". The conflict of interest between management and settlers is one of the major causes of decline in productivity of the project and of other similar major agricultural development projects in the semi-arid areas of Sudan.

Population of the project area in 1980 was about 350,000, including some 150,000 tenants and their families, and 200,000 nomads. About 30% of the tenants are Nubians. Butana nomads constitute over 60% of total tenants and 80% of nomadic tenants. The Nubians, who were relocated from the Aswan reservoir area, live in twenty-five villages with all services. The nomads, whose grazing lands were taken for the Khashm el Girba resettlement project, live in fifty-two villages poorly equipped with services. The three main crops are cotton, wheat and groundnuts.

Annual rainfall in the area ranges from 250 to 300 mm. Losses from evaporation and seepage are about 17%, and field application losses are about 14%, so that less than 70% of available water is actually used for irrigation.

In the first five years of project operation, the project appeared to be successful, but in recent years, problems of management, water shortages, lack of spare parts, etc., have caused a decline in productivity and a decrease in area cultivated. Animal husbandry dominated the area prior to implementation of the irrigation project, and with the decline in yields from cash crops, the nomads are turning again to pastoralism.

Formulation of the irrigation project not only neglected the settlers' perceptions of development and change, but also appeared to have neglected such factors as declining fertility, weeds and water availability.

Table 8.3 Reasons for satisfactory performance of agricultural projects

	Percentage or projects affected by this factor		Percentage of projects where this factor was					
			The most important		Second most important		Third most important	
Contributory Factor	1985	1984	1985	1984	1985	1984	1985	1984
Design merits: Appropriate project content (simplicity, sufficient local resources, or suitable technology	73	54	30	29	8	17	24	6
Design merits: Appropriate institutional arrangements	65	51	14	9	27	29	11	9
Strong borrower support (for project goal and, during implementation, for providing adequate local finance, input supplies)	73	57	30	26	22	14	3	11
Successful procurement	8	6	5	-	8	-	-	-
Successful execution of civil works	35	23	5	6	3	6	8	3
Good institutional performance	51	37	5	11	16	6	16	11
Good performance by consultants or technical assistance	14	14	-	3	5	3	3	3
Favourable economic conditions	8	17	3	3	3	6	-	3
Favourable support of pricing and other government policies	11	6	3	3	-	-	-	-

Figures relate to agricultural projects reviewed in 1984/5 (The World Bank, 1987).

8.8.2 Irrigation technology

Procedures for designing dams, canals and diversion works have been established through colonial experience in India and Egypt, for example (see Houck, 1952), but these are generally for large scale works.

Modern irrigation systems are considerably more efficient than the older methods. Traditionally irrigation has been by flood in developing countries. Sprinklers however are reputed to be much more water efficient, although they are more capital intensive. Another technological advancement is the automatic type of lateral valves and schedulers. They again have proved efficient in advanced countries but may not be understood or may be mismanaged in less appropriate environments. Travelling irrigation sprinklers and even centre-pivot type schemes require careful operator training before they are viable. In general, the use of the smaller drip type nozzles appears to be less technological but requires more maintenance in the way of cleaning of nozzles, etc. Computer crop scheduling is also obviously not appropriate unless permanent contract type managers are brought in to assist with the operation, i.e. a back up economic input is required for many years before such projects can be taken over by the local people in many cases.

Methods of distributing water have been developed in third world countries which are most appropriate to the circumstances. For instance, the surjan which was developed in Java (Pickford, 1987).

The surjan is a system of parallel furrows through which the water is diverted and more particularly rainwater is caught and held. Different crops are planted in the furrows and on the ridges. For instance, trees, which have deeper roots, may be planted on the ridges and even rice could be planted in the furrows.

The Food and Agricultural Organization of the United Nations has started a review of existing irrigation schemes. They are concentrating on small-scale methods such as the construction of wells. They are also looking at resettlement and provision of roads and infrastructure to ensure the projects are viable. They consider the necessity for continuing with livestock despite the problems due to poor grazing practices, realizing that keeping livestock is a basic part of life in many situations, either acting as wealth indices or means of transport. Often the use of livestock for food is of small consideration as it is expensive to rear when considering the amount of grazing land required.

There are also many mission stations which are improving agricultural practices. By forming a focus, the awareness of standards of living and appreciation of the values in life are instilled in the people. A sense of permanence and, therefore, awareness of the environment and improvement in living standards results. This in turn leads to a desire to improve crops and therefore irrigate and fertilize and manage well.

Considering the number of people in the world involved in agriculture and irrigation, and in fact only in these fields, then the amount of money spent on research and even understanding their problems is small and probably at the moment insufficient to hope to raise the living standards of the majority of the world's population.

In underdeveloped areas the competition for water is not likely to be fierce. It may be more appropriate to design a low efficiency system with resulting savings in costs. Unlined canals around flood irrigation may be most appropriate for small-scale plots. Where the plots are scattered saturation of the subsoil may not be a problem.

The net result is that more area and hence more people, can be supplied for a fixed budget. Operation manpower is likely to be higher but less skilled than the sprinklers or drips. That is, however, one of the objectives, i.e. to provide employment for rural people and reduce use of pumps, pipework and complex application systems to a minimum.

Land classification is an important but often neglected aspect in planning irrigation. Soil composition is important from crop growth point of view, texture for drainage and moisture retention, salinity for affecting the transpiration process, slope for drainage and access, and depth for ploughing and drainage. Stone content, weather, ease of initial cleaning, aspect for sun and elevation for supplying water require further consideration. Then ownership, previous use, social attitudes and present cover will influence the decisions as well as economics of water supply, land preparation, fertilizing, crop suitability, weeding, reaping and marketing. Finally, human resources, training and management must be available.

Salts in the soil and water can affect productivity. Alkalis can be brought to the surface by capillary action and cake as water evaporates. They also affect the transpiration process and yield. In such soils, overhead sprays may be preferable to flood irrigation.

Losses due to seepage and soil evaporation can be greater with flooding and unlined ditches, and the water table can rise causing saturation and flooding.

8.8.3 Water requirements

The amount of irrigation water depends on:

- Method of irrigation
- Rainfall
- Temperature
- Wind
- Humidity
- Groundwater
- Effectiveness of water
- Increased yield
- Salinity
- Crop
- Crop density and foliage
- Number of crops per year
- Time of year
- Losses in conveyance, storage and soil

Generally the amount of water is measured as an equivalent depth over the area to be irrigated. The rate of application can be 20–40mm per day, depending on the soil type and method of application, but the period between applications will depend on crop requirements. As a rule of thumb, 0.5m to 1m of water is required per year, including field losses which can be between 10 and 50 percent. Transmission losses can be equally large (see Doorenbos and Kassam, 1979).

Crop requirements can be calculated using a formula such as that of Olivier (1961) based on a modified Penman approach (1948). An alternative formula is that of the U.S. Department of Agriculture (1974):

Daily potential evapotranspiration in mm:

$$E_{tp} = 0.000673 \times 25.4\,[C_1\,(R_n - G) + 15.36C_2\,(1.1 + 0.017 \times 0.625W)(e_s - e_d)] \quad (8.8)$$

Where:

C_1 and C_2	=	mean air temperature weighing factors ($C_1 + C_2 = 1$)
e_s	=	mean saturation vapour pressure in mb
e_d	=	saturation vapour pressure at mean dewpoint temperature
W	=	total daily wind movement, km
R_n	=	daily net radiation in $\text{ca}\ell/\text{cm}^2$
G	=	daily soil heat flux in $\text{ca}\ell/\text{cm}^2$
c_2	=	$0.959 - 0.0125\Gamma + 0.000045\Gamma^2$
Γ	=	mean daily air temp in $^{\circ}$F = $^{\circ}$C x 1.8 + 32
e(T)	=	$-0.7 + 0.295T - 0.0052T^2 + 89 \times 10^{-6}T^3$
G	=	$5[\Gamma - (\Gamma_{-1} + \Gamma_{-2} + \Gamma_{-3})/3]$

T_{-i} = mean air temp for ith previous day

$$R_n = 0.77R_s\left[\frac{0.9R_s}{R_{bo}}+0.1\right]R_{bo} \quad (8.9)$$

$$R_{so} = \text{solar radiation on a clear day} = 760\ \exp\left[\frac{\text{Day1}-107}{157}\right]^2 \quad (8.10)$$

Where: Day 1 = March 1 in Northern Hemisphere, or September 1 in Southern Hemisphere.

R_{bo} = net outgoing long wave radiation on a clear day

$$\left(\frac{0.37-0.044}{e_d}\right)\left(11.7x10^{-6}\right)\left[\frac{T_a^{\ 4}+T_b^{\ 4}}{2}\right] \quad (8.11)$$

T_a and T_b = max and min daily temperature in ^oK

As an approximation, the potential evapotranspiration can equal reservoir evaporation, which is some 20% less than "A" pan evaporation. With dormant periods, the evapotranspiration can be considerably less than free surface evaporation; however, field and conveyance losses can add up to 100 percent on to net requirements.

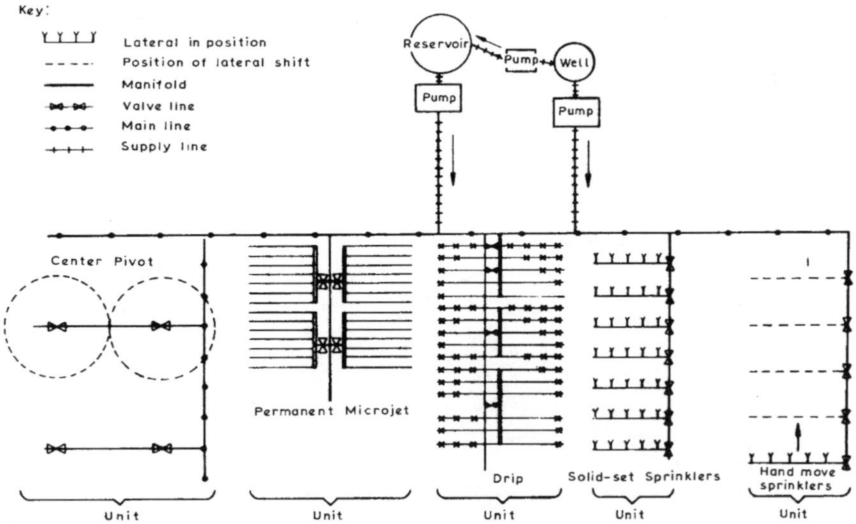

Figure 8.12 Types of irrigation water distribution (Karmeli et al., 1985)

8.8.4 Factors and objectives in the selection of emitters

The selection of a given type of emitter (drip, sprayer or sprinkler) for low intensity irrigation, is based on a number of factors as follows:

- The nominal emitter discharge
- The nominal emitter operating pressure
- The relationships of emitter discharge and pressure
- The size of flow cross section (nozzle size in sprayers and in sprinklers; orifice or flow path size in drippers)
- The vertical angle of water jet (for sprayers and sprinklers)
- The wetting diameter of a single emitter
- The wetting pattern of a single and/or a group of emitters
- The spacing and position of emitters along and between laterals

The selection of the emitter based on the determination of the factors listed above, is carried out by simultaneously satisfying a set of objectives which are directly affected by the emitter characteristics. The basic parameters in each case:

1. The application rate of the irrigation water is

$$I = \frac{qE}{bxr} x1000 \tag{8.12}$$

Where: I = application rate - mm/hr
qE = nominal emitter discharge - m^3/hr
b,r = emitter spacings - m x m
AGD = gross application depth - mm

2. Time of application

The time required for the desired depth of application is given by Tapp = AGD/1

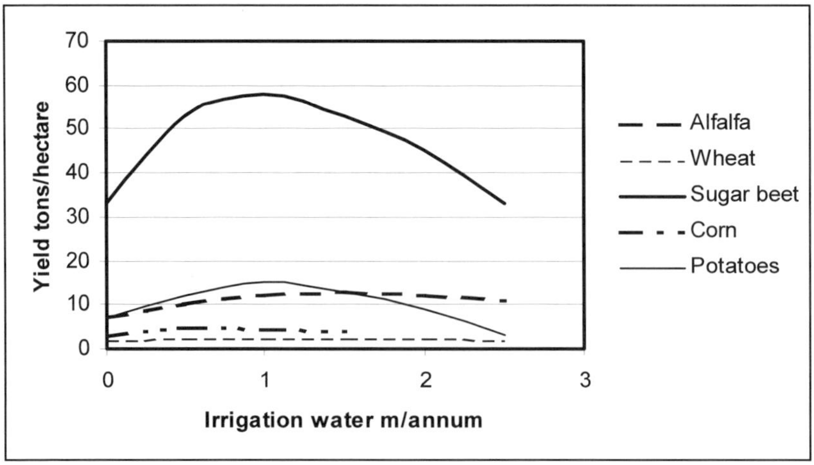

Figure 8.13 Crop yields as a function of water application

8.9 PASSIVE USE

Irrigation is a large consumer of water. The water is lost by evaporation from channels, dams and wet soil, as well as by transpiration from crop leaves. The total evaporation is higher than from natural vegetation because farming intensifies the growing rate and production. There may also be a fallow period with natural vegetation. The gross evaporation can range from 200mm per year for sparsely vegetated arid areas to over 2000mm per year for intensively irrigated land, plus transmission losses. But natural vegetation also uses rain and runoff and groundwater. Forests in particular are high evaporative loss areas. Forestry may have to pay a premium on land use because of the high abstraction rate.

Industry and municipal users return a high proportion of water used to water courses but contribute to the pollution load. Over 75% is generally returned to water courses or the ground. Hydro power does not consume water but diverts it.

There remains a tertiary type of user of water resources. That is, navigation and recreation. They do not divert, or cycle or evaporate water but rely on the water being where it is. Fish life and other aquatic resources also depend on the water without detracting from its availability. But a reserve is required for their use. This may have to be considered in planning storage or discharge facilities. And facilities may have to be built to facilitate their passage, i.e. navigation locks and fish passes.

REFERENCES

Abu Sin, M.E. (1985). Planners and Participants Perceptions of Development in the semi arid lands of Sudan. In *Case Studies in Sudan,* H.R.J Davies, The United Nations University.

Balance, A., Stephenson, D., Chapman, R.A. and Muller, J. (2000). A geographic information systems analysis of hydro power potential in South Africa. *Jnl. Hydroinformatics*, 02.4, IWA, 247-254

Doorenbos, J. and Kassam, A.H. (1979). *Yield response to water*. FAO Irrigation and Drainage paper no. 33, Rome.

Dutkiewicz, R.K. (1986). *The potential for small-scale hydro plant*. Energy Research Institute, University of Cape Town.

Garrad Hassan and Partners Ltd (2001). *New Solutions for Energy Supply – Renewable energy for rural electrification in Eastern Cape*. EU Energy series document 2614/GR/01.

Houck, I. (1952). Irrigation. Chap. 17 in Davis, C.V. *Handbook of Applied Hydraulics*, McGraw Hill.

Karmeli, D., Peri, G. and Todes, M. (1985). *Topics in irrigation systems design and operation.* Course in Continuing Eng. Educn., University of the Witwatersrand

Kubečka, J., Matena, J. and Hartvich, P. (1997). Adverse effects of small hydropower stations in the Czech Republic: 1. Bypass plants. Regulated Rivers: *Res. Management*., 13, 101-113.

Meadows, D.H., Meadows, D.L., Randes, J. and Bejrers, W.W. (1972). *The Limit of Growth*, for the Club of Rome, Pan Books, London.

Olivier, H. (1961). *Irrigation and Climate*. Edward Arnold, London.

Penman, H.L. (1948). Natural evaporation from open water bare soil and grass. *Proc. Royal Soc.* No. 1032, 193/

Pickford, J. (Ed) (1987). *Developing World Water*. Grosman Press, p260-309.

Saparta, D. and Munoz, M. (1995). Water consumption in distribution networks, short term demand forecast. In *Improving Efficiency and Reliability in Water Distribution* Systems. Ed. Cabrera, E. and Vela, A.F., Kluwer Academic Publications, Dordrecht.

Stephenson, D. and Turner, K. (1997). Water consumption patterns in Gauteng. *Proc. Inst. Munic. Engrs*. S.A., Jan.

Stephenson, D. (1999). Hydraulic problems associated with small hydro power. *Proc. IAHR Conf.*, Graz.

The World Bank (1987). *12th Annual Review of Project Performance Results*. Washington.

Twort, A.C., Law, F.M., Crowley, F.W. and Tatnayaka, D.D. (1994). *Water Supply*, 4th Ed., Edward Arnold, London.

University of Salford (1983). *Report on The Development of Small Scale Hydro-Electric Power Plants,* Dept. Civil Eng.

U.S. Dept. of Agriculture (1974). *Scheduling irrigations using a programmable calculator*. Publicn. ARS-NC-12.

CHAPTER 9

Demand Management

9.1 BALANCING SUPPLY AND DEMAND

Although there are theoretically water resources sufficient to meet world requirements now and for the foreseeable future, the increasing distances required to pump water, and the storage required to meet droughts, make the cost of exploiting new sources higher and higher. The price of further water is therefore likely to increase. In addition, the expected increasing living standard of many of the population will mean that greater volumes of water will be sought, even though consumption may start at a minimal level. Resources are being depleted or polluted. This affects not only availability for human consumption but also the amount of water in rivers and for nature. A balance will therefore have to be achieved between consumption and new supplies (see e.g. Rademeyer et al., 1997). This cannot be achieved except by considering marginal costs and increasing tariffs. The occasions when water tariffs need to be considered will also affect the instrument used to control usage. During crises (e.g. drought), short-term tariff increases may be applied, whereas in the long-term, the average tariff will depend on historical costs and the cost of new sources.

Water consumption can be limited by physical, sociological or economic means (instruments). Physical means include cutoffs or pressure control by reduced pumping or constrictions in pipes, e.g. orifices or washers. The latter costs money in waste of energy and cost of installations. On the other hand, it may even out the water drawoff variations by making consumers take water uniformly over more hours per day and provide in-house storage to meet peak consumption. The former (curtailing supply over periods of hours), could result in higher peaks when supply is resumed, but this will in turn reduce pressure and therefore peak drawoff. Demand control by pressure reduction could result in different drawoff patterns. Roof tanks could be filled at night. This will save distribution pipe costs but not necessarily reduce total volume of use. It may also be possible to reduce supplies to uneconomical, no longer valued consumers with compensation, in preference to newer consumers. In the long-term, water-saving plumbing devices could be installed. These include small and double action cisterns, low-volume showers, and automatic tap closers. Invention of water savings devices such as reuse of basin water for toilet flushing, not only saves water in that situation, but they make people aware of water scarcity.

Sociological methods include appeals, way of living or legal action. Appeals, through the media or on monthly accounts, rarely last long before consumers forget the urgency. Long-term changes in ways of life to reduce water consumption will generally be caused by increasing water costs, together with public relations campaigns. Legal enforcement of water restrictions, in associated with fines, can be effective but costly to apply. It may mean inspectors checking consumers, or relying on spying neighbours. Then fines would have to be imposed by courts unless incorporated in water accounts. Such methods include prohibiting use of water on gardens on specified days, banning filling swimming pools or use of hosepipes for flushing drives. Consumer awareness can encourage local reuse of grey water, e.g. wash-water for gardening.

Economic methods include water tariffs, metering or charges on discharges. Theoretically, the best system would be to charge prices which reduce the usage to meet availability. This is, however, an unknown equation since the true value of water may not be known to the supplier or even the consumer. It may also involve tiered tariffs. That is, successively increasing consumption will be charged at higher rates so that the basic requirements of consumers, particularly domestic consumers, are met and more luxurious uses are charged at higher rates. This assumes there will be no trading between consumers (Moore, 1989). It may also encourage consumers to seek alternative sources which, although they may be more costly in total supply, may be cheaper to individual consumers.

Apart from the socio-economic objectives of providing water, there is a long-term value of water. If the world population and standards of living continue to increase, water will become scarcer. It may also occur that climatic change requires more careful use of water owing to reduced availability or greater variability in rainfall.

The traditional approach to supply management is to meet demands with successively more expensive schemes until the demand balances the supply.

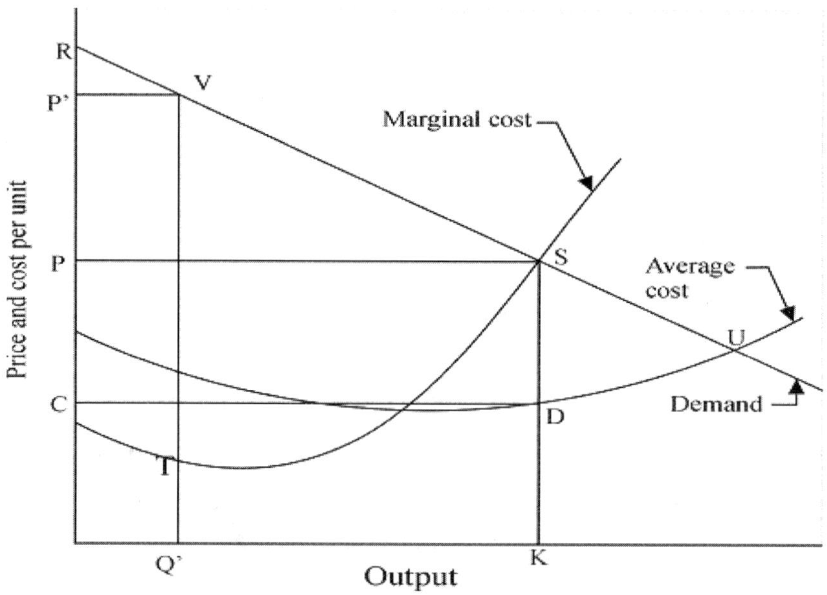

Figure 9.1 Supply and demand with different price structures

However, unless marginal pricing is applied, the average supply cost will always be less than the marginal additional cost of water, so that the demand will continue to increase asymptotically.

9.2 ECONOMIC THEORY OF SUPPLY AND DEMAND

A fundamental concept in economics is the law of supply and demand. Figure 9.1 shows theoretical supply and demand for water. At higher prices, producers would be willing to supply more but consumer demand would decrease; at lower prices, consumers would demand more but producers would cut back on supply. Figure 9.1 shows the theoretical equilibrium condition between the price and the quantity supplied and demanded for average costing and marginal costing (Hirschleifer et al., 1960).

With increasing price, the reduction decreases because further reductions may require changes in behaviour that are inconvenient or contrary to personal or social norms. And at even higher prices, there will be no reduction at all if it means cutting into essential uses like cooking and waste disposal. On Figure 9.2, this relationship is shown by an increasingly steep demand curve as prices increase on the left side of the graph. At lower prices, people will buy and use more water, but there is a limit on how much water anyone can use, even if it is free. So again as price falls, demand eventually drops off as well.

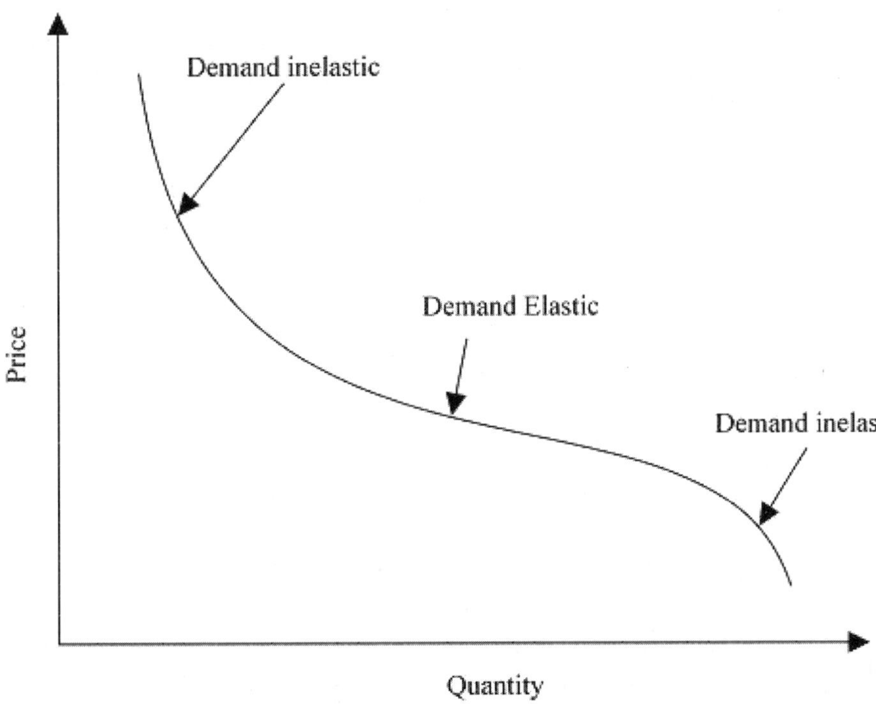

Figure 9.2 Showing how elasticity changes at different points along the water demand curve

The rate at which demand changes as price changes is called the *price elasticity of demand.* (Similarly, there is a *price elasticity of supply.*) Conceptually, when demand changes a great deal for a given change in price demand is said to be *elastic*. When demand does not change very much compared to the change in price, demand is said to be *inelastic*. Economist calculate the elasticity of demand e as:

$$e = \frac{dQ / Q}{dP / P} \tag{9.1}$$

Where P is price and Q is quantity.

As new water schemes are commissioned, the average cost per unit (long-run average cost or LRAC) is likely to increase due to more expensive projects succeeding cheaper projects. On the other hand, over the life-span of each project the short-run average costs (SRAC) may reduce as consumption increases and more efficient use of facilities occurs (Fig. 9.3).

The short run cost changes are due to construction of new schemes or abnormal one-off costs. For example when a new project such as a dam is constructed it may be able to meet demands for another 20 years but initially the cost (probably annual interest and loan redemption) have to be met. Without a balancing fund this may result in a price hike for the water. This could in fact be a compound effect, for the increase in cost of water could reduce demand temporally, resulting in a further price increase.

A water company should anticipate these reactions to smooth to price increases. And new schemes could be delayed by applying restrictions for a few years to reduce costs and enable the new project to come in at a higher demand base. If there is more than one source of water the short term cost in Figure 9.3 may not rise to the marginal cost as costs could be averaged over old and new schemes.

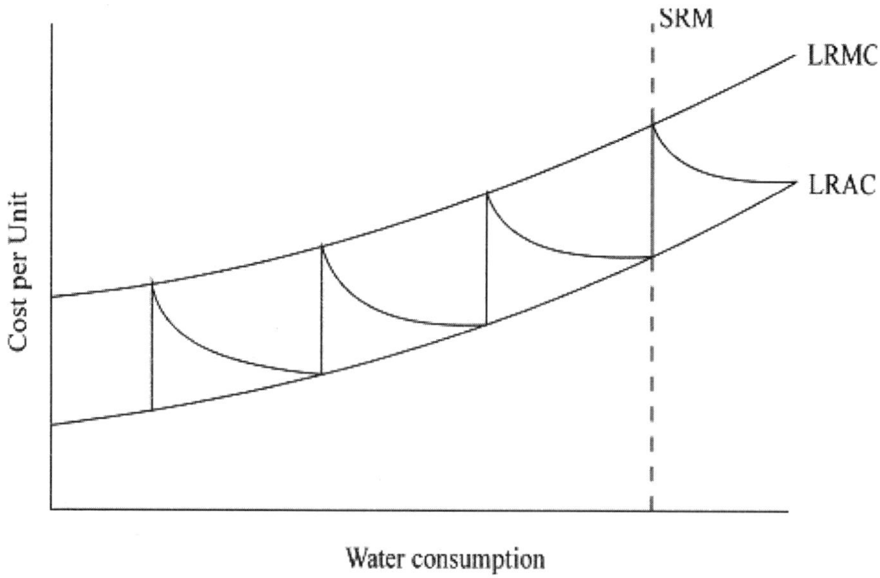

Figure 9.3 Short-run average costs of water supply

9.2.1 Effect of metering

Those consumers who pay average cost will tend to use more than those paying higher marginal cost. If the water is metered, it is the (long range) marginal cost to the consumer which influences the consumption (Fig. 9.4) since the supplier can observe who is using water excessively and charge them a higher marginal tariff. If it is unmetered, only public responsibility, which is related to LRAC, controls consumption (Q_1), or the water company can only charge an average tariff which will encourage greater use.

The marginal cost of not metering is area ABQ_1Q_2. This may be compared with the cost of metering.

Actually, the marginal cost varies slightly with metering, so the comparison is a bit more complicated (see Henderson Sellers, 1979).

9.3 MANAGEMENT BY USE OF TARIFFS

If the true value of water to consumers could be assessed, it may be possible to charge a limiting tariff. This method could be applied on a long-term basis or less effectively for short-term (crisis) demand reduction. However, one must be careful of applying crisis criteria persistently. Somc consumers may locate their organization based on indicate water tariffs but the use of variable tariffs to manage water during drought, for example, must be explained and incorporated within the overall tariff system.

The level of consumption could be decided at the planning stage, if the cost of assured water is balanced against the cost to the economy of rationing. However, the operational basis will be from a different perspective.

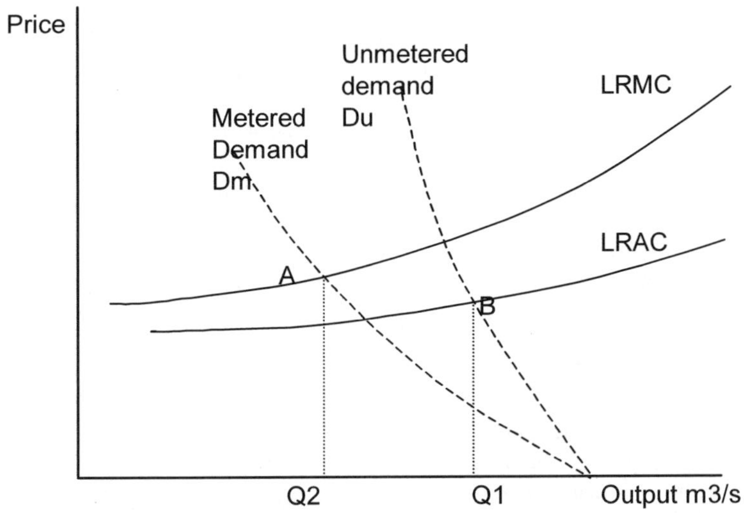

Figure 9.4 Demand curves with and without metering

Unfortunately, a uniform tariff cannot be applied in this way to restrict the use of water, for the poorest sectors of the economy may not be able to meet the tariffs which would be imposed on industry in order to force them to restrict water. Therefore, a percentage reduction, or a differential tariff or shadow value may have to be incorporated. The shadow value may not be paid by the poorer sectors but it should be added onto the cost of water to others. The alternative would be to charge a tiered tariff, i.e. the first volume would be at the original tariff and above a lifeline supply rate the tariff would be successively increased as a function of the percent of the lifeline supply rate. In this way, poorer consumers will only pay marginally more for excess consumption, whereas richer or industrial consumers would pay considerably more. The tariffs would have to be based on the economic value to all consumers. Dandy and Connarty (1994) indicate increased tariffs reduce consumption but to a limit.

Hong Kong's experience with tiered tariffs (Chan, 1997) is that the resulting demand management is limited. But they were limited by having to keep charge levels within inflation. Their most successful experiment in saving water was to use sea-water for flushing. Tucson's experiments with block rates also failed due to the politicians' control on maxima (Agthe and Billings, 1997), but their summer rate differential reduced consumption. Australia is also experimenting with demand management (Duncan, 1991).

In fact, water pricing experiences throughout the world (Dinar and Subramanian, 1997) show that external objectives of politicians or administrators can destroy the efficiency of water use control through tariffs. Increasing prices can instead be intended for many purposes, e.g. financing new schemes, becoming financially self-sufficient or cross-subsidization.

In the long run, it may also be that the consumer could find alternatives to being restricted in water usage or paying higher tariffs. He may seek alternative sources such as groundwater. These sources may have a higher operating cost but as they are intermittent it may not be as severe as long-term usage. This results in efficient conjunctive use of alternative resources.

Consumers may also elect to reuse water and if necessary purify the effluent reused. Again this may be a higher operating cost alternative but, owing to the limited duration effect, could be ameliorated.

The effectiveness of economic methods to control use will vary with the consumer. Industry may be most sensitive to price increases, whereas poor people will hardly be able to adjust their consumption even though they may find it difficult to pay. The richer domestic consumer is likely to have most elasticity in demand, but this is likely to constitute a decreasing proportion of the total.

In order to put objectiveness into water tariffs, Bahl and Linn (1992) suggest a five-part tariff based on:

Variable costs:	Consumption
	Maintenance
Fixed costs:	Connection
	Development
	Upgrading

The above basis is, however, not sufficiently detailed to control use or obtain a method of cost allocation. There are other factors which affect water tariffs, e.g.:

- Capital and operating costs

- Opportunity cost
- Time-of-use or peak-load basis (Eskom, 1994)
- Size of property (e.g. Lumgair, 1994)
- Size of connection
- Zoning of district or purpose of use
- Timing of application
- Investment reserve
- Conservation
- Environmental
- Foundation consumers
- Insurance to ensure continuity during shortfalls
- Capacity allocation (Dudley, 1990)
- Tiered
- Cross-subsidization of income groups
- Location

9.4 TIMING

There are three stages during which the tariff for water needs consideration. (Table 9.1 summarizes which methods of demand management arc applicable to which occasion.).

9.4.1 Long-term (planning and design)

Before a water scheme is constructed, the capital cost of the project is likely to be the most serious economic consideration. Average running costs will be added to discounted capital cost of dams and conduits for alternative schemes in order to select the most economical alternative. If rationing is to be considered at this stage as an alternative to larger resource schemes, the true economic cost to the consumers due to shortfall also needs to be included. (This is not the same as the income to the water supplier which may even increase due to punitive tariffs during shortfall.)

Table 9.1 Demand management methods and their use

Method	Crisis management (Drought, non-payment)	Operational time-frame	Long-term (Planning and design)
Technical	Pressure reduction Scheduled supply Valve closure	Flow control Orifices	Metering Loss control Plumbing devices
Social	Appeal Social persuasion Advertisements	Legislation Cross subsidies	Consumer awareness Education
Economic	Fines Punitive measures	Differential tariffs Trade	Supply and demand economics. Marginal prices

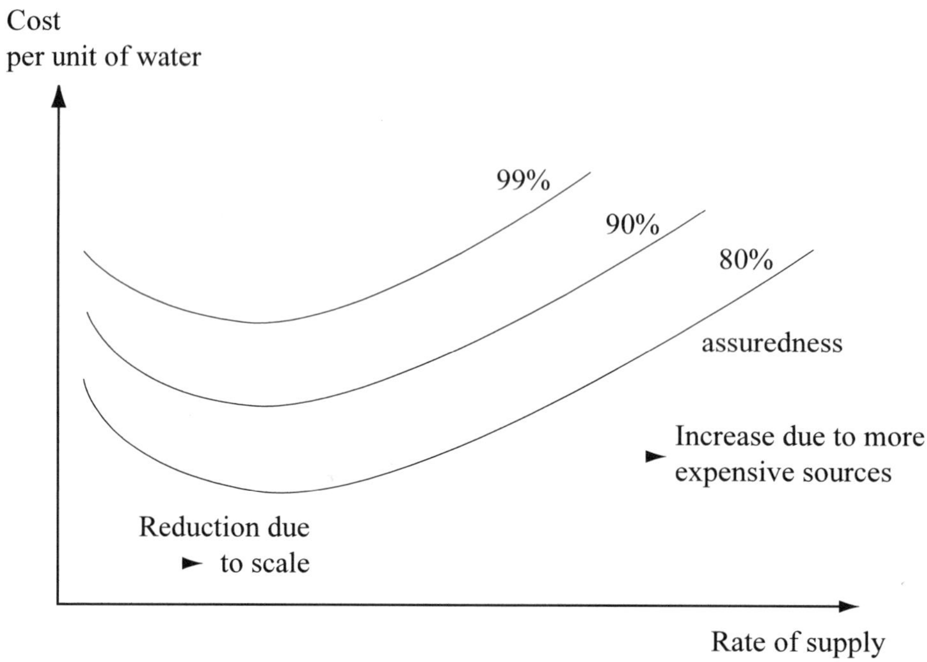

Figure 9.5 Effect of assuredness on cost of water

When new water schemes are being considered, the cost of the scheme and consequently the average cost of water to consumers is the prime criterion. Alternative sources and levels of assuredness will be compared (Berthouex, 1971). This section is concerned with the reliability of supply during drought, and typically the more reliable the surface source, the greater the cost will be (see Fig. 9.5).

9.4.2 Operational time-frame

Once the scheme (e.g. dam and waterworks) is built, its cost does not feature in operational optimization. The object of the new optimization exercise is to minimize economic loss due to restrictions. This may mean shuffling the available water around to minimize total economic loss. The result will be an operating policy for a reservoir.

After a water scheme is commissioned, the perspective changes and day-to-day, as well as annual, supply rates change. Each year, the tariff may be revised as the supply rate increases, and hence the tariff could be reduced if it were solely to meet fixed repayment costs. However, funds for future more expensive schemes also have to be raised so it rarely happens that the tariff drops over the years. An operational policy for reservoirs may be designed to enable water to be conserved during drought. The control of usage could be by tariffs. The tariff may be consumer orientated, i.e. lower tariffs for the poor, higher for the rich, or industry. A tiered or sliding tariff structure generally results. (Fig. 9.6 shows the resulting effect on consumption).

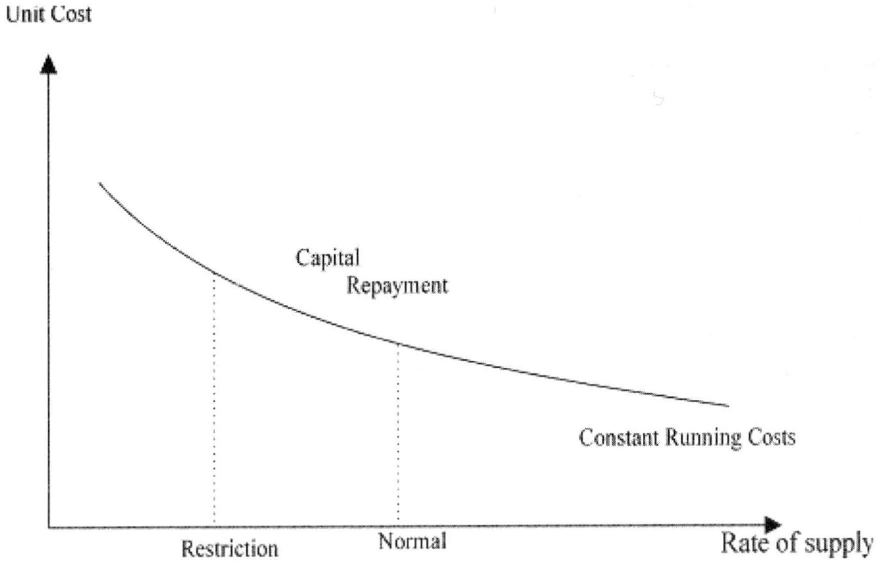

Figure 9.6 Effect of tariff on consumption

The objective of the water works should be to cover costs. They should not unduly be enriched, by charging marginal costs or basing tariffs on what the market will pay. There may also be a planning component and a stabilizing component in the tariff.

The consumer on the other hand is entitled to minimize his costs. He may seek alternative sources of water, or insure himself against shortfalls. Industry and agriculture would suffer real losses if water was restricted or charged at an uneconomic rate. He may store water, or trade it, or purify and reuse wastewater. The trading or reselling or buying water could affect the water works efforts in the short term so the supplier needs to think of similar saving measures. Generally an operating cost intensive source will be retained as a standby, as the costs may not be incurred. Such schemes could include recycling or boreholes.

Capacity allocation is not a tariff-based method of controlling water usage provided there is some other way of controlling the volumes used. There are a number of other methods for controlling water use during periods of shortfall or crisis. For example, public appeal has been resorted to with limited success. There are also methods of physically restricting supply of water by control valves, orifices and pressure reduction. The latter have been employed with roof tanks so that consumers can draw at peak rates while inflow is restricted. It may also be that the consumer could find alternatives to restrictions or paying higher tariffs. He may seek alternatives such as groundwater. These sources may have a higher operating cost, but as they are intermittent it may not be a severe penalty. This is efficient conjunctive use of alternative resources.

9.4.3 Crisis management

When there is a shortage at the source, e.g. during a drought, then there could be rationing of water, but at the same time the authorities have to meet fixed costs. The tariff may have to be increased (see Fig. 9.7).

Assuming that an emergency has arisen in the way of drought or some other reason for inability to supply water, then the method of restricting water consumption could be based on an economic system as follows:

Penalties or punitive tariffs

Higher tariffs could be charged for total consumption if consumption is above a set figure (Davis, 1995). Alternatively, a marginal penalty could be applied for consumption above a certain figure. This method is not guaranteed to reduce consumption correctly because the supplier has not necessarily estimated the value of water to the consumer.

Purchase system

If there were a free market, then consumers could bargain amongst themselves to purchase different allocations of water.

Shortfall surcharge

Due to lower sales figures by the water authority, they may have to increase tariffs in some way to meet their costs which cannot all be reduced in proportion to the amount supplied.

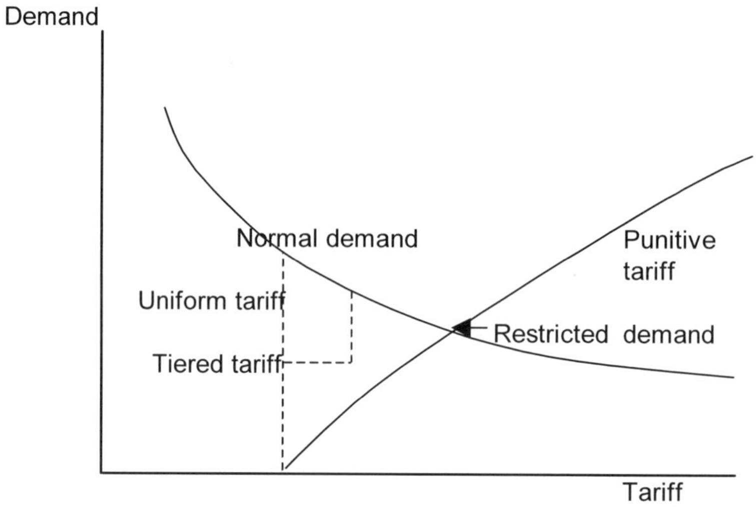

Figure 9.7 Effect of restrictions on cost of water supply

The problem of time lag arises with crisis management by means of tariffs. Following the establishment and promulgation of punitive tiered tariffs to meet a certain requirement, it may be months before the tariff is charged, detected and evaluated by a consumer. He will then change his consumption, but possibly not by the amount desired by the biller. So the process may be iterative.

9.4.4 Notes on management by use of tariffs

If the true value of water to consumers could be assessed, there is likely to be a wide range of tariffs and the supplier may unduly benefit from overcharging. One must be careful of applying long-term criteria during crisis. Some consumers may locate their organization based on indicated water tariffs, but the use of variable tariffs to manage water during drought must be explained and incorporated within the overall tariff system. The level of rationing can be decided at planning stage if the cost of assured water is balanced against the cost to the economy of rationing. However, the operational basis during drought will be from a different perspective.

A drought would be identified if the water level in the reservoir is low and the probability of refilling the reservoir during the current operational season is remote. The objective is to minimize the probable economic damage by applying water restrictions. The fact that water restrictions may be implemented by use of tariffs is incidental, but it has the advantage that the tariff can be more easily decided if the level at which the tariff will influence the consumption is known, i.e. the economic value of water to the consumer is known. Unfortunately, this may result in recuperation of excess income or possibly under-recovery of income by the water supply authority and therefore a balancing fund would have to be built up by the water supply authority to ensure he does not make a profit or loss, if it is an autonomous non-profit-making organization. After ranking all consumers, a relationship between minimum damage and level of restriction could be established. Then the objective would be to minimize the probable damage or economic loss due to restrictions. In order to apply the restrictions, the cost of water must be increased to its perceived economic value.

There is some optimum tariff which may be charged for water at which a compromise between consumer and supplier satisfaction is achieved (Riley and Scherer, 1979). This level may never be achieved, because of conflicting objectives, political intervention or the complexity in achieving it. Political objectives could be towards achieving socially acceptable levels of supply (Triebel, 1994).

9.5 THE COST OF WATER

To control use of water by means of tariffs requires estimating the marginal value of water as well as the marginal cost. The components which make up the supply cost of water include (see also Table 9.2):

- Capital costs
- Operating costs
- Quality control, purification, pressure maintenance, supply rate including back-up for droughts
- Funding of indirect projects such as redistribution of wealth or national improvement in health and economy

- Deterrents for conserving resources such as a premium to reduce usage of water
- Components to pay for environmental protection or reclamation
- Community funding including training
- Reserves for future expansion and to ensure continuity of supply or jobs
- To cross-fund, e.g. other department's shortfalls, or redistribution of charges.

The historical basis on which tariffs are calculated is generally the cost of supplying the water (Stephenson, 1995). However, there is the possibility of charging for water before it has been controlled or tapped by man. This is a form of funding as the real cost is zero, seeing it is a renewable source. If the resource is mined, such as the use of groundwater at a rate greater than the natural replenishment rate, then there may be a long-term cost to the environment.

The historical cost has been the one most commonly used for establishing water tariffs (Palmer Development Group, 1994). The income from water tariffs is used to meet the costs of repaying loans, operation, maintenance, fuel, management and often a fund for future expansion. Based on average cost, the water authority will charge a tariff which could be the total expenditure divided by the total sales of water.

A deviation from this method of costing is the marginal cost basis. Based on the fact that additional augmentation costs more than the original source of water, new users may have to pay more. Alternatively, all users may have to meet the additional cost. An alternative marginal effect may be the reduced cost due to bulk supply since the cost per unit delivered from a source decreases the larger the pipeline or the supply system.

If the total income from tariffs is only to meet average costs, then it is purely a financial calculation. However, there are invariably economic components which make the historical or average cost basis rather academic. For example, the non-technical components described above may be added onto the total cost.

The cost of water is not static even though historical costs may be constant, until the loans are repaid. Invariably, there is no reduction in average tariffs when costs are paid off, since expansion increases expenditure faster than the reduction in loan repayments over years.

Costs increase because supplies have to be augmented and these augmentation schemes are invariably from more and more costly sources. There is also inflation of prices causing the unit cost to increase. Policy factors may also cause increasing cost to some of the consumers. For example, subsidization or redistribution of resources may mean more acceptable costs to some, but others will have to pay more to meet total costs. There may also be cost increases of a temporary nature due to limited sales, for example during drought, which means that the unit price must be increased to meet certain fixed costs.

Historical water costs vary enormously throughout the world, and it is difficult to compare them internationally. They depend on the cost of installations at the time, inflation since then, the standards of supply and the ability of the consumer and government or authority to meet costs.

The cost of municipal water in Europe and North America is of the order of \$5/m^3. In South Africa, it is less than R3/m^3 (US\$ 0.50c), and in some regions in Africa, it is sometimes free. An affordability of 1 to 2% of income is a yardstick in developed countries, but in some developing communities they may pay up to 10% of their income.

Table 9.2 Factors affecting water prices

Supply costs	Charges	Controls
DIRECT	Sale of natural resource for income	Differential tariffs
Dams	Prevention of over-exploitation	Subsidization
Pumping	Cost of alternative sources	- communities
Pipelines	Cost of depletion	- localities
Reservoirs	Fines	- relocation
Purification	Pollution – cost of purification	- use type
Administration	Environmental restitution	- higher marginal cost
Repairs and maintenance	Control of usage	Drought rationing
Upgrading	Economic benefits	
Land	- health	
INDIRECT	- time	
Financing	- education	
Risk minimization	- commercial	
Standby equipment	Taxes	
Monitoring	Affordability	
Future more expensive sources	Permits	
Commissions	Willingness	
Mismanagement	Bearability	
Inefficiency		
HIDDEN (NOT CHARGED)		
Labour disruption during construction		
Rerouting communications		
Loss of land surface		
Loss of future potential		
Alternative uses of water		
Environmental impact		
Wastewater disposal		
Siltation		

The methods developed for justifying water resource projects, particularly in the United States, in the mid-20th century, were based on comparing benefits and costs of projects before ranking them, or deciding on the scale or priority of development. Whether these techniques can be applied to water supply is doubtful. In particular, the evaluation of benefits which cannot be cashed in could distort the market. It could result in over-expenditure or power-building in government centres which fund water supply projects. At the most for water supply, it should be used for ranking projects but the social impact needs to be evaluated for inclusion in the decision-making process.

When trying to assess the value of water to a user with regard to curtailing supply, the true long-term value may not be the applicable figure. The user will only consider his operating benefits minus costs, since capital expenses cannot be avoided. He will also consider primarily cash benefits, since intangible benefits, e.g. education, are long-term. So, it is important to distinguish between long-term and short-term benefits as well as tariffs.

Table 9.3 The benefits of water supply

Benefits
Income
Health
Improved quality of life
Time - education
- leisure
- economically productive
Commercial and industrial development
Agricultural
Power generation
Environmental

The principles of economics, however, should be used for comparing projects and optimizing supplies. Thus the possibility of alternative sources or inter-basin transfers may have to be compared in some fashion.

The benefits and costs of water supply are not easy to evaluate (see Gibson, 1987). Tables 9.2 and 9.3 list some of these. The costs can vary not only for the direct installation costs but also the social impact costs. These could be as obscure as changing social customs due to different methods of water collection. There are also changing population demographics which are difficult to evaluate, and the interruption of the economy by providing temporary construction employment. The river patterns may change if the water is dammed. This may affect agriculture. The environment is affected whether it is due to burying pipelines or construction of structures. More particularly, it is affected by the change in hydrology if the demand is for surface water. Groundwater is also obviously affected and the effects are not as readily seen in the short-term, but in the long-term, it could have severe environmental implications.

There are also opportunities lost as the water cannot be used for other purposes and also future planning will change owing to the lesser availability of water. Costs of planning also need to be considered in the total system cost, and if all direct, indirect and hidden costs were included, it is likely that the level of water supply would be reduced in many countries.

On the other hand, the benefits of providing water are many. Not only are they those listed below but also they have a multiplying effect in parallel with many services. That is, money is injected into the economy, the level of development increases, standards of living increase, expectations increase and therefore the entire economy is provided with an injection. Of course, there is also the effect of increasing price leading to lower consumption (Postel, 1985).

The human rights issue means that if water is to be provided to all, then it must be marketed at affordable rates, which vary considerably. It therefore appears that some form of differential tariff system would be required whereby the richer subsidized the poor. This could be disguised in various ways. For example, incremental water consumption would be charged at a higher and higher tariff. This assumes that the full cost is to be recovered by the water supplier. Subsidization by the government could further complicate the issue. This in fact may be necessary if the policy is set by the government.

An alternative to the cost recovery pricing system would be the production cost pricing system. This would imply that prices were pushed up to reflect the value of water to the consumer, whereas the price may not be pushed to the limit of affordability, it would reflect some value to the consumer (see Mirrelees et al., 1994).

The third alternative is the water scarcity pricing system whereby the price of water is increased to reflect its value (see Berk, 1981). This may be on a permanent basis or temporarily during drought. Unless a thorough understanding of the affordability of water is obtained then the price to limit consumption during scarcity maybe a matter of trial and error.

The problems of setting affordable tariffs, particularly to poorer communities, will draw in the following considerations:

- Adequate quality of service, that is pressure and flow
- The possibility of upgrading the system as living standards or affordability improve
- Labour-orientated construction to inject money into the community
- Flexibility to ensure that various levels of demand are satisfied to their standard
- Charging for services to recover what is possible, but also to instill a sense of value
- This may involve prepayment systems or flat rate systems to simplify collection of rates
- Speed of delivery which is a function of financial resources and technical resources

The problem of non-payment for water complicates the issues – the cost must be borne by others until pressure is sufficient to right the problems causing non-payment.

Methods of subsidizing water costs vary. If the subsidizer does not want to become involved in the politics, he may subsidize the water supplier and this could be by means of direct payments or reducing taxes or cost of raw water. The alternative of payment to the consumer is complicated not only by administration or the need to appear equitable and just, but also in the method of payment. It would appear more logical to subsidize indirectly, that is by reducing taxes or providing other services to reduce expenditure. Donors often subsidize the capital cost of the system, particularly rural water supply schemes. It is also not easy to decide how to discriminate between recipients subject to different levels of subsidization. In many cases, abuse of the systems needs consideration (misappropriation or resale).

The value of water to a consumer is influenced by risk (Cortruvo, 1989). If there are frequent interruptions (due to breakdowns) or lengthy rationing (drought) or pressure drops or pollution, or high tariff increases, the value is diminished. Unfortunately, supply authorities generally give no indication of these or the associated probability of occurrence. Some are catered for, e.g. emergency storage, and others may be completely unknown, e.g. future price increases.

9.5.1 Future trends

The future is likely to see increasing water costs. This will automatically reduce consumption. The theoretical correct way to control consumption would be to charge marginal costs on the top consumption, but the administration and lifeline requirements make this difficult.

Conflicting objectives make economic methods impractical for accurate day-to-day control, but economics can be used in the longer term.

Physical ways of limiting consumption (pressure reduction, cutoff) are only applicable in periods of crisis, and long-term education of consumers is seen as a necessity.

9.6 ECONOMIC VALUE OF WATER

Whether water is a 'free good' or an 'economic good' is a paradox which has been the subject of much debate (see Janusz, 1998). The International Conference on Water and Environment in Dublin, 1992, listed the following guiding principles:

- Water has an economic value in all its competing uses.
- There is a basic right of all human beings to have access to water and sanitation, at an affordable price.
- Managing water as an economic good is an important way of achieving efficient and equitable use and of encouraging conservation and protection of water resources.
- The value of water can be measured in terms of:
  (a) its utility, which results in an economic benefit;
  (b) its exchange value;
  (c) its scarcity. The scarcer the resource, the greater its value.

The price is not the same as the value of water. Public organizations use one or more of three prices (Aswathanarayama, 2001):

1. The market value based on supply and demand.
2. The administered price based on cost recovery or political decisions.
3. The accounting price, which could be the shadow value of the water, or marginal value.

The shadow value is a function of the value in alternatives uses. For example it could be the replacement value (not possible for domestic use), or the value for alternative consumers.

9.7 LOSS CONTROL

Losses of water can be up to 50% in older reticulation pipe systems. And in irrigation systems it can be equally alarming bearing in mind the large quantities used. Many cities quote losses of 20 to 30% but the figures are seldom less than 15%. The losses may not only be the fault of the supply system, there may be consumer plumbing leaks or requirement of water which cannot be charged for. This is not strictly water lost, but could result in a loss of revenue.

The term unaccounted for water is also used, but some losses can be measured but not avoided e.g. flushing out pipelines or reservoirs, or water used for street washing. If a proper interdepartmental charging system could be developed it may reduce some of these losses. And it is not only water which is lost. It could be revenue, or energy or customer relations which is lost, all of which represent financial loss to the water company.

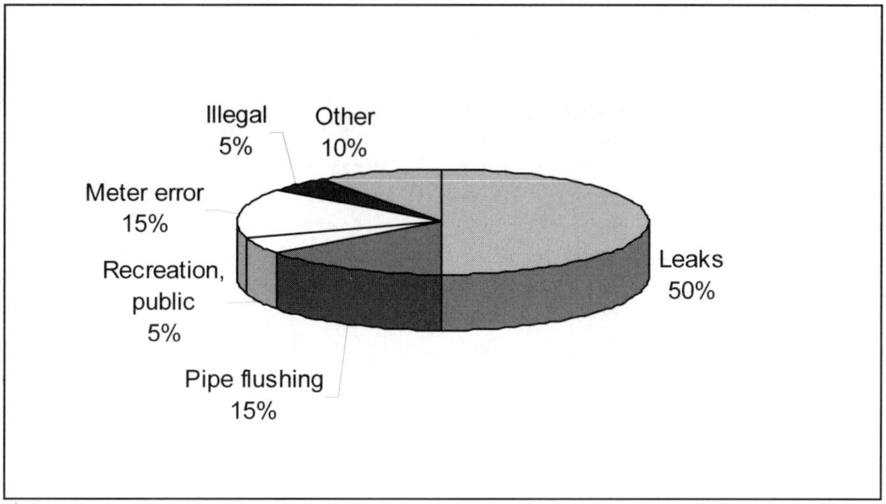

Figure 9.8 Components of water loss in reticulation systems

Whereas in the past losses used to be taken for granted, with improvement in business sense the cost of losses, as well as the waste of resources is realized and appropriate steps are taken to minimize losses. The following methods are used to reduce losses of water (Stephenson, 1998):

- Passive maintenance (repairs when notified of leaks)
- Active maintenance (vigilant inspection programme)
- Water audits
- Monitoring
- Zoning to limit maximum pressures
- Pressure reduction by valves
- Metering
- Targets e.g. 5 ℓ/hs/hr or 500 ℓ/km/hr
- Education of consumers

Table 9.4 Types of losses in water supply

Loss of:	Water	Revenue	Other
To Supplier	Leaks, Bursts	Non metered, Meter slip	Pilfering
	Flushing streets,	Illegal connections	Office inefficiency
	reservoirs	Public use, Fire fighting	Fraud
	Overflows	Bad accounting	Energy, friction
	Backwash filters	Shortage of water	Data
To Consumer	Demand management	Plumbing leaks	Confidence in supplier
	Stolen	Meter misreading	Pressure
	Leaking fittings	Corruption	Damage to property
	Wastage	Ineptitude	Water quality

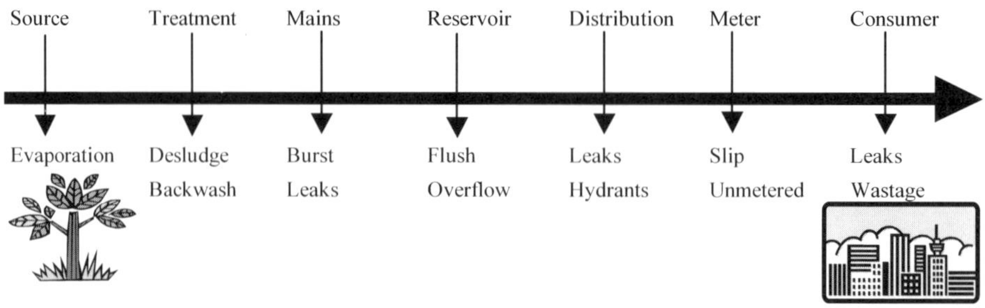

Figure 9.9 Location of losses in a water supply system

Apart from the inertia of supply organizations, there are physical factors affecting the loss rate. The following factors affect losses:

- Pipe age
- Pipe material
- Corrosion protection, internal and external
- Wall thickness, pressure class, standard
- Jointing system
- Changes in pressure
- Soil type, moisture
- External damage
- Number of consumers, connections
- Maintenance

Many water supply companies now take active steps to find losses and reduce them. The following leak detection methods are applied:

- Night flow measurement
- Step pressure/flow measurement
- Remote sensing using infra red waves
- Visual inspection
- Gas/tracer injection and detection
- Seismic refraction
- Resistivity surveys
- Noise frequency detection
- Noise correlation

In all cases cost effective loss control should be aimed at. That implies careful management to ensure the correct level of maintenance. Apart from revenue loss the supplier should conserve resources as they are a national asset and of latent value in their natural state. In fact national monitoring of resources and if necessary imposition of control measures from the resource and pollution point of view are desirable.

Loss reduction improves the economics of supply as the costs are reduced. This assumes cost effective loss control. But there are also hidden benefits. Less resources are consumed. Older systems can have their life prolonged. Structural damage at leaks is minimized. And if losses are due to theft, better consumer response is achieved. Rehabilitation of pipelines and reservoirs is a specialized task which has benefits beyond the saving in water. Relaying of pipes by plastic sleeves or in-situ applied mortar reduces friction and energy losses. The capacity of pipes is increased. So the pipes may supply more consumers and disruption of roadways by new pipes is avoided.

Rehabilitation of water works forms a component of loss control. Pipeline relining, repair or replacement needs to be considered in the alternatives. The subjects of asset management and optimum level of maintenance are covered in Chapter 12.

9.8 WATER HARVESTING

The interception of water in its various locations on a small scale is often referred to as water harvesting. This may be by means of tanks below the gutters or roof. Then it is called rainwater harvesting. It could be done on a larger or commercial scale with plastic sheets to collect the water.

In arid areas and on the coast, e.g. Namibia, dew harvesting has been attempted. There are winds carrying moist air and large temperature drops at night. During the night, moisture condenses on vertical mesh screens and is led into containers.

Runoff can be harvested by catching runoff in furrows. This is done on agricultural plots and the water is permitted to infiltrate or is diverted to dams.

Forestry harvests water by catching precipitation on leaves or on the mulched ground. Trees use a lot of water and forests reduce runoff. The question arises as to whether the water may legitimately be intercepted to reduce runoff downstream. In some instances the forester may be charged for abstracting from the catchment runoff. The original runoff has to be measured previously or modelled. And the possibility of use downstream should be established.

Similarly there may be attempts by authorities to charge for use of boreholes. Aquifers contribute to total runoff and by abstracting, the water table drops and lateral discharge diminishes. Greater infiltration may occur so there is less surface runoff too. However, there is no record of considered compensation for recharge of groundwater by discharging water or waste water onto the ground. There are also possible charges for discharging polluted water into sewers so this makes up for otherwise disposing of wastes, but quality control is a problem which will not easily be solved by tariffs.

REFERENCES

Agthe, D.E. and Billings, R.B. (1997). Equity and conservation pricing policy for a government run water utility. *J. Water SRT-Aqua*, Vol. 46 (5), 252-60.

Aswathanarayama, V. (2001). *Water resources management and the environment.* Balkema. Lisse.

Bahl, R.W and Linn, J.F. (1992). *Urban Public Finance in Developing Countries*. Oxford Univ. Press.

Berk, R.A. (1981). *Water Shortage*. Alot Books, Cambridge, Mass.

Berthouex, P.M. (1971). Accommodating uncertain forecasts. *J. Am. Water Works Assoc.*, Vol. 66 (1), 14.

Chan, W.S. (1997). Demand management. *Water Supply*, Vol. 15 (1), 35-39, Blackwell Science.

Cortruvo, J.A. (1989). Drinking water standards and risk assessment. J. *Inst. Water and Environ. Management*, Vol. 3 (Feb), 6-12.

Dandy, G.C. and Connarty, M.C. (1994). Interactions between water pricing, demand managing and sequencing water projects. *Proc. Conf. Water Down Under*, Adelaide, Austr. Inst. Eng., 219-224/

Davis A. (1995). Rand Water's response to the drought. *IMIESA*, Vol. 20 (9).

Dinar, A. and Subramanian, A. (Eds) (1997). *Water Pricing Experiences*. World Bank Tech. Paper 386, 164 pp.

Dudley, N. (1990). Alternative institutional arrangements for water supply probabilities and transfers. *Proc. Seminar Transferability of Water Entitlement*. Univ. New England, Amidal.

Duncan, H.P. (1991). Water demand management in Melbourne. *Water* (June) 26-28.

ESKOM (1994). *Time-of-Use Tariffs*. Megawatt Park, South Africa.

Gibson, D.C. (1987). The economic value of water. *Resources for the Future*, Washington, D.C.

Henderson-Sellers, B. (1979). *Reservoirs*. MacMillan, London.

Hirschleifer, J., de Haven, J.C. and Milliman, J.W. (1960). *Water Supply Economics Technology and Policy*. Univ. Chicago Press.

Janusz, K. (1998). Economic value of water. In H. Zebidi (Ed*) Water, a looming crisis?* IHP-V. Tech., Dec 19\8, Paris, Unesco, 407-416.

Lumgair, G. (1994). *Water Supply Tariff Formulae for Local Authorities*. MSc(Eng) report, Univ. Witwatersrand, Johannesburg.

Mirrelees, R.I., Forster, S.F. and Williams, C.J. (1994). *The Application of Economics to Water Management in South Africa*. Water Research Commission, Report No. 415/1/94, Pretoria.

Moore, J.W. (1989). *Balancing the Needs of Water Use*. Springer-Verlag, New York.

Palmer Development Group (1994). *Water and Simulation in Urban Areas. Financial and Institutional Review. Report 1: Overview.* Water Research Commission, Pretoria.

Postel, S. (1985). *Conserving Water: The Untapped Alternative*. Worldwatch. 67, Washington.

Rademeyer, I.J., Van Rooyen, P.G. and McKenzie, R.S. (1997). Water demand and demand management, *S.A. Civil Eng*. (Sept), 13-14.

Riley, J.G. and Scherer, C.R. (1979*).* Optimal water pricing and storage with cyclical supply and demand. *Water Res. Research*, 15 (2), 253-259.

Stephenson, D. (1995). Factors affecting cost of water supply to Gauteng. *Water S.A.*, 21(4), 275-280.

Stephenson, D. (1998). *Water Supply Management*. Kluwer Academic Press, Dordrecht.

Triebel, C. (1994). Tariffs for water supply and sanitation. *Proc. Natl. Water Supply and Sanitation Policy Conf.*, Kempton Park, SA.

CHAPTER 10

Hydraulic Structures

10.1 THE PURPOSE OF HYDRAULIC STRUCTURES

Without hydraulic structures we could not control water. Something is needed to contain or transport water. Usually a man made structure is needed to meet human requirements. Despite objections, without structures in or across rivers we would not be able to access water except by primitive means of traveling to the source at the correct time. Our purpose should be to reach a balance between utility and acceptability. Structures are built on rivers and at water sources for many purposes. These may include:

- Measuring structures such as weirs, flumes, orifices or installation of measurement devices for measuring water depth or flow.
- Storage structures such dams, barrages for storing water from periods of excess to periods of shortage.
- Transfer conduits. These can include tunnels, pipelines and channels.
- Control structures. Gates, valves and constrictions to control flow rate, pressure or volume.
- Abstraction or diversion. Intakes on rivers or wells are for drawing out water.
- Protection or rehabilitation such as erosion control.

In constructing a hydraulic structure, we usually change the flow pattern. This may be local in the case of measurement or more extensive in space and time in the case of storage and conveyance. It may be temporary in the case of construction works or more permanent in the case of dams, canals etc. Even if the natural flow pattern were irregular, the fact that water is stored or diverted will change the flow pattern in time and space. There is therefore an effect on the immediate construction area as well as downstream of the structure. The effect could be positive or negative and it needs a proper impact study in order to assess the benefits and adverse effects of a structure.

By storing and diverting water we change the energy level at points in the system. Naturally flowing water tends to a short-term equilibrium by reaching a flow velocity whereby the friction drag equals the potential energy loss along the river. If we slow the water down, it will decrease the kinetic energy and increase the potential energy. A problem can then arise in that when the water is released downstream of the structure, such as a dam, it has a high kinetic energy or high velocity which can cause erosion. Similarly, upstream of the dam where the kinetic energy of the water is reduced, it may deposit sediment.

10.2 MEASUREMENT STRUCTURES

The most common form of flow measurement in a natural river is a weir. The flow rate can be accurately assessed by measuring the depth of water upstream of the crest of the weir. In the case of a simple horizontal crest weir, the flow rate is given by the equation (in SI units)

$$Q = CBH^{3/2} \tag{10.1}$$

Where:the flow rate Q is in cubic metres a second,

C is a coefficient depending on the shape of the crest, for example a sharp crest has a coefficient of 1.8 (in S.I. units),

B is the width of the weir,

H is the height of the water surface above the crest in metres assuming there is negligible flow velocity at the point where H is measured.

If there is a flow velocity owing to the shallow depth of the weir, then H must be corrected by the velocity head $v^2/2g$. The coefficient C can be affected by the pool depth upstream of the weir and the width of weir because side contractions can reduce the value of C. Since Equ. 10.1 is not dimensionless, the value of C depends on units.

Where the flow rate can be very low, then a V-notch may be more accurate because it causes a greater depth of flow which can be measured more accurately for small flows. As the flow rate increases, so the width increases and there is a non-linear relationship as indicated by the equation

$$Q = \frac{8}{15} Cd\sqrt{2g}\ \tan(\theta/2)h^{5/2} \tag{10.2}$$

Where:g = gravitational acceleration (9.8 m/s^2)

C_d is the discharge coefficient, approximately 0.585, and

θ = angle of the notch

For example, for a 90° notch, $Q = 1.38h^{5/2}$ in SI units.

Sharp crested weirs are prone to siltation in some environments and a weir with a gradual slope upstream and downstream was developed called the Crump weir. This has similar types of discharge relationships.

Weirs are usually designed to measure up to a certain regular high flow rate. However, there may be floods when the entire weir structure and the banks of the river are inundated, in which case a calibration is required. In addition, backwater from downstream should be avoided but at very high floods the weir could be inundated and again requires calibration.

It is also possible to calibrate natural channels and measure the depth and to calculate the flow rate from a uniform flow equation such as that of Manning:

$$Q = \frac{A}{n} R^{2/3} S^{1/2} \qquad (10.3)$$

Where: A is the cross sectional area in square metres
n is Manning roughness, 0.013 for concrete, 0.03 for grass, 0.1 for vegetation
R is the hydraulic radius, A/P
P is the wetted perimeter in metres, and
S is the energy gradient in metres per metre.

Sometimes the bed gradient is measured and if the flow can be assumed to be uniform, this is used instead of the energy gradient. Flumes may also be used for flow measurement and by constricting the width of the channel it is possible to increase the flow velocity such that the flow depth passes through critical depth. Thus, weirs and flumes work on the principle of upstream measurement and not a differential measurement of water depth. They both result in a higher flow velocity downstream than in a normal channel and as a result erosion can occur.

The flow in conduits is associated with energy loss due to friction and turbulence and the potential energy is what drives the flows through the conduit. When flow is measured, it is often by converting the kinetic energy into potential energy upstream of the measuring device so that a pressure difference can be measured. This is much simpler than measuring the flow velocity. Indeed the flow velocity varies across the cross-section of a conduit and needs to be integrated to get an exact total measurement.

The equation of Bernoulli is used to relate potential and kinetic energy of flow:

$$(Z + p/\gamma + v^2/2g)_1 = (Z + p/\gamma + v^2/2g)_2 + h \qquad (10.4)$$

Where: Z is the elevation of the bed of the water containing structure,
p is the water pressure at the bed,
γ is the unit weight of water so that p/γ is the depth of water at that point,
v is the water velocity, and
h is the head loss between two sections due to friction and turbulence.

Figure 10.1 Katse dam in Lesotho, 180m high.

10.3 DAMS

Dams are the biggest structures which man builds to control water. They may range from a few metres high in the case of storage weirs or barrages, to hundreds of metres high for some structures. Dams may be used to store water for use during dry periods. They can be used to create head sufficient to generate hydro electric power. They can also raise the level of the water to enable to be diverted to agricultural lands or through pipelines to cities or other consumers.

The safety of dams is paramount. A number of catastrophes have occurred due to failure of dams, such as in Vaiont in Italy where two hundred people were drowned. This was a concrete arch dam which failed when the side of the mountain slipped into the reservoir. A number of embankment dams have failed due to water seeping through underneath, i.e. piping, and eventually opening a large gap in the wall. This type of failure can be rapid, i.e. over a few hours, and can cause a high surge of water to rush down the valley owing to the large volume of water behind the dam and the high head. This danger should be considered in assessing the disadvantages of dams.

The structural strength of a dam may be due to self-weight of the embankment or concrete. This type of dam is called a gravity dam. Alternatively, the concrete may be arched in one or two directions to transmit the thrust to the embankments of the river. Buttress dams take the water pressure on upstream concrete slabs and transmit it to buttresses which take the thrust down to the foundation of the river. The abutments and foundations of concrete dams, therefore, need to be firm and strong. Extensive testing needs to be done before a dam can be built.

Water pressure upstream of a dam at depth will be high owing to the great water depth and the water can seep through the foundations towards the downstream side of the dam. There will therefore be pore water pressures in the foundations which can tilt the dam wall upwards and add to its instability. Uplift pressures, therefore, need to be considered in analyzing the stability of the dam against overturning or sliding.

Figure 10.2 A dam spillway

Sedimentation is a problem which many dams experience, particularly in developing areas where soil erosion is not under control. The volume of sediment running off the catchment can be hundreds or even thousands of tonnes per square kilometre per annum. The sediment is deposited in the dam when the water velocity of the river is reduced. If the capacity of the reservoir is small relative to the average flow of the river, the reservoir can rapidly silt up. Eventually an equilibrium may be reached whereby the cross-sectional area of the reservoir is so reduced that the water velocity is high enough to maintain sediment in suspension. Little can be done to erode the deposited sediment once it has settled. This is because, even with a large gate through the dam wall, the water velocity is only high in the region of the gate and within a few metres back, the water velocity is so low that it does not erode the sediment when the gates are opened. One solution may be to have large barrage-type gates which open up when sediment-laden water flows into the reservoir. The water then passes through the reservoir at a high flow velocity.

This is expensive as metal or inflatable gates are costly. The alternatives of dredging or off-channel storage are also expensive so the entire problem needs careful consideration at the planning stage. Bear in mind also that once the sediment has been deposited, then the water flowing over the crest of the dam or through the outlets is clean and has a potential for further erosion downstream, particularly as it is at a higher velocity than would be the case without the dam.

Releases should be made from the dam to sustain the environment downstream and also to meet riparian demands and the needs of communities that have become dependent on natural river flow.

By reducing floods with dams and dropping the sediment behind the dam wall, the outflow can have a considerably different pattern than before the dam was built. This is the case for the Nile river downstream of the Aswan dam where agriculture was dependent on flood levels moistening the soil and the deposit of silt to enable crops to be grown.

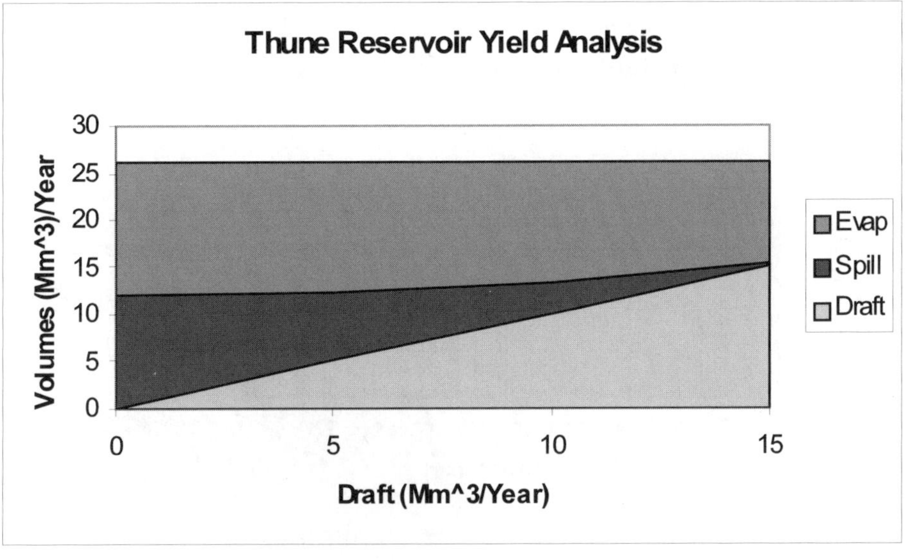

Figure 10.3 Dam evaporation as a function of yield

The evaporation from a dam is also increased over that of the natural river so there is a net loss of water from the system even greater than the abstraction from the dam. Downstream users therefore suffer because of the reduced flow and in particular no flow during the low dam levels unless releases are made (Fig. 10.3).

The structure to pass floods through a reservoir is another expensive component. The spillway needs to take the full flood of the river apart from the attenuation effect that storage may have. This flood needs to pass without damage to the dam or downstream valley as the dam wall could otherwise be endangered. Various types of spillways have been developed to reduce erosion downstream. A stilling basin may be constructed downstream of the dam wall to create a hydraulic jump and to enable the water to flow out at sub-critical velocity. Alternatively, flip buckets spray the water into the air. Chutes may reduce the water velocity particularly if they are stepped. Large steps are common with rolcrete-type dams and offer potential energy dissipation to reduce the downstream water velocities. The normal ogee-type spillway does not offer much in the way of energy dissipation but it is relatively erosion resistant and minimizes the variation in upstream water level.

Siphon spillways are notable for passing a high discharge with a low head rise upstream. This is because the full depth of the dam is utilized in creating the velocity through this siphon spillway. In the case of crest-type spillways, it is only the head above the crest which is used to generate water velocity, so the backup is generally higher than for a siphon-type spillway.

The problem with a siphon spillway is that its capacity is limited and if an excessive flood over the design capacity of the siphon occurs, it can overtop the dam wall. Thus the siphon needs to be fairly large in capacity, with the result that even smaller floods are magnified by discharging through the siphon once it is primed. There is therefore a high erosion potential downstream unless a stilling basin is constructed. Successively larger parallel siphons can reduce this problem but they will require more space.

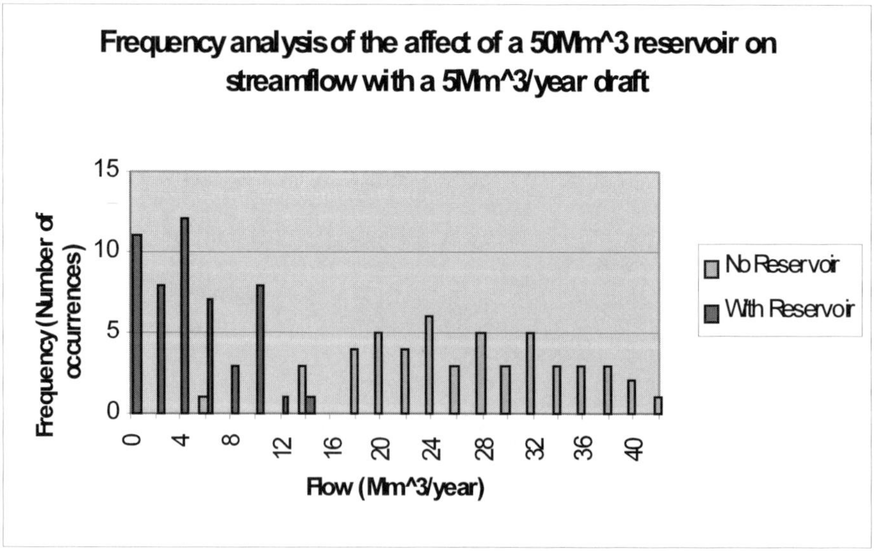

Figure 10.4 Effect of dam on downstream flows

Underground storage dams is a subject confined to aquifers with specialized properties. In arid areas with sandy soils, water can be stored in the aquifer as it would in a dam. In Botswana, many of the rivers have deep sand beds and old dams which have silted up can store up to 30% of their capacity as groundwater. Evaporation is considerably reduced compared with a surface water reservoir, but the cost of abstracting the water is higher.

Frequently, sandy aquifers which have the highest permeability and storage coefficient are close to the coast or other water sources. By drawing down the water level, the water from the sea or polluted sources may be drawn towards the aquifer with a deterioration in the quality. In addition, lowering of water tables can cause subsidence of the ground.

10.4 EFFECTS OF CONSTRUCTION OF DAMS

Over 45 000 dams have been built over the last century. Dams were initially built to provide water for irrigated agriculture, to control floods and for domestic and industrial use. More recently though, dams have been providing hydropower. Despite all the benefits that the construction of a dam brings, there are also negative impacts, such as, diverting the flow of rivers, this in turn affects certain communities' rights and access to water. This results in a significant impact on their livelihood and the environment. Decisions to build dams are being contested increasingly as human knowledge and experience expand.

The World Commission on Dams considers that the completion of any dam project must be the sustainable improvement of human welfare, i.e. a significant advance of human development that is economically viable, socially equitable and environmentally sustainable. However, if there are other options which offer better solutions, we should favour them over large dams. As populations grow, an increase in economic development leads to a higher consumption and demand for water, therefore water is becoming a more and more valuable resource. What better way to protect that resource than to store it in a dam for a drought? But the debate is not related to the way the water is stored. It is related to what the dam will do to the river flow, to the rights of access to water and river resources, to whether it will uproot existing human settlements, disrupt the culture and sources of livelihood of local communities and deplete and degrade environmental resources. A dam built in one country could affect downstream countries availability, so international rights need consideration. Conflicts over dams are more the conflicts over water. They are conflicts over human development and life itself.

10.4.1 Large dams

There are various definitions of large dams. The International Commission on Large Dams (ICOLD), established in 1928, defines a large dam as a dam with a height of 15m or more from the foundation.

Figure 10.5 Three Gorges in China, set to become the site of the largest dam in the world

If dams are between 5-15m high they are classified as small, 15-30m they are classified as medium, and greater than 30m or with a reservoir volume of more than 3 million m^3, they are classified as large dams. Using this definition, there are over 45 000 large dams around the world

The two main categories of large dams are reservoir type storage projects and run-of-river dams that often have no storage reservoir and may have limited daily pondage. Within these general classifications, there is considerable diversity in scale, design, operation and potential for adverse impacts.

Reservoir projects impound water behind the dam for seasonal, annual and, in some cases, multi-annual storage and regulation of the river. Run-of-river dams (weirs and barrages, and run-of-river diversion dams) create a hydraulic head in the river to divert some portion of the river flows to a canal or power station.

Dams have been seen as an important means of meeting the needs for water and energy services as well as a long-term strategic investment with the ability to deliver multiple benefits. When building a dam, additional benefits must also be thought of, such as, regional development, job creation and fostering an industry base which has an export capability. Other goals which create income from dams include the income from export earnings, either through direct sales of electricity or by the processed products from electricity intensive industry such as aluminium refining.

In countries where water is plentiful, such as Canada, Norway or Brazil, dams have been developed solely for hydropower. Whereas in countries where water is scarce, such as South Africa, Spain or Australia, dams have been built with large storage capacities to match water demand with storage supply, and for security against the risk of drought. In most of these countries, dams have been constructed to capture and store water during wet seasons for release during dry seasons.

Dams have been constructed to a lesser extent to improve river transportation and many dams are also used for recreation, tourism and aquaculture.

Table 10.1 Estimated annual investment in dams in the 1990s ($US billion per year)

	Developing countries	Developed countries	Total
Dams for hydro power	12-18	7-10	19-28
Dams for irrigation	8-11		
Dams for water supply	1.5	3-5	13-18
Dams for flood control	0.5-1.0		
Total	22-31	10-15	32-46

Dams offer a large economy of scale. Many factors are relatively independent of dam height, e.g. geophysical exploration and remediation, spillway size, bank protection and outlet works. Therefore it is more profitable to build large dams, until the yield approaches the flow rate of the river. Other advantages of large dams are:

- Instruments of development
- Water used for irrigation can be maximized
- Water used for industrial and urban centers which continually grow
- Electricity generated for the national grid can replace fuel plants
- Flood protection requires large volumes
- Multi purpose development improves economic viability..

10.4.2 Problems associated with large dams

Although dams have contributed to economic growth, the advantages they bring have come at a cost. Large dams have transformed the world's rivers. The World Resources Institute (WRI) found that large dams modify 46% of the primary watersheds. This means that certain ecosystems will not receive the required water to keep them alive. These dams fragment and transform aquatic and terrestrial ecosystems with a range of effects that vary in duration, scale and degree of reversibility. These watersheds are the home of 40% of the world's fish species. Ecosystem transformation also impacts on river estuaries, often closing mouths of major rivers, salt intrusion, destruction of mangroves and loss of wetlands are some of the disadvantages of large dams. But the environmental ecosystems are not the only systems that have been negatively affected. While many people have benefited from the services that large dams provide, their construction and operation have led to many significant negative social and human impacts. These negative impacts include displaced families, riverine communities, especially those downstream of dams whose livelihoods and access to the resource of the river have been altered. More broadly, whole societies have lost access to natural resources and cultural heritage that were submerged by reservoirs or rivers transformed by dams.

Negative effects of large dams include:

- Destruction of ecosystems
- Negative social impacts
- Large investment
- High risk

From this, it can be seen that large dams have both positive and negative impacts, but because dams will always be with us certain positive and negative aspects need to be addressed.

Critics of dams point to:

- The need for more sustainable and appropriate alternatives to dams
- The imperative for improved transparency, accountability and public participation in the planning of water and energy projects
- The importance of prior project approval by potentially effected groups
- The need for protecting and promoting the rights of potentially effected people, and for setting in place measures to reduce inequities
- The necessity of reparation measures to address the legacy of unfulfilled commitments and unresolved problems.

Dam proponents underline:

- The evolution and change in practices over time
- The recognized need for social and environmental concerns to be elevated to the same level as safety concern
- The importance of ensuring the affected people are better off as a result of dam development.

If these steps are taken into proper consideration then according to the World Bank, these dams will be complete. How many incomplete dams are there?

Dams provide essential services and benefits, including the supply of water (for irrigation, drinking water, sanitation or manufacturing), electricity and flood control. These are inputs or intermediate goods and services and economic activities that have the potential to deliver widespread improvements in income, employment and food security.

In many countries (e.g. India, Pakistan, China), a significant proportion of food production is dependent on irrigation, and that proportion is likely to increase markedly globally through the twenty-first century. Dams can provide an essential means of sustaining and expanding irrigated agricultural production, and hence food security, food independence and the potential for agricultural produce trade. Population growth will increase national and global food demand and increase the need for dams to supply water.

Dams are designed to alter the distribution of water in space and time, and social impacts on those using the water before or after construction are therefore inevitable. However, these impacts can be positive (e.g. supply of water to new users, reduced flood risk) as well as negative (e.g. reduced opportunity for floodplain agriculture, involuntary resettlement).

Negative impacts can be fully compensated. If the dam project as a whole generates net benefits; careful planning and good design can avoid or minimize adverse impacts. Development planning involves making choices about where and how to invest, and there are likely to be winners and losers from all such decisions. Dams are no exception to this rule.

Alternatives to dams have significant environmental, health and social costs (e.g. in the case of hydropower, the impacts of coal burning or nuclear power stations). The critical question about dams is not the severity of negative impacts in isolation losses, but the overall balance of costs and benefits, and the adequacy and efficiency of measures to compensate losers. It is important to look at the larger picture of distribution of costs and benefits across all major groups, first to confirm if in fact dams deliver more benefits than costs, and second because the natural sources of funds to compensate the potential losers are the gains of the potential winners.

Social impacts:
Economic and social costs may be considerable, and often have not been taken fully into account in the appraisal of dam projects.

Some costs are tangible (e.g. lost houses, fields or forests), but others are intangible (e.g. loss of heritage, culture). Both are important. Costs are borne by the particular groups of people, particularly the poorest and least powerful (tribal groups, women, remote rural dwellers); these people are unable to enjoy benefits of the dam (e.g. irrigation or cheap power). There is therefore an unequal distribution of costs and benefits both socially and spatially.

Although these problems are recognized by governments, donors and professional groups, they are not being overcome effectively because of the inherent rigidities of the dam process and the inherent complexities of the impacts (e.g. in remote downstream locations). The surveys necessary to predict impacts talk place too late in the design process, and are often not extensive enough to influence the decision to build, they can therefore do no more than provide a basis for compensation or mitigation. The costs of full compensation or mitigation can be so great that they are often not met in full.

The complexities of the problems created by dams are such that they cannot always be fully mitigated or compensated. Dams therefore impose net costs on certain groups or categories of people. Those who lose often lack an effective channel for seeking justice, and certainly lack influence on the decision to construct the dam at any stage.

Calculations of benefit-cost ratios or net present value tend to overestimate direct or primary benefits and fail to take full account of the magnitude of costs, and the distribution of all costs and benefits, particularly those that are secondary or indirect.

Not all impacts can be expressed in economic terms, and these are not effectively captured by any economic approach to the calculation of costs and benefits. For example, an attempt to put a money value on a flooded sacred site or ancestor's grave is both theoretically problematic and likely to lack legitimacy among local stakeholders. Legitimate grievances can therefore remain even when there has been a careful compensatory process using a 'cost-benefit' framework.

Actual practice lags ways behind best practice in many parts of the world, particularly where the institutions of formal civil society (media, access to law, openness of government) are weak.

10.5 DAM CONSTRUCTION

There are various stages in the planning, building and operation of a large dam. Each stage has its own advantages and disadvantages.

Different features of a dam's possible purpose and operation are:

- Planning a large dam
- Building a large dam
- Building power lines/access roads/irrigation canals/other infrastructure
- Impounding/flooding
- Managing the reservoir

- Supplying electricity from hydropower
- Supplying irrigation water from the reservoir
- Supplying water from the reservoir
- Managing floods
- Permitting access to the reservoir for recreation, navigation and relaxation.
- Refurbishment/upgrading
- Decommissioning

10.5.1 Impacts during planning and construction

Both positive and negative impacts of the planning phase of dams are minor compared to those that come later. The major positive impacts of dams at the planning stage are enjoyed by the companies and individuals employed to plan, design and construction them. The interested and affected parties include not only contractors, consultants and bankers, but the many workers employed on all aspects of the project, and those who are sustained by the business generated by the planning and construction process. These organizations form a powerful interest group acutely aware of the possible wider economic benefits of a possible new dam, and they wield considerable influence in debates that shape the future of dams. Note however the risks associated with dam construction can be large. That is from an economic and safety point of view.

The chief negative impacts of the planning phase relate to the fear and uncertainty created in the possible project area (or series of project areas, if alternative projects are being considered). Economically, this can cause two problems, land speculation and lack of investment. With the announcement of dam construction, people with the means to employ legal mechanisms to claim land, take a legal hold of areas, which are likely to increase their land values. This affects small landholders or others (e.g. women) who have restricted access to legal mechanisms.

Dams are often discussed years before project development is seriously considered, and once identified in this way, a form of 'planning blight' descends, making governments and businesses reluctant to invest in infrastructure or business facilities that might be flooded. Communities can live for decades starved of investment in this way. A related problem is the fear than many people living in a possible reservoir area feel. Such psychological stress cannot be effectively quantified in economic terms. Such fear can cause socio-cultural impacts, for example loss of respect for traditional leaders.

Resettlement:
The most serious negative impacts of impoundment are due to the trauma of resettlement, or the socio-economic and cultural costs to displaced people who are not resettled.

Substantial numbers of people have been moved to make way for reservoirs. The human cost of dam construction varies greatly between countries, but globally it is significant. Locally it can be devastating. For example, in India about four million people have been displaced by reservoirs and irrigation schemes. The Sardar Sarovar Dam on the Narmada River will flood 265 000 people, and the Narmada Sagar Dam 170 000. A disproportionate number of oustees are people from tribal or landless people (43% of the outstees from the Narmada Sagar Dam, for example, are landless).

Evacuees face the greatest relative costs in the relocation stage, particularly where it is rushed. There are a disturbing number of instances where involuntary resettlement of reservoir evacuees has been enforced and accompanied by violence. Some of these examples are very recent, for example in the Narmada Valley. It is a suggested principle of this study that no large dam should be constructed with the use of coercion or force. Even if the adverse impact is low, it stirs up a resistance which can obstruct progress.

The positive impacts of resettlement, if any, can only be felt after the initial trauma of displacement has receded. Usually, it is the second generation of displaced communities that is in a position to use the resources available to them. Sometimes, this does result in increased social and economic mobility for those members of the community who are in a position to use the increased access to the market or other job opportunities. However, this is more dependent on the agency vested in people, rather than as a deliberate policy. Program, policy and outcome cannot be directly linked. Communities act upon programmes in complex ways and the consequences can be positive as well as negative.

Impacts at the dam site:

There are a distinct series of impacts of dam construction at the dam site. The most obvious of these are associated with the construction activity itself. As civil engineering projects, dams demand large amounts of unskilled labour and smaller but significant amounts of skilled labour. New jobs are therefore created both for skilled trades persons (most of whom are drawn from the national or even international labour market), and for unskilled workers. Some of the latter may come from surrounding communities, including sometimes those flooded by the reservoir. However, it is common for reservoirs to be located in areas whose population lack key skills (e.g. literacy, numeracy) and the necessary contact to obtain even menial jobs, unless a positive recruitment programme is in place. Experience in Latin America, for example, showed that there is little recruitment of local labour except in certain high-profile (and high budget) projects such as Itaipu

Dam design and construction is often done by private corporations, often from outside the country. The most senior and best-paid posts will often go to non-nationals, and often nationals, and often non-locals (mostly men). A high proportion of salaries, particularly of highest-paid staff, may be remitted out of the dam area, to distant relatives or even internationally in salary remittances. Only a proportion of the investment in the dam that is paid locally stays within the local economy.

The presence of a large dam construction labour force creates a demand for a wide range of other products. These range from specialist sub-contractors for a dam itself, a local construction industry for the creation and maintenance of housing and office facilities, and a large service industry for the construction operation and the people building it. Economic activities range from transportation or building to garden maintenance of personal services. Often a large camp or town builds up at the dam site, gaining a life of its own on the back of the construction activity. However, the barrack-like communities that result are dysfunctional, and (as at the Yacyretá Dam) unsuitable for family accommodation because of problems of alcoholism, drug abuse and prostitution.

The existence of a construction town can be taken as a major economic benefit of dam construction. Both the WAC Bennett and Peace Canyon dams in Canada are located within 25km of the town Hudson's Hope, on the Peace River and are the biggest employers in the town. Similarly, when questioned about benefits additional to electricity production at the Churchill Falls Dam in Canada, the example given was "A permanent settlement with population about six hundred was established near the dam to service it. This town, [Churchill Falls] still exists".

Very often, the positive impacts of reservoir creation are concentrated in or near the dam site because of the existence of infrastructure (road access, health and education facilities) or of a market (initially among construction workers). New economic activities such as fishing often demand capital (e.g. deep-water boats and nets). These activities can attract specialist migrants from outside the area (e.g. Ewe fishermen moving to the Volta Lake) or in-migrants able to recycle formal or informal sector earnings from work on the dam or associated service industry. Dam towns such as Kariba in Zimbabwe have secured the lion's share of the economic benefits of both tourism (game fishing, water sports and hotels) and commercial fishing. The companies owning these facilities are not owned by people from the dam or reservoir area, and if profitable can attract foreign investment.

Dam construction towns are a magnet for evacuees, although these may lack the skills and contacts necessary to obtain skill-enhancing formal sector jobs. Even if those without compensation, re-housing and establishment in new livelihoods are able to find a niche in the dam site urban economy, this is likely to lack security. Living conditions in dam towns can be poor, and problems such as sexually transmitted disease rife.

10.5.2 Impacts in the catchment

Dam construction can have significant negative impacts on land use in the catchment far above the reservoir, particularly due to land-use controls imposed to reduce soil erosion or maintain water yield. A typical side effect of a dam is a ban on agricultural activity in the catchment, sometimes linked with a proposal to create a protected area (e.g. a National Park) in the catchment. The latter may create certain benefit streams (e.g. from wildlife tourism) but such developments can impose severe costs on subsistence resource users, particularly indigenous people.

10.5.3 Building power lines, canals, roads, etc.

Engineering works that are an inherent part of a dam project can have their own impacts. Such works include the construction of power lines, irrigation canals or access roads. Positive impacts centre on the work created in the construction process. Beneficiaries here include companies engaged in construction, materials or equipment supply, and their employees and shareholders. Roads and power lines also allow access to previously inaccessible areas, allowing pioneer farmers, hunters and vacationers to enter and exploit their various resources. Negative impacts are, in a sense, the mirror image of these positive impacts. Secondary displacement is a problem. People with land or holding resource rights traversed by utility corridors suffer in much the same way as those losing land to the reservoir, especially if they are not adequately warned and compensated. These can be speculation in land and resources. In-migrants, whether permanent cultivators or itinerant hunters bring a range of threats from economic competition, disease, and challenges to established cultural norms and practices.

10.5.4 Impacts of reservoir management

There are relatively few direct positive impacts of reservoir management in the reservoir itself. Those that exist relate chiefly to the possibility of the development of new open-water fisheries, as happened with the Volta Lake in Ghana and Lake Kariba on the Zambezi river. These can generate economic benefits, and need to be considered in an assessment of wider social impacts. The productivity of reservoir depends on many factors, including the chemistry, turbidity and temperature of inflowing waters, as well as the nature of the land and vegetation flooded. The development of the fish and fauna in tropical reservoirs is complex, and may involve an initial peak of population as nutrients from flooded areas feed into the ecosystem, followed by a slump to a lower level. Reservoir fish populations may require very different fishing techniques to those used by previous fishing inhabitants of a river floodplain. Invasive floating plants such as the Nile cabbage or the water hyacinth can create problems for fish-catchers (and hydroelectric turbines) as for example at Kariba on the Zambezi river.

10.5.5 Impacts of supply of hydropower

The main positive impact of hydroelectric power generation is the increase in power availability. Hydropower is relatively clean in comparison with fossil fuels, and particularly coal (avoiding pollution from SO_2, NO_2 and particulates, and the greenhouse gases released by fossil fuel burning). It avoids the long-term costs and risks of radioactive fuel storage. It is also relatively cheap over long time periods. Foreign currency can be earned from electricity export. Secondary effects of hydropower generation include increased economic activity (industry, commerce, household), and reduced domestic drudgery in electrified households (light, cooking). There can be health benefits from reduced growth in mortality from respiratory disease, and environmental benefits from reduced growth in acid rain precursor emissions and climate change precursor emissions.

These benefits are experienced by a range of people, including all users of electricity, and particularly women (who bear the burden of domestic work). Hydropower utilities and their owners and employees gain, while industries dependent on rival sources of energy lose: there are negative impacts on residents in areas of fossil fuel extraction (jobs lost) or burning (health benefits but jobs lost). Although hard to measure, there may be global benefits and benefits to future generations from reduced anthropogenic climate change.

Negative impacts of hydroelectric power generation include the impact on non-hydropower electricity producers, the problem of methane from flooded vegetation, and the sometimes impacts of hydro dams on downstream people such as the impacts of the Mananatali Dam in Senegal. Downstream impacts include effects on downstream floodplain economies (fishing, agriculture, navigation), effects in estuary or delta resource fishers, and effects on offshore fisheries. These impacts are similar to those caused by other kinds of dams, but the continuous releases, and the possibility of unseasonal flood releases in generating arrangements designed to meet peak power surges, give hydroelectric dams distinct downstream impacts. Negative impacts are felt by the owners of enterprises and the workers (and their dependents) in the oil, gas and nuclear industries, by all people in downstream communities and by downstream industry and commerce.

10.6 WATER CONDUITS

Most pipelines and tunnels are beneath the surface of the ground and offer little obstruction and affect the environment in very small ways. The construction activity can have the main impact. However, surface channels or canals can prevent the migration of wild life and are aesthetically undesirable in many locations. Although surface channels are usually more economical than closed pipes, they have disadvantages such as loss of water by evaporation, abstraction by illegal consumers, and the above-mentioned environmental problems.

The carrying capacity of a conduit depends on the surface roughness and the cross-sectional area and slope. Whereas closed conduits can often operate under pressure and therefore follow the ground undulations, open channels have to more or less follow the contours with a gradual slope downwards towards the discharge point. They therefore often wander and have a considerably longer length than a direct pipeline, for example.

The carrying capacity of a channel is frequently calculated using the Manning equation 10.3, whereas the carrying capacity of a pipeline or tunnel is usually obtained from the Darcy-Weisbach equation as follows:

$$h = \lambda L\, v^2/2gd \tag{10.5}$$

Where: h is the head loss or drop in water level from one point to another over length L
λ is the Darcy friction factor
v is the water velocity
d is the pipeline diameter (equal to 4R in non-circular conduits), and
g is gravitational acceleration

The hydraulic radius R is the ratio of cross sectional area to wetted perimeter and is equal to d/4 for a full circle. The Darcy roughness coefficient λ is a function of the physical size of roughness particles on the surface of the conduit as well as a measure of turbulence or Reynolds number. The Reynolds number, vd/ν is a measure of the difference between laminar and turbulent flow. For large conduits, fully developed turbulent flow results and the Darcy friction factor as well as the Manning number n are independent of the flow. For smaller Reynolds numbers, the Reynolds number does affect the friction factor and it increases the lower the Reynolds number. The Darcy coefficient can range from 0.011 for smooth surfaces to 0.02 for old rough pipes, and it is also higher for low Reynolds numbers, i.e. small bores.

Pipelines operate at velocities between 1 and 2 m/s while canals operate at lower velocities. High velocities can cause erosion and surges, and if the pressure head is reduced due to high velocity there may be cavitation damage. This takes place in outlet pipes and over dam spillways where the downstream pressure is low. Abrasions due to suspended particles or deposition, are also of concern for transport of raw water.

10.6.1 Pipeline design (Stephenson, 1989)

Pipelines are an efficient and environmentally acceptable means of transporting water. The circular cross section is efficient hydraulically and economical. It takes the water pressure stresses around the circumference hence it is structurally efficient. External loads from the soil above do however require special design in the case of thin wall pipes. To resist internal pressures, the pipe wall thickness for a thin pipe should be calculated from the equation

$$PD = 2FT \tag{10.6}$$

where P is internal pressure, D is diameter, F is permissible wall stress and T is wall thickness, all in consistent units, e.g. N/m^2 for stress and m for dimensions.

Internal pressures are made up of hydrostatic pressure, friction pressure and transient pressures due to water hammer and pump startup. Pipeline walls can only withstand limited stress and pumping lifts over about 300m (representing 3000 kN/m^2) and above this the lift may have to be split with booster stations. Although factory and field test pressures are 25 to 100% above operating pressure there are many uncertainties accommodated in the margin, such as material deterioration and laying stresses. High velocities in closed pipes can lead to water hammer overpressures on closing valves or surges in open shafts. An hydraulic analysis of such systems is needed to determine design pressures or to provide pressure relief. As an elementary estimate the water hammer head rise due to sudden valve closure is $H=(c/g)V$, where c is the wave celerity e.g. 1000m/s for rigid pipe, g is gravitational acceleration, 9.8 m/s^2 and V is the original water velocity, in m/s.

Pipe materials have changed in the last century from cast iron through steel, concrete, and glass reinforced polyester (GRP) for larger bores, to Polyethylene (PE) and Polyvinylchloride (PVCs) for small bores. Research into composites and structured walls will no doubt make further changes in the future. Thin wall pipes have to be designed together with the compacted trench soil backfill as an engineered system to resist collapse under external loading. The deflection of a ring under external load is given by an equation of the form

$$d = 0.1WD^3/(8EI+0.06E_sD^3) \tag{10.7}$$

where d is deflection or distortion in vertical diameter, W is the superimposed loading in kN/m of pipe, E is the pipe elastic modulus, I is the wall section moment of inertia, E_s is the soil modulus and D is the pipe diameter.

Since the friction head loss is a function of the pipe diameter to the power of 5 and flow rate squared, the cost per unit of water delivered reduces by the diameter to the power of 2.5 for any fixed head loss gradient. This results in an economy of scale unparalleled in other transport systems. The big disadvantage is that pipelines still represent a big capital outlay, the same as for dams. Therefore bad estimates in flow forecasts can result in overexpenditure and unduly high water costs. However with modern electrical pumps drives the power input can be minimized and pipe capacity can be upgraded to an extent as demands increase. Owing to the high capital cost it pays to operate pipelines at a high load factor, and meeting peak drawoffs from reservoir storage close to the consumer.

Pipelines are fitted with many types of valves, including isolating valves, non-return valves, scour valves, pressure or flow control and air release valves. Special valves have been designed for each purpose as erosion and cavitation can occur in badly designed systems. Access is difficult once a system is in operation so failsafe systems should be sought.

Computer modelling of pipeline systems is now common for a number of purposes:

- Analysis of flows and pressures in complex systems.
- Simulation of pumping systems.
- Optimization of pipe and reservoir sizes.
- Water hammer and transient pressure calculation

10.6.2 Canals

Canals are not as flexible in operation as pipelines. Flow cannot be stopped simultaneously along a canal length as it flows under gravity. So large open reservoirs or overflows are required to take diverted water. The cost of canals is generally less per metre than the equivalent capacity pipeline. But canals follow contours with a gentle gradient and may therefore be inconveniently situated and longer than a direct route which a pipeline could take. Maintenance and patrol of canals is required. The problem of traversing a canal and its effect on animal paths is a problem. Similarly surface water runoff transverse to the canal is obstructed as well as groundwater interflow. For the latter reason uplift can be a problem when a canal is emptied.

10.7 ENVIRONMENTAL STRUCTURES

There is a category of hydraulic structure designed to protect the environment. This includes erosion protection of land and rivers. And stabilization of river banks and beds subject to seepage pressures or other natural changes. We are forced to stabilize what was once a naturally changing landscape, because of man's pressure for space, safety and economic efficiency. Stone pitching can be used for many purposes in erosion control, provided adequate stone size and filter layers have been selected (Stephenson, 1985). Dumped or placed rockfill can be used to:

- Protect bed and banks of erosive channels
- Fill gabions (wire beskets) for walls and reno-mattresses for linings
- Dry stone walling
- Riprap on embankments for wave protection
- Closure of channels
- For stabilizing dams
- Groynes and walls to direct flows away from banks
- As drainage layers
- As energy dissipation in high water velocities

The stable stone size is given approximately by the equation

$$d = \frac{0.25v^2}{g(S-1)\cos\sigma(\tan\phi - \tan\sigma)} \tag{10.8}$$

where d is mean stone size, v is water velocity, g is gravity, S is stone specific gravity, σ is the slope angle in the flow direction, and ϕ is the angle of incline of dumped rock, e.g. 30°.

Graded filters are required under rock layers to avoid erosion of the bed material. This can involve filter fabrics or successively smaller stones in layers.

Riprap is dumped rockfill lining an embankment. The drag due to wave action on rocks lining a shore is a function of the roughness of the exposed surface. Large loose boulders reduce runup but incur higher forces, and smooth well packed linings have less drag, and therefore more freeboard is needed. Concrete units such as tetrapods also reduce wave action but require careful design.

Apart from erosion protection there are other applications of structures for environmental enhancement. Fish bypass channels around dams facilitate movement of fish to breeding grounds. Pollution control works protect our streams and water bodies.

Yet another series of structure of use for navigation are harbours, and locks. Locks control water level, enabling ships to be lifted or lowered from one water level to another using only water power.

10.8 HYDRAULIC MODELS

Owing to the empirical nature of dam design and the many unknowns, hydraulic models are frequently constructed. These are used to determine loads on dam wall, to optimize spillway size and energy dissipation downstream, and to design outlets. Outlets are often gated with screens, and may be high or low level. Water velocities can be high and may be abrasive, cavitational or fluctuating. Air entrainment and vortex formation are problems it remains difficult to predict theoretically.

Hydraulic models are built to scales of 1/10 to 1/100 for example, depending on the detail required. Vertical scales may be distorted to achieve representative grades or hydraulic suspension of sediments. The flow is scaled in accordance with the Froude law where the velocity and time scale are equal to the square root of the depth scale.

To illustrate some aspects which require modeling and the thought process in designing a model, the modelling of a morning glory spillway and de-aeration shaft is described (Stephenson and Metcalf, 1991).

10.8.1 Model study of a drop inlet with air entrainment

Model tests were carried out on the intake to a 38km long, 4.5m diameter tunnel forming a link in the Lesotho Highlands Water Project which was designed to meet shortfalls in water requirements in South Africa. Upstream of the tunnel across Lesotho/South Africa border is the Muela hydropower plant, the intake of which draws from a tailpond and a 48km long, 4.3m diameter transfer tunnel leading from a 170m high storage dam at Katse. The system is designed to ultimately provide 77 m^3/s.

The delivery tunnel is designed to operate without control valves. Water is abstracted from the Muela hydro power plant tailpond by way of a high level intake operating as a 'morning glory' spillway, and there is a drop to tunnel level. The tunnel itself flows full at all flows as there is outlet control, created by the level of the outlet and tunnel friction.

A high level intake is necessary to maximize the silt storage capacity of the on-catchment tailpond and to minimize changes in tailpond water level during power generation cycles. The minimum operating level is at 1760m above sea level.

At the outfall into the Ash River, a control weir is set at level 1734m above sea level. Therefore, under low flow conditions, the tunnel is fully submerged, while under maximum flow, the Muela tailpond operating level is 1773.5m above sea level. Figure 10.6 shows the delivery tunnel in longitudinal section.

The intake works comprise eight components (see Fig. 10.7) which are:

(a) a 24m diameter outer weir 0.5m below minimum operating level designed to minimize approach velocities and tailpond rise during power generation
(b) a 15m diameter screen ring to exclude large debris
(c) a 7.6m nominal diameter inner weir set at 1759.0m above sea level, with an ogee (morning glory) crest profile intended to
  (i) minimize air entrainment under free flow
  (ii) act as a low head loss orifice when drowned out
  (iii) maintain water depth to limit the screen velocities to 0.6 m/s.
(d) a 5m diameter drop shaft, 30m deep, with a 90° elbow at its lower end
(e) a de-aeration chamber 75m long and of 50 m^2 sectional area at downstream end
(f) an air shaft to remove air released in the chamber
(g) a bifurcation serving the phase I and phase II tunnels
(h) a gate shaft for each tunnel.

When the project is complete and the tunnel linings are aged, the intake will run completely submerged. By contrast, at early stages an alternation between free flow (upstream crest control and drop) and drowned out (tunnel conductance control) working states was anticipated. Under the first state (drop-inlet), air will be entrained by the free fall from tailpond level to the tunnel hydraulic grade line.

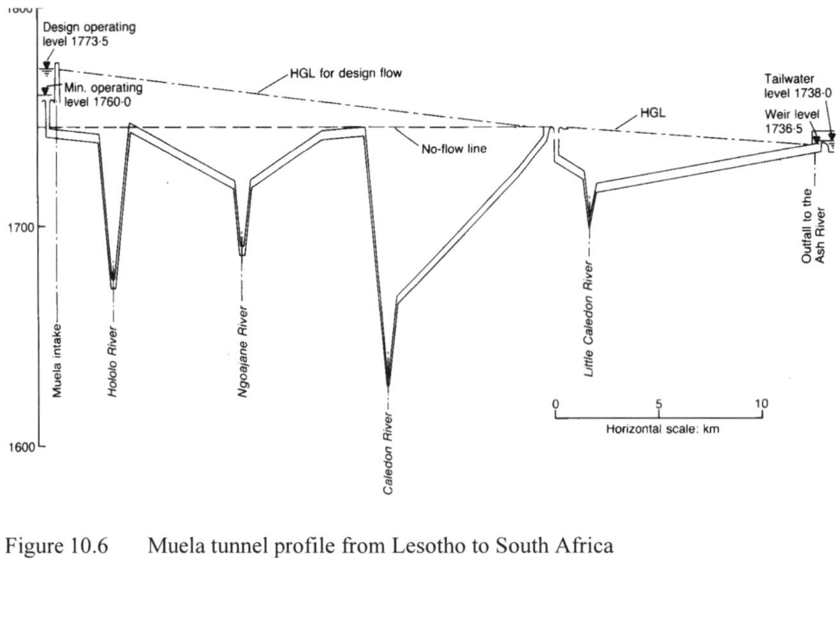

Figure 10.6 Muela tunnel profile from Lesotho to South Africa

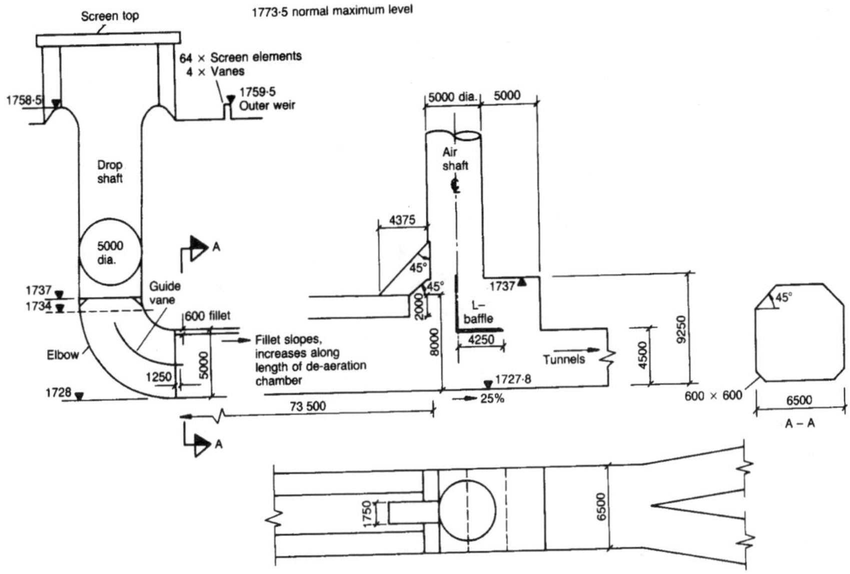

Figure 10.7 Intake components: general arrangement

Any air drawn into the tunnel could affect conveyance capacity and cause unstable flow conditions because the vent shafts are limited to 200mm diameter. Model tests were therefore commissioned to study the air entrainment in the drop shaft and to ensure that air was effectively released upstream of the gate shafts by way of the de-aeration chamber and air shaft. Preliminary estimates of likely air entrainment and necessary de-aeration chamber length and air shaft size were made by the designers on the basis of Haindl's theory (1982) relating air flow to Froude number. It was predicted that the air volume to be handled in the de-aeration chamber would be at least 8 m^3/s at a water flow of about 40 m^3/s.

A de-aeration chamber length of 75m was calculated assuming a minimum bubble rise velocity of 0.15 m/s, a throughput per metre width of 8.5 m^3/s, and an air-water ratio of 12%.

10.8.2 Model objectives

The two objectives of the model testing were; firstly, to minimize air entrainment below the drop while simultaneously maximizing air release efficiency; secondly, to predict the air entrainment rate on the prototype. Scaling-up of air entrainment rate is open to question, and extrapolation cannot be done solely by means of the Froude law of dynamic similitude. The Froude law requires similarity between dynamic and gravitational forces on model and prototype. While the first objective could be met with a single suitably scaled model, for the scaling-up of results at least two (and preferably more) models were considered desirable. Air entrainment in the drop shaft was considered to be function of:

(a) free fall
(b) flow rate
(c) separation from walls
(d) breaking up of jet
(e) counterflow of air

While it was possible to investigate the effect of some of these factors on the model, others could not be studied, e.g. counterflow of air up drop shaft. Other concerns were:

(f) avoidance of negative pressures in the drop shaft at the ogee transition
(g) avoidance of erosion by high velocities
(h) necessity for a guide vane in the elbow and its shape
(i) optimization of tunnel and de-aeration chamber , which influenced inlet cost
(j) quantification of confidence band regarding experimental error and scaling-up.

10.8.3 Construction and testing of model

A 1/25 scale model was used to optimize the intake design. The model scale resulted in a 4m long de-aeration chamber based on the original tunnel length, and a drop shaft and forebay height of approximately 2m. The original tender called for a model to a scale of 1 in 12, which would have resulted in model dimensions about twice those selected. The alternative proposal was for a 1/25 scale model for optimizing design, followed by a half-width 1/12½ scale model to confirm results and to extrapolate aeration rates to the prototype. The smaller model facilitated alterations, reduced costs and construction time, and provided additional information for extrapolating model results. The 1/25 scale model was much lighter than the 1/12½ model. In the end, two further models at 1/50 and 1/100 scale were built to assist extrapolation.

The first model scale (1/25) was calculated to be sufficiently large to avoid surface tension and laminar flow effects: i.e. the nappe thickness was maintained above 10mm and the Reynolds (VD/N) numbers above 2000. A remaining problem was what criterion to use in scaling-up air entrainment and release rates: i.e. would the air entrainment and removal ratios be the same on the model and the prototype? It was recognized that there is a scale effect a solution was to be to build a number of models to different scales and extrapolate.

The drop shaft and structure beyond it were made of Perspex. The crest was made of plaster.

10.8.4 Air entrainment down drop shaft

Air is entrained when the falling jet impinged into the headwater of the tunnel. The air entrainment rate for the worst condition was at a maximum at 45–55 m^3/s (prototype flow) and decreased for low flows or higher tunnel headwater levels.

A device which reduced air transport to the vent shaft was a central guide vane around the elbow below the drop shaft. The vane appeared to trap air bubbles beneath it, before they were washed around into the de-aeration chamber. The bubbles coalesced under the guide and escaped back up the drop shaft.

The most troublesome aspect of the air elimination proved to be associated with surging in the de-aeration chamber, created by release of large pockets of air up the vent shaft. The majority of the air bubbles collected on the soffit of the de-aeration chamber within the initial 5m length. They coalesced into larger bubbles which surged along the chamber to the vent shaft and accelerated up the shaft. A corresponding volume of water returned back down the shaft, carrying fine bubbles with it. These bubbles were projected into the forward flowing water and carried into the tunnels. Numerous baffle systems to divert the air pockets that flowed in the vent shaft were investigated with little effect. The surging was inherently associated with large air pockets, and all that was achieved was to direct the air pockets of surges elsewhere.

In the end, the symptoms of the problem were eliminated by retarding and obstructing the fine bubbles being transported downwards by the surging water. This was achieved with an enlarged chamber beneath the vent shaft and an L-shaped baffle to stop the bubbles. The shaft was increased in size until no improvement in surging was apparent. In practice, the 5m diameter shaft will be reduced in diameter above the water level, for economy of construction.

Early tests indicated higher air entrainment rates than had been expected, particularly at high flows (50–60 m^3/s). The theory of Haindl (1982) indicated air/water ratios of up to 40% at 10 m^3/s, dropping to 7% at 60 m^3/s. In initial tests, corresponding rates of 9–11% were obtained. However, as the testing optimized the model, air rates dropped to 6-4% respectively (Fig. 10.8). The latter results were encouraging but had to be confirmed before extrapolation to the prototype, using the larger (and later, smaller) models to extrapolate entrainment.

Peterka (1956) indicated that prototype air flow can be up to four times that indicated by a model, whereas Mateos (1988) indicated that the model underestimates by 12% in the case of a 1/8 model, but that a 1/6 model is acceptable. Muller and Stephenson (1984) found that a low head siphon model underestimated air entrainment compared with the prototype. This effect was borne out by Westrich and Barczewski (1982) in tests on a cone valve. In general, the Froude number is found to be the most significant variable (Sharma, 1976; Ahmed, 1984). Ervine and Himmo (1984) studied air entrainment in long shaft, which is of most relevance here.

Koschitzky et al (1984) state that a large model (1/8 scale) produced 10% more air than a smaller one (1/30 scale). Par and Saho (1984) however, indicate that scale is important only for Reynolds numbers less than 3×10^6 for aeration down ramps. Straub and Anderson (1958) indicate that air entrainment is a function largely of y/y_c, where y is water depth and y_c is critical depth.

De-aeration in the horizontal chamber was considered to be dependent on the overflow rate, i.e. the ratio of flow rate to surface area of the de-aeration conduit or chamber. This is suggested by an analogy with simple settling theory, and is confirmed by Ervine and Ahmed (1982). This means that the velocities have to be similar in the model and prototype. An attempt was made, therefore, to reduce the flow rate in the de-aeration chamber by a factor of y^2, where y is the geometric scale.

Ellis et al (1978) and Morgan and Reid-Thomas (1961) indicate also that comparable velocities are necessary in the model and prototype to reproduce air entrainment more accurately. However, such a scaling law (overflow similitude) would increase velocities by a factor of 5 for a 1/25 scale model, creating very high turbulence and flows which would be impractical to handle in the model.

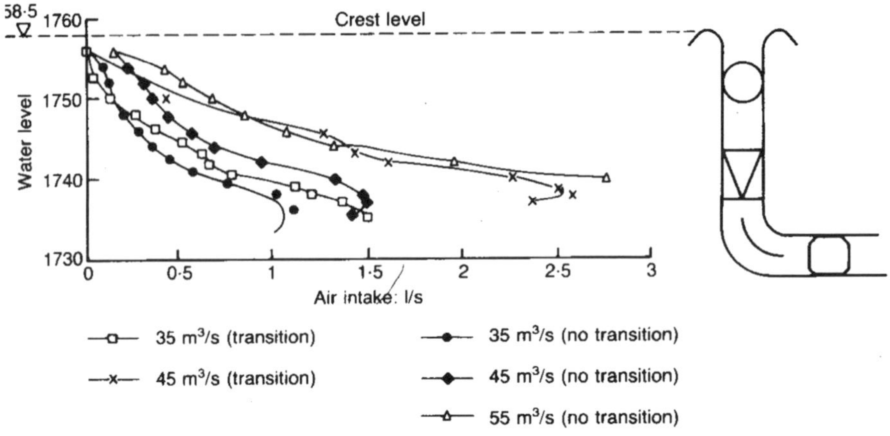

Figure 10.8 Model air entrainment rates at different water levels

For the overflow theory, the model velocity is equal to the prototype velocity which is five times larger than for the Froude law. The Froude number is therefore also five times larger, which results in abnormally high air entrainment in the drop shaft and around the elbow.

At a model flow of 63 ℓ/s, which would represent 40 m^3/s prototype flow on the overflow scale, considerable air entrainment appeared, and air bubbles occurred in the entire cross-section of the chamber for most of its length. Most of the large volume of air, however, escaped up the vent and little air escaped into the tunnels. Despite the fact that the volume of air appeared higher, the water flow was also high, so that the net percentage of air was marginally lower than for the Froude scaling tests.

It appears from observations that the following problems influence the theory.

(a) The size spectra of air bubbles are not exactly the same in both models or prototype: hence the overflow rate should not be exactly the same.

(b) Bubbles expand as they rise, as bubble volume is proportional to the inverse of the absolute pressure to a power less than one, and the atmospheric pressure of 10m water remains even when the bubble is on the surface.

(c) Atmospheric pressure is not scaled. Tests were conducted in Johannesburg (altitude 1600m). It may be possible to reduce atmospheric pressure more on a smaller model by creating a vacuum over the intake and vent shaft. The model cannot be too small, however, or it is difficult to create sufficient vacuum. A 1/5 scale model would require nearly 8m vacuum which is very difficult to obtain, and anything smaller would make it practically impossible to scale by vacuum.

(d) Air release could possibly conform with the overflow theory, whereas air entrainment does not, and it is a function of the Froude number. A hydraulic jump is one of the most effective air entrainment mechanisms There was no hydraulic jump on the Froude model, one did appear on the overflow model.

(e) Turbulence is not modelled correctly, especially with the outflow theory, and re-entrainment of bubbles is highly dependent on turbulence.

(f) Coalescence and interference of bubbles and walls, was not scaled correctly.

10.8.5 Comparison of air entrainment rate with theory

There have been many attempts to predict air entrainment rates. Early correlations were with Froude number (see Kalinske and Robertson, 1943). Later, Wisner (1965) indicated a higher air entrainment rate based on prototype measurements. Thomas and Goldring (1989) corrected for bubble transport rate. Ervine and Ahmed (1982) clarified the effect of model scale . They suggest an equation for total air entrainment rate

$$q_{at} = 0.00045(v_i - 0.8)^3 \tag{10.7}$$

where q_{at} is the total air entrainment rate per unit width (m^2/s) and v_i is the water velocity entering the plunge pool. This equation indicated the strong effect of scale, as a model based on the Froude law must have velocity proportional to the square root of the scale. The air entrainment ratio will therefore decrease for smaller models. The number 0.8 is taken as the critical transport velocity but it could be less than this. The rise velocity of bubbles was also taken as 0.25 m/s.

The theoretical prototype fall velocity (20 m/s) was used in the calculations. The actual velocity will be less due to friction and obstruction of the drop shaft by bubbles rising (which also raise the effective water level in the shaft). The latter may be the reason that observed air entrainment rates were less than practical, using Ervines equ.

10.8.6 Model scale effect

The doubts emanating from the overflow study prompted a more empirical approach. Two further models were constructed: one at a scale of 1/50, the other at a scale of 1/100. These models were simplistic representations: i.e. the drop shaft and de-aeration chambers were maintained at the same diameter as the drop shaft. The length of the drop shaft and the de-aeration chambers were scaled correctly, as well as crest shape.

The results depicted the predicted trend in air entrainment rate: the air entrainment ratio decreased the smaller the scale. In each model, the peak aeration rate occurred at about 50 m^3/s (all flow rates converted to prototype values using the Froude law of similitude). Table 10.2 summarizes maximum air entrainment rate of four models.

The effect of model scale on air entrainment was that of reducing significantly the air entrainment ratio for small models. The air/water ratio for the 1/12.5 scale model (6%) was four times as large as for the 1/100 model (1.5%). Geometric projection on log paper would indicate even bigger rates (13%) on the prototype. Theory indicates, however, that this is not likely on the prototype.

Table 10.2 Maximum measured air entrainment rate of four models

Scale	Air entrainment: %
1/100	1.4
1/50	2.5
1/25	4.3
1/12.5	7.4

REFERENCES

Ahmed, A.A. (1984). The process of aeration in closed conduit hydraulic structure. *Proc. Symp. on Scale effects in modelling hydraulic structures*. Int. Assoc. of Hydraulic Research, Esselingen, 1-5.

Ellis, J. et al. (1978). Surge protection of the Kielder water transfer works. *Proc. Instn. Civ. Engrs*., Part 1, Vol. 64, May 227-246.

Ervine, D.A. and Ahmed, A.A. (1982). A scaling relationship for 2-dimensional vertical drop shaft. Prof. *Int. Conf. on Hydraulic modelling of civil engineering structures*, Coventry, BHRA, Cranfield, 195-214.

Ervine, D.A. and Himmo, S.K. (1984). Modelling the behaviour of air pockets in closed conduit hydraulic systems. *Proc. Symp. on Scale effects in modelling hydraulic structures*. Int. Assoc. of Hydraulic Research, Esselingen, 1-10.

Haindl, K. (1982). Aeration of hydraulic structures. In *Developments in Hydraulic Engineering*, ed. P. Novak, Macmillan, Vol. 2, Ch. 3.

Kalinske, A.A. and Robertson, J.M. (1943). Closed conduit flow. *Trans. Am. Soc. Civ. Engrs*., Vol. 108, 2205, 1453-1516.

Koschitzky, H.P. et al (1984). Effects of model configuration, flow conditions and scale modelling spillway aeration grooves. *Proc. Symp. on Scale effects in modelling hydraulic structures*. Internatl Assoc.Hydraulic Resrch, Esselingen, 1-5.

Mateos, C. (1988). Personal communication with J. Zwamborn, Cedex.

Morgan, H.D. and Reid-Thomas, D.D. (1961). Discussion on maintenance of hydroelectric schemes and the development of sidestream intakes. *Proc. Instn. Civ. Engrs.*, Vol. 20, Nov., 345-370.

Muller, J.R. and Stephenson, D. (1984). Air regulation of a siphon spillway. *Civ. Engr. S. Africa*, Vol. 26, Oct, 507-512.

Par, S. and Shao, Y. (1984). Scale effects on modelling air demand by a ramp slot. *Proc. Symp. Scale effects in modelling hydraulic structures*. IAHR, Esselingen, 1-4.

Peterka, A.J. (1956). Performance test on prototype and model. *Shaft Spillway Symp. ASCE*, 121.

Stephenson, D. (1985). *Rockfill in Hydraulic Engineering*, Elsevier, Amsterdam, 206pp.

Stephenson, D. (1989). *Pipeline Design for Water Engineers*, Elsevier, Amsterdam, 256pp.

Stephenson, D. and Metcalf, J.R. (1991). Model studies of air entrainment in the Muela drop shaft. *Proc. Instn. Civ. Engrs*., Part 2, Vol. 91, Sept, 417-434.

Sharma, H.R. (1976). Air entrainment in high head gated conduits. *Jnl. Hydraul. Div. Am. Soc. Civ. Engrs*., 102, HY11, Nov., 1629-1646.

Straub, L.G. and Anderson, A.G. (1958). Experiments on self-aerated flow in open channels. *Jnl. Hydraul. Div. Am. Soc. Civ. Engrs*., Vol. 84, Dec, HY7, 901-35.

Thomas, N.H. and Goldring, B.T. (1989). Planar plunge-zone flow patterns and entrained bubble transport. *Jnl. Hydraul, Res*., Vol. 27, (1.3), 363-383.

Westrich, B. and Barczewski, B. (1982). Model and field measurements of the air demand of a cone valve in a closed conduit. *Proc. Symp. Scale effects in modelling hydraulic structures*, Int. Assoc. of Hydraulic Research, Esselingen, 1-5.

Wisner, P (1965). Role of Froude number in study of air entrainment. *Proc. 11th Cong. IAHR*, Leningrad.

www.dams.org. (2002). Dams and Development: New Framework for Decision-Making.

CHAPTER 11

Economics of Water Resources Development

11.1 ECONOMIC ANALYSIS

Economics is generally used as the basis for decision-making in engineering projects. It is also used for planning resources development and making investment decisions. There are other criteria which may also influence the decisions such as politics, conservation or aesthetics, but economics remains the favourite as it can be evaluated in numerical terms and therefore alternatives can be compared or ranked.

The way of comparing costs and benefits varies depending on the analyst, but the following standard methods can be employed. In general, the procedure is to compare economic benefits with costs. Invariably there is a time lag and the question of how to compare cash flows at different periods of time arises. That is, $1 derived in the future may be worth less than $1 today. We therefore discount cash flows to common time periods in order to make a comparison.

11.1.1 Definitions

Capital is the money which is set aside or invested as an earning resource. It may be in fixed assets such as buildings or plant, or it could be as floating assets such as stock or equity.

Revenue is income including earnings from assets.

Interest is the remuneration on capital or the payment for the use of the capital.

Redemption or *amortization* is the provision made from revenue for the repayment of capital indebtedness.

Sinking fund is a fund composed of regular amounts set aside out of revenue at regular intervals to provide at a stated rate of compound interest, specified capital sums in a defined period.

Depreciation is the wastage of an asset.

Present value is a sum which if invested now at compound interest will produce a specific amount at the end of a defined period. Alternatively it is the sum which with accumulated interest creates a fund out of which equal annual instalments can be paid over a defined period, the funds being exhausted at the end of that period.

Life : The physical life is the period during which assets remain functional and this could be typically twenty to a hundred years depending on whether it is a mechanical or a civil installation.

The economic life is the period during which assets have a positive economic value.

Loan life is the period over which a loan is borrowed and this may not equal economic life.

Maintenance is the keeping of the asset in serviceable condition by repairing and servicing it.

Running costs include services, labour, materials and other commitments to operate an asset.

Finance is the manipulation of money such as raising loans, paying, taxing, and is usually handled by an accountant.

Economics is concerned with the planning and evaluation of alternatives.

Risk is the chance that something will pay off or not.

Internal rate of return is the effective interest or discount rate at which benefits equal costs on the same time base.

11.2 RESOURCE EVALUATION

Resources such as water and land are limited so there is competition for them. The engineer or accountant has to make a decision as to which project should be built and which are preferable, as well as the scale of construction and the timing. All these decisions can be made using economic evaluation methods (e.g. King, 1967; ICE, 1980).

On the other hand, financial planning is concerned with cash flows, such as income and expenditure, taxation, foreign exchange and raising loans, and this is generally handled by an accountant. Fiscal planning is done by the national financial controllers and involves policy making. It affects inflation, ability to trade internationally, interest rates, exchange rates and duties and taxation.

In completely free economies, the laws of supply and demand will result in a price for resources. But scarce resources may cost more. The bigger the competition, the higher the bidding price. In the case of public works, the price is usually based on cost as there is no profit incentive. Bach (1977) discusses equilibrium between supply and demand in a free economy. Whether there exists a free market for water is debatable.

The interest on a loan to finance a capital project can be regarded as a cost together with operating costs, for example fuel, maintenance, etc., and distribution or selling of the commodity. Benefits may not be evaluated by a public organization but have to be compared with costs for a financially viable project. Thus the present value of sales of water or power, need to be compared with the capital cost plus the present value of the operating costs. Alternatively, the approach may be to minimize costs, for example flood control, and there may not be a direct income, particularly if it is a public project. Alternatively, there may be savings in costs. For example, if a more expensive source is eliminated. The methods described below generally narrow down to comparing benefits and costs on a common time basis.

11.3 PRESENT VALUE ANALYSIS

A method of comparing values at different times is developed below. \$1.00 this year will yield \$1.05 at the end of one year if invested at 5% p.a. interest, and $\$[1 + (r/100)]^n$ if invested at r% interest over n years. It is probably only worth

$$\$\left(1+\frac{r}{100}\right)^n \Big/ \left(1+\frac{f}{100}\right)^n \approx \$\left(1+\frac{r-f}{100}\right)^n \tag{11.1}$$

if inflation is f% p.a. But let us ignore inflation to start with. Thus we cannot compare absolute amounts at different times. They should all be converted to a common time basis.

11.3.1 Discount rate

We refer to the interest rate as the discount rate, which is not always strictly correct. Other discount rates which could be used are:

1. The time rate of preference, or
2. Current bank interest rate
3. Long term bond rate
4. International loan rate where exchange rate features
5. The interest rate less rate of inflation, or more accurately $R = \frac{1+i}{1+f} - 1$ (11.2)

where r is effective rate, i is interest rate and f is inflation rate(all fractions here). E.g. which is preferable assuming a discount rate of 6%; (a) \$1 000 now, or (b) \$400 at end of year 1 plus \$500 at end of year 2 plus \$300 at end of year 3?

b).Present value(PV) = $\$400/(1 + 0.06) + \$500/(1 + 0.06)^2 + \$300/(1 + 0.06)^3$ (11.3)
= \$37.14 + \$445.0 + \$251.9 (PV of \$300 in 3 years time)
= \$1 074.3, which is more than \$1 000

Therefore alternative (b) is preferable. In other words a bank would need \$1 000 now to make a loan of \$1 000 now, but would need \$1 074.3 now to lend us \$400 after 1 year, \$500 after 2 years and \$300 after 3 years.

11.3.2 Inflation

Inflation also needs to be accounted for. A sum of \$377.4 invested now would yield \$400 after 1 year, but it would only buy goods then costing $\$400/(1 + f)$ now. So, if f is greater than r, it pays to spend as much as possible now, borrowing as much as possible. But this adds to inflation and makes the inflation worse as, if there are more buyers than sellers, prices are forced up.

To be viable,

1. A project must earn enough to repay costs, and

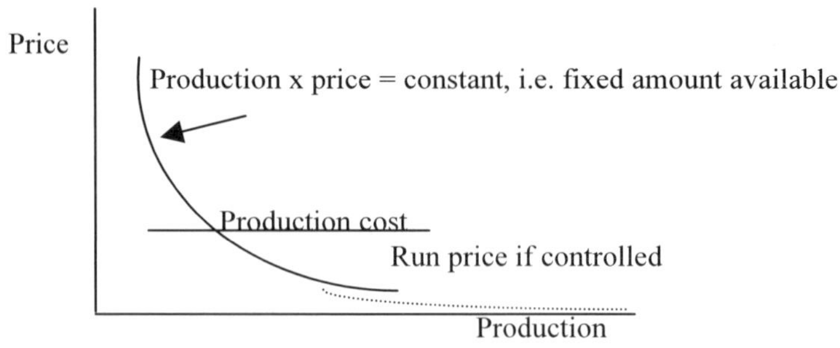

Figure 11.1 Price versus Production

2. The return must be greater than the 'opportunity' cost, which is the return of the best project displaced by the one in question, especially if the projects are mutually exclusive, i.e. capital is rationed.

Instead of maximizing the net present value, we may calculate the discount rate which would result in zero net present value. This is the internal rate of return, and if the interest rate is less than this, the project is profitable.

Inflation is the process whereby the purchasing power of money declines with time. The rate varies with the product, and indices are published by departments of statistics for various types of goods.

Causes of inflation:

1. If there is a constant quantity of goods available each year (production) and there is more and more money available in people's pockets, they will spend the surplus money on whatever goods are available (unless the additional money in circulation is saved).
2. If there is no scope for increased production, prices will rise due to shortages (in free markets).
3. If there is an overproduction, then, in a price-controlled market, the costs of all producers have to be met, so prices are put up on the items sold, to pay for the costs of the unsold items.
4. When trade is booming and there is full employment, people can ask for higher salaries. The employer just increases his prices to meet his costs.
5. If factories are underproducing, then overheads increase.

11.3.3 Taxation

Tax is a negative cash flow and must be treated as a cost. It must be considered in estimating the rate of return. Taxation is payable out of profits indicated by accounts. The accountant is entitled to reduce taxation to a minimum (e.g. spreading profits over time, or spreading losses between subsidiaries, since losses do not qualify for tax refunds). National projects can usually ignore company tax.

11.3.4 Public projects

Repayment of loans by public bodies is pretty secure, so interest rates are lower than for private enterprise, (just enough is charged to repay loans for perfectly secure investments). One could take interest rate as rate of return of the best private project, which will be sacrificed. In the case of public works often comparisons of costs and benefits are not done, especially as benefits are difficult to estimate, e.g. water supply, although Hirschleifer et al. (1960) maintain it should be evaluated. The least cost scheme is selected to meet objectives.

The method of financing public projects can differ from private, e.g.:

1. External, e.g. public loans, local or overseas, or taxes or rates.
2. Internal, i.e. out of current income. Public utilities may try to do this, hence there could be an increase in tariffs.

The second method is wrong. The present should not be paying for posterity's benefits (which may never occur, or posterity would be more able to pay for it allowing for inflation and greater population). Nationalized industries may hope to stabilize the economy, provide employment, foreign exchange, but it often lead to empire building or vote catching. They also cause monopolies, stifle private enterprise and competition.

11.4 PLANNING HORIZON AND PROJECT LIFE

In the case of public works, it is generally necessary to construct successive projects as demands increase. Thus industrial or urban areas expand with increasing population. This is still being felt at a rapid rate in developing countries but the opposite is being experienced in many developed countries. Soon in the latter case the question may be which projects to close down whereas for the developing areas the water demand will be met by successively further or more expensive projects, thus increasing the marginal cost each time additional sources are tapped.

The scale of the project as well as its location on nature will influence the costs. Generally the larger the scale the more economical the water is and this will be indicated by a benefit cost analysis.

Civil engineering works generally have effective lives considerably longer than the loan period. Some dams and waterworks can have indefinite lives. Maintenance and even decommissioning costs should however be added when evaluating a project, although they would be discounted to some future time.

Mechanical and electrical parts will have shorter lives. In fact the fiscal life of engineering works depends a lot on the level of maintenance. Rehabilitation and selected replacement will also extend the life and could be regarded as investments at future years.

It may however be uneconomical to extend the life of a project beyond a certain time because of its high maintenance and operating costs or because of obsolescence.

The question arises of what loan life should be used. Ideally, it should be the life of the project but this is impossible to determine and it is not easy to borrow for such a long period. The interest rate will also be affected by the loan period. Markets are more likely to offer short-term loans because the risk is less, particularly in the case of developing countries. It may therefore be necessary to take a succession of loans over the life of a project and these may be at different interest rates. The cash flow is often weak during initial stages after commissioning so it is important to extend the loan as long as possible. The entire series could however be brought to a present value using spreadsheet methods. It will however be noted during calculations that the sensitivity to the loan period is not as high as to the interest rate for high interest rates.

11.5 RISK AND UNCERTAINTY

The various risks a water resource project could be exposed to include:

- Hydrological – Flood, Drought
- Geophysical – Landslide
- Structural – Cracks
- Demographic – Population decrease
- Economic – Slump
- Financial – Interest rate

To minimize cost of risk one or more of the following strategies could be adopted:

- Design conservatively for economic risk
- Assume high interest
- Design for shorter term
- Underdesign to reduce exposure
- Overdesign to reduce hydrological risk
- Use probability methods
- Plan carefully using economic principles

In Chapter 12, some of the methods of minimizing the cost of risk are described.

It is difficult to estimate what future demands and economic conditions will be like. We often assume conditions in the future, but should really do marginal analyses or use probability theory (which is what life assurance actuaries do).

The costs cannot always be evaluated and one cannot always work with averages. A company may go bankrupt if it experiences a certain loss, therefore it must remove the possibility.

Table 11.1 Calculation of optimum project risk

Q flow m^3/s	D River depth	T Recurrence Interval yrs	E Probability of exceedence	R Q range	P Probability of being in range	C Average cost of damage over range	P x D	Average damage over range
36	2	2	.5	0-36	.50	\$1m	1.00	\$0.5m
53	3	5	.2	36-53	.30	\$3m	.90	\$0.9m
80	5	25	.04	53-80	.16	\$6m	.80	\$0.96m
120	6	100	.01	80-120	.03	\$8m	.18	\$0.24m
180	7	1000	0	120+	.01	\$10m	.07	\$0.1m
					1.00		Σ=2.95	\$2.7m

Note: (i) ΣP x D (2.95) = average depth

(ii) Average damage over range (\$2.7m) is probable insurance payout, therefore one should adjust premiums to pay this

As an example the average damage cost of a river embankment is estimated below. A rule of thumb for allowing for uncertainty is to underdesign a system by about 5-10% depending on the uncertainty, or to adjust discount rate, e.g. increasing the rate favours a low capital cost scheme, which is preferable if future demand is uncertain to save operating costs. This is why private enterprises often prefer the rate of return method.

11.6 FORMULA RELATING ANNUAL CASH FLOW TO PRESENT VALUE

If the interest rate as a fraction is r and the number of years that the loan is taken is n, then the present value of a single amount in year n is $1/(1 + r)^n$. Conversely, if \$1 is invested this year, it will be worth $\$(1 + r)^n$ after n years. This formula is used to handle single cash flows.

On the other hand, if the capital sum is borrowed, this has to be paid back over a number of years n with an interest rate r. The following section derives the relationship between the capital cost or present value and the annual repayment of interest and redemption.

$1/(1 + r)$ will amount to \$1 in one year according to the compound interest formula.
$1/(1 + r)^2$ will amount to 1 in two years.
$1/(1 + r)^n$ will amount to 1 in n years.

The total is $c = 1/(1 + r) + 1/(1 + r)^2 +\ldots 1/(1 + r)^n = ((1 + r)^n - 1)/r(1 + r)^n$. (11.4)

This is the term by which annual payments must be multiplied to obtain the present value. Alternatively the annual repayments on a capital loan c will be

$$cr(1 + r)^n/((1 + r)^n - 1) \qquad (11.5)$$

For example, the present value of \$1 a year at 12% interest over 20 years is \$7.47. Alternatively, if you invest \$7.47 now, it will produce \$1 per year for 20 years, the capital and interest being repaid over that period.

Since yearly payments usually vary from year to year owing to changing conditions, for example, changing maintenance costs, replacement, etc., it is easier to do all the calculations on a spreadsheet. Thus each row could represent a year and the columns provided for capital cost, inflated cost, and cost discounted to present value at whatever year it is to be considered. Then any column brought to a common time period could be summated.

11.7 METHODS OF PROJECT COMPARISON

The following methods have been used for comparing costs and benefits for different schemes:

- Annual worth method (comparison on an annual cost and benefit basis)
- Present value method (everything discounted to today's value)
- Future value method (everything discounted to some future date)
- Internal rate of return (effective interest rate where costs equal benefits)
- Capital recovery cost

The following methods have been used by accountants but are not necessarily recommended:

- External rate of return
- Explicit reinvestment rate
- Pay out period method

The discount rate and the loan duration can both affect the viability of projects. Some of the methods are orientated to sensitivity studies of factors such as interest rate, period of loan or the internal rate of return.

Projects may be selected, rejected or sized using a number of alternative methods:

Selection of the project with *maximum difference between benefits and costs* would generally result in the maximum net benefit. This method may not however consider limitations on capital availability or the demand for water. To a large extent it may be undesirable as it may under-utilize capacity and waste money. Public projects may also not have benefits to evaluate numerically and would tend to plan on a selected time horizon, although this could be optimized. This method may be used for selecting an optimum size or scale of a project, such as a dam, but the above-mentioned factors may play a part in the final decision.

The *ratio of benefit to cost* is another possible criteria for selecting projects. It will be observed in Figure 11.2 that the ratio of benefit to cost does not occur on the same project as the maximum difference between benefits and costs. In fact the maximum ratio could occur for a zero sized project so it is no use for optimizing the project scale.

It may therefore be wiser simply to use the benefit-cost ratio to rank projects once they have been sized using the difference between benefits and costs.

It is not an easy task to decide what benefits to include in a project. Whereas some benefits may be direct, for example sale of water or sale of hydroelectric power generated at a dam, there are also secondary, indirect and intangible benefits which may not be evaluated easily. Secondary benefits could include sale of fish. Indirect benefits may include the value of the reservoir for recreation. Intangible benefits could include the improved living standards of communities receiving water and electricity.

Intangible benefits are even more difficult to pinpoint, let alone evaluate. For example, employment creation by constructing a waterworks or by operating it, is a definite benefit to the economy. What the figure should be is not easy to decide. Often shadow values are used. The shadow value of employment may be the entire payment to the labour force if there is complete unemployment, but it should be the opportunity cost, i.e. difference between pay-out and alternative labour income. Even this may be multiplied by a factor less than 1 as it is somewhat of a simplification.

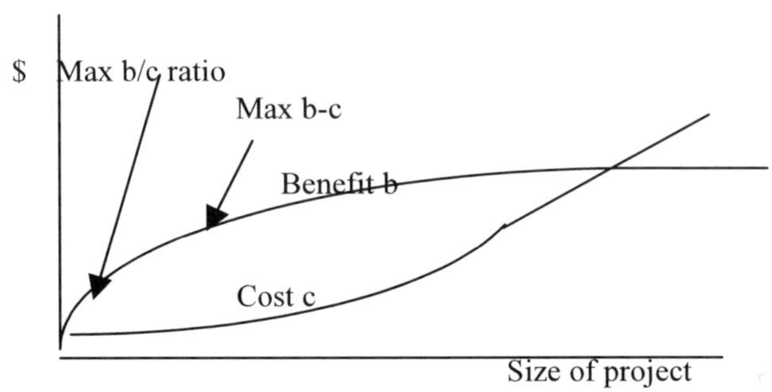

Figure 11.2 Relationship between benefits and costs

The latter types of benefit, i.e. indirect and intangible benefits, would probably only be accounted for in public works. Private enterprise would only include direct benefits as it works on a cash-in-pocket basis. Similarly, taxes may be included as a cost by private enterprise. Customs and other duties would be regarded as a cost by quasi-government organizations but probably not by the nation because those duties are ploughed back into the national economy.

The following list is a summary of *benefit types*:

- Direct, e.g. sale of water
- Secondary, e.g. sale of byproducts
- Indirect, e.g. recreation
- Intangible, e.g. more wetlands
- Shadow, e.g. employment creation
- Social

The following list gives some of the benefits which have been considered for water resources projects:

Benefits:

- Water
- Irrigation
- Power
- Fishing
- Navigation
- Recreation
- Health
- Services and infrastructure
- Access to finance
- Education and training
- Employment
- Wealth generation

Disadvantages:

- Inundation of land
- Relocation of communities
- Culture and heritage loss
- Crop loss
- Exposure to new diseases
- Temporary nature of employment
- Debt
- Crime
- Risk
- Environment
- Earthquakes due to weight of reservoir water
- Changes in flow regime
- Import of labour with new social habits, diseases
- Pollution due to construction

Some of the disadvantages are compared with advantages in Chapter 10.

11.8 SYSTEMS ANALYSIS

The systemization of projects for catchments is a good way of getting a grasp of the factors involved and their relationships. It is therefore a useful exercise to set the system down in constraint form or equations after depicting them on a plan with interactions indicated by arrows or other means. Systems analysis methods such as linear or non-linear programming, can be used once the system has been formulated in mathematical terms. Many spreadsheets have the facilities for optimizing constraints in numerical form subject to optimizing an objective function which can be set out in terms of the variables.

Complex systems can be handled by decomposition but again the systematic method of describing them is useful to the planner, even though sometimes simplifications have to be made to obtain a numerical answer. Multiple objective optimization and conflicts can be handled in similar ways.

An example of multiple-objective planning occurs in stock market analysis. Although the prime purpose for investing on the stock market is to optimize returns or a combination of growth and income depending on the tax situation, the volatility of shares can be assessed in terms of some statistics such as the coefficient of variation in the price. This could be allocated a weight and the corresponding economic return also allocated a weight so that the combined function could be obtained. Other objectives which may be introduced into water resources planning could be according to quality of water, risk of drought or flood, or public image as seen by politicians who may prefer short-term benefits to longer-term benefits, and safety or value of life.

There is considerable concern about the effect on populations and communities of large-scale water projects. Thus the relocation of communities displaced by lakes has been a subject of investigation by the World Commission on Dams. These costs are now added to the cost of the project. The costs of degradation of the environment or improving the environment around waterworks are now also included due to environmental impact analyses which have to accompany most water resource development projects.. Much has been written on water resources planning methodology (e.g. Goodman, 1984; James and Lee, 1971).

The decomposition of problems in to sectorial or regional problems using linear programming was developed by Dantzig (1963) and applied to water resources problems by Stephenson and Petersen (1991).

11.9 FINANCING WATER RESOURCES PROJECTS

It is generally easier for national authorities to borrow money for water resource development from international funders at attractive interest rates than for small private projects. This may be because organizations such as the World Bank (1980) regard development of water resources as a prime way of developing national economies and communities. Water is associated with quality of life, irrigation is valuable for food production and hydro power for renewable energy, and the project provides local employment and this income can be circulated in the economy. Sometimes there may be hidden reasons for favourable international loans, e.g. clauses requiring the donor country to provide design or installation. Or international loans may have secondary costs, e.g. increased repayments in local currency where there is high inflation.

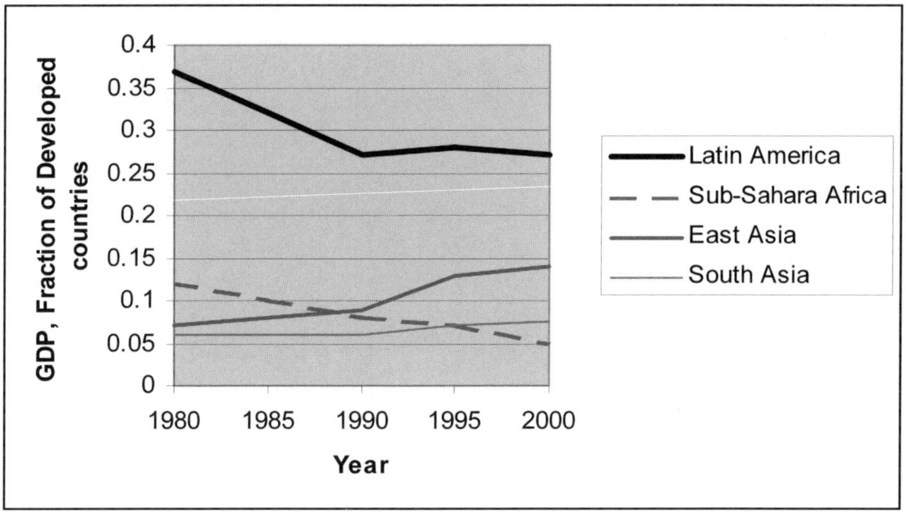

Figure 11.3 International trends in Gross Domestic Product

It has been realized in recent decades that money cannot always be spent effectively on water resources projects and a number of internationally-funded projects have not met expectations.

11.10 INTERNATIONAL FUNDING AGENCIES

Apart from internal budgetary provisions, which in developing countries are relatively small, external financing of the development of the water supply sector in developing countries is now provided from many institutional sources. The principal lending agency to the sector is the World Bank, whilst the regional development banks, namely, the African, Asian, Caribbean and Inter-American Development Banks, now provide significant inputs in terms of loans and technical assistance.

Bilateral sources such as the United States of America Agency for International Development (AID), Australian, German, Dutch, Nordic countries, Canadian International Development Agency (CIDA), Overseas Economic Co-operation Fund (OECF) of Japan, and the UK Overseas Development Administration (ODA), are also important contributors in financing projects in the sector.

Other financing agencies, such as the Organization of Petroleum Exporting Countries (OPEC), the Kuwait Fund for Development and JICA (Japan International Cooperation Agency), have co-financed projects with the development banks. Several of the United Nations (UN) agencies play an important role in the development of the sector. The principal UN agencies involved in sector development, particularly in water supply are:

- the World Health Organization (WHO), which provides technical assistance to developing countries;
- the United Nations Development Programme (UNDP), which is the co-ordinator of the International drinking Water Supply and Sanitation Decade (DWSSD) and

which provides grant funds for technical assistance related to the sector in general (e.g. training) and preparation of projects;
- the United Nations Children's Fund (UNICEF), which has financed and implemented rural water supply and sanitation schemes in many developing countries.

All these agencies play an active role in the development of the sector, and although there have been cases of overlapping, the various agencies, by and large, have defined distinct and useful roles for themselves (IWES 1983, Ch. 4; Hollingworth, 1988).

Each of these agencies has a different approach and criteria in the selection of projects for which financing is being considered. However, there is a general theme in that each agency does, to a varying degree, subject a project proposed for financing to an appraisal (UN, 1968).

11.10.1 The World Bank

A general statement of the nature and objectives of the World Bank follows plus the annual lending programme as in World Bank Report (1987). It is a characteristic of all the above-mentioned institutions that their developmental role is not confined to project lending. Particularly attention should be paid to the diverse nature of the lending instruments.

The expression "The World Bank", as used in their Annual Report, means both the International Bank for Reconstruction and Development (IBRD) and its affiliate, the International Development Association (IDA). The IBRD has a second affiliate, the International Finance Corporation (IFC). The common objective of these institutions is to help raise standards of living in developing countries by channeling financial resources from developed countries to the developing world.

The IBRD's charter spells out certain basic rules that govern its operations. It must lend only for productive purposes and must stimulate economic growth in the developing countries where it lends. It must pay due regard to the prospect of repayment. Each loan is made to a government or must be guaranteed by the government concerned. The use of loans cannot be restricted to purchases in any particular member country. The IBRD's decisions to lend must be based on economic considerations.

The International Development Association was established in 1960 to provide assistance for the same purposes as the IBRD, but primarily in the poorer developing countries and on terms that would bear less heavily on their balance of payments than IBRD loans. IDA's assistance is, therefore, concentrated on the very poor countries – those with an annual per capita gross national product of less than $791 (in 1983 dollars). More than fifty countries are eligible under this criterion.

The International Finance Corporation was established in 1956. Its function is to assist the economic development of less-developed countries by promoting growth in the private sector of their economies and helping to mobilize domestic and foreign capital for this purpose. Membership in the IBRD is a prerequisite for membership in the IFC, which totals 127 countries. Legally and financially, the IFC and the IBRD are separate entities. The IFC has its own operating and legal staff, but draws upon the Bank for administrative and other services.

While the World Bank has traditionally financed all kinds of capital infrastructure such as roads and railways, telecommunications, and ports and power facilities, its development strategy also places an emphasis on investments that can directly affect the well-being of the masses of poor people of developing countries by making them more productive and by integrating them as active partners in the development process.

In 1990, the Bank, in response to the deteriorating prospects for the developing countries, inaugurated a program of structural-adjustment lending. This lending supports programs of specific policy changes and institutional reforms in developing countries designed to achieve a more efficient use of resources and thereby:

1. contribute to a more sustainable balance of payments in the medium and long term and to the maintenance of growth in the face of severe constraints; and
2. lay the basis for regaining momentum for future growth.

11.10.2 Investment experience

Economic, political and social conditions throughout the world have changed dramatically during the existence of The World Bank, and much has been learned about the process of economic development, its complexity and the pitfalls involved. We understand better today than we did in the 1950s that development is a long, slow process, and one that is often painful. Lessons learned from projects the Bank has financed can be invaluable in planning for future water resource development projects, no matter how they are funded.

Water resources planning has become a very broad discipline in the past twenty years, but it is only recently that we have recognized the importance of considering from the earliest stage of water resource planning the strong inter-relationship between:

- Water resource development.
- Agricultural development (particularly shift from subsistence farming to cash crops).
- Energy requirements
- Infrastructure development.
- Modification of environmental and social systems.
- Governance

In the past, planning studies addressed technical and financial factors in detail, with little attention given to cultural variables. It has become obvious that cultural and social factors must be given particular attention from initiation of planning studies.

In 1973, it was estimated that almost 40% of those living in developing countries lacked basic human necessities and suffered from malnutrition, disease and illiteracy. Since then the focus of the World Bank's programs has shifted from comprehensive economic planning toward safe water, health care and education. These social components have become an integral part of many projects. Since most of the poor in developing countries are rural, the Bank's priorities have shifted to more agricultural projects. And in more recent years Institutional Capacity building and good Governance have featured.

Governments frequently provide irrigation water and electricity for agriculture at little or no cost, at fees that do not cover the cost of even operation and maintenance. Traditionally, governments also provide infrastructure, agricultural extension, health and educational facilities. A large portion of funding for agriculture often goes for irrigation development, but once farming becomes commercialized, infrastructure (such as roads and electricity) is needed. Rural road construction is usually the first requirement, since access to markets is needed for commercial agriculture. Improved roads also encourage those involved in health and education programs to live in villages and provide access to markets. Low agricultural productivity in many developing areas reflects limited investment in rural roads, water, electricity, and so on, as well as agricultural technology. Investments in infrastructure in developing countries have been concentrated in urban areas where large numbers of people can be served at low unit cost.

Most governments in developing countries would agree with the five objectives of agricultural development identified by the World Bank (Baum and Tolbert, 1985): growth, sustainability, stability, equity and efficiency. These are integral to the broad national goal of "food self-sufficiency" of many countries.

- Growth in agriculture is often the primary objective because more food is needed for a growing population.
- Sustainability is maintaining adequate levels of production into the future.
- Stability is important because farm policy must even out the inherent variation of agricultural production due to fluctuating weather patterns and trade cycles.
- Equity involves fair distribution of agricultural benefits among all those involved.
- Efficiency is of particular concern to developing areas because any waste of resources is a real loss to the overall economy, and in agricultural societies such losses can be very large.

Many factors determine the success or failure of a project. The World Bank's review indicated that those factors include the following:

- Project formulation was considered to be critically important, especially the clarity and acceptance of objectives, the technical, administrative and financial feasibility of the project, and the thoroughness with which the project was developed and appraised. Over one-third of the 189 projects were judged to have been adversely affected by deficiencies in project design or appraisal.
- Institutional capacity of borrowers is vital in determining the success of projects and the extent to which benefits can be sustained in the future.
- Strong borrower support is required. Many projects were delayed or scaled down because of a shortage of local funds; others were adversely affected by constraints beyond the control of the project agency, such as manpower shortages and high staff turnover.
- External events adversely affected many projects reviewed in 1985. The world recession after 1979 depressed world markets, and in some cases prices were much lower than projected in planning studies. Also a large number of projects were severely affected by prolonged drought, political unrest, wars and administrative changes.

The Bank also concluded that some issues identified in prior years continued to be important factors in project success and need continued attention, including:

- Encouraging environmental impact assessment by project-related dialogue and support.
- Data on social impact indicators are incomplete.
- The quality and continuity of senior management is vital to project success.

- Training to develop local skilled and managerial personnel is a significant factor in project success.
- Funding for management, operation and maintenance is as important as funding for project construction.

Projects were categorized in three groups in assessing economic benefits:
1. Projects for which the economic rate of return (ERR) captures the economic benefit of the investment, principally agricultural, industrial and transport projects. Comparison of the ERRs estimated in project planning with evaluation shows that prior to 1978 re-evaluation values tended to exceed original planning estimates, but that since 1978 the re-evaluation values are less than original estimates.
2. Projects in the utility sector where revenues are used as proxy for economic benefits in estimating ERRs.
3. Projects for which no quantifiable indicator of economic return was available at evaluation were assessed subjectively as to likely economic outcome as well as for their achievement of other objectives.

11.10.3 Social impact

The bank has been giving increasing attention to social impacts of projects, focusing on reducing socio-economic imbalances, improving quality of life, and enhancing skills and income-earning capability. Most projects with specific social objects concentrated on education, agriculture, urban water supply, population, health and nutrition. Project objectives have been designed to enhance living standards by:
- Improving access to and quality of such services as water supply and waste disposal, transportation infrastructure, health facilities, and so on.
- Raising productivity and income through improved education, credit, agricultural inputs, small-scale industrial employment.

Educational components appended to projects without careful preparation and coordination were not fully successful.

Conclusions on social impacts from the review included:
- Assessing the social impact of projects raises difficult questions of methodology and data collection.
- Social impacts, especially those involving basic cultural change, take time; such objectives are probably best addressed through a series of projects.
- Equity and efficiency considerations sometimes suggest alternative approaches; in such cases, both the social and economic costs and benefits must be weighted carefully.

11.10.4 Technological impact

Technology transfer has been an important objective for many projects. Of the 1985 agricultural projects reviewed, 44 out of 55 sought to use improved technology to increase production and yield. The technology was generally not new, but was adopted from experience in other areas and was new to the project area.

The introduction of sophisticated technology takes time, and it is important to allow time for the learning process. In some cases, local agencies are slow to recognize the complexity of the process. There also may be difficulties with spare parts. Even where projects included financing for parts, problems were encountered. Also, in some education programs, special workshops were underutilized because of inadequate store or parts. There is always the risk that projects may not be adequately maintained or be abandoned after a short time. The problem of finding skilled personnel to sustain the new technology is similar. Experience has shown the importance of thorough training to ensure that the technology will continue to be used. Conclusions regarding technology change were:

- To be successfully adopted, the technology must be proven to be both sound and appropriate to the needs of the user.
- It is important to ensure that the borrower can absorb and sustain the improved technology.
- Due attention must be given to cultural and social factors that might inhibit acceptance of new technology, especially in agriculture.

11.10.5 Environmental concerns

Projects faced a number of environmental concerns typical for power and water supply projects. There were two successful resettlement programs in Ghana and Thailand as part of construction of hydroelectric projects. In Zambia, changes in natural stream flow due to construction of a dam and reservoir required augmentation of low flows by special releases from the reservoir. In Thailand, improved downstream conditions due to construction of a hydro project resulted in investment in a controlled water supply system for irrigation.

The Bombay water supply and sewerage project had an environmental impact typical of many similar large-scale projects of this type. Priority was given to alleviating a critical water shortage, and improvement of the wastewater disposal system became increasingly more costly and was considered to be of a lower priority. While it had been recognized that the sewerage components of the project could not keep pace with increased wastewater production and that the already extreme pollution in receiving water would worsen, what happened was worse than expected because some of the sewerage components had to be cancelled during implementation due to funding limitations.

11.10.6 Sustainability of project benefits

The 1985 review indicated that sustaining the estimated benefits of power and water projects appears to require as much attention as implementation of a project. A number of projects had shortages of spare parts (even lack of fuel) because of foreign exchange shortages. This led to maintenance deficiencies and loss of production and distribution capacity. It is at least as important to provide for resources (including foreign exchange) for management, operation and maintenance, as it is for constructing a project.

One of the major concerns regarding sustainability is excessive losses of energy and water in transmission and distribution systems by theft through illegal connections, leakage or faulty metering. Objectives of 25% of the power projects and 45% of the water projects included reducing losses, but in most cases results were not satisfactory.

11.11 SOCIO-ECONOMICS

Health and wellbeing generally go hand in hand with water supply. This is the reason the World Health Organization and other international organization push to establish potable water projects in developing countries. However, there are a number of non-technical factors to be considered in making such projects.

Since the early 1980s, it has become increasingly evident to the international assistance and lending agencies that social issues have not been adequately addressed in project planning for water resources development and that major unforeseen social and environmental problems have resulted from such programs in less developed countries. The problems have been primarily related to:

1. Resettlement of people living in the project area due to construction.
2. Acceptance of the project and its responsibilities by local people.
3. Inadequate sanitation measures.
4. Water-related diseases.
5. Food production and supply.
6. Ecological change.

The nature and importance of social and environmental impacts in developing areas are different from those in the United States and other industrial countries, and the social and environmental effects are often so closely inter-related as to be inseparable. In some cases, such impacts have been so acute that projected benefits have not been realized.

The emphasis of impact studies related to water resources in developing areas has been on economic and, more recently, environmental effects have rarely been thoroughly addressed. When social impacts have been studied, the focus has been on populations as a whole, without giving attention to the special role of women as users and conveyors of water. In recent years, increasing emphasis has been given to the crucial role of women in planning agricultural development programs because it has become evident that national objectives for food self-sufficiency cannot be met unless the role of women in agricultural production and food processing, preservation and marketing is taken into account. However, the role of women in relation to water resources development has been largely ignored.

11.11.1 Health and wellbeing

Water is a key health factor in developing areas. The World Health Organization estimated in 1980 that about 32% of rural population and 73% of urban population in developing countries had access to safe water. About two-thirds of the total population of 4 billion (600 million in urban areas and 2 billion in rural areas) were without safe drinking water and waste disposal. In several countries, only a very small percent of the rural people have safe water: 2% in Kenya; 3% in Gambia and 5% in Zaire.

Water resource development programs have the potential to improve the health and socio-economic wellbeing of people in developing areas. Lack of a reliable adequate supply of safe drinking water is probably the greatest cause of disease in developing countries, and disease hinders productivity and, therefore, economic development.

Traditionally in many rural areas, women and girls spend several hours each day fetching the household water supply from natural sources, often from great distances. Access to water in rural areas in many parts of the world is difficult; supplies are frequently polluted; supplies are often limited (sometimes seasonally, sometimes for extended periods of drought); and natural sources are often a considerable distance.

Typically most developing countries have high rates of population growth accompanied by increasing migration from the countryside to cities as people seek to improve the quality of their lives. The growing urban centers will require greatly increased supplies of safe water in the near future, and in semi-arid areas potential supplies are often severely limited. Also, as an area develops and becomes industrialized, there is increased demand for water, and the associated waste discharge usually leads to increased pollution of water supply sources (Stephenson, 2001).

Economic growth and development imply improved living standards for all the people, including better nutrition, better health and health services, better educational opportunities, higher income, and better housing. Few development alternatives have greater potential for improving the health and social well being of people than water supply projects. However, it is often difficult to show project economic justification on the basis of improved health.

The 1980s were designated as the International Drinking Water and Sanitation Decade, with the objective of providing safe water and adequate sanitation to all people by the year 2000, but this goal was not met. In less developed countries, two-thirds of the people still do not have reasonable access to adequate supplies of safe water, and the World Health Organization estimates that 80% of all diseases in developing areas is related to unsafe water supplies and inadequate sanitation measures. In such areas, water-related diseases contribute to high infant mortality, low life expectancy, and a poor quality of life. Undernutrition and malnutrition clearly reduce the resistance of children to disease and the productivity of adults.

11.11.2 Assessing Social Impacts

The following types of impacts related to health and social well-being should be considered when examining potential impacts of alternative water resource programs and recommending programs in developing areas:

1. *Impact on those living in a project area*:
   - Changes in communicable disease patterns.
   - Local sanitation problems.
   - Deterioration of water quality (surface and ground water).
   - Adverse impacts on fish and wildlife populations.
   - Lowered nutrition due to decrease in per capita food supplies during construction
   - Increased employment opportunities with labour-intensive project.
2. *Impacts on immigrant project workers*:
   - Impaired health due to endemic disease, exposure to chemicals, physical hazards.
   - Lowered nutrition levels because local food supplies are insufficient.

3. *Impacts on those relocated from project area:*
   - Problems of ethnicity.
   - Safe water and sanitary measures in relocation areas.
   - Compensation for land "in kind". (Land prepared for farming when resettled.)
   - Soil conditions appropriate for same crops as project area.
   - Adequate food supplies during resettlement period and until first harvest.
   - Access to other towns and health centres.
   - Fair compensation for lands.
   - Timely relocation.
4. *Impacts on health services*:
   - Greatly increased need for local health services.
   - Increased need may be too costly for local resources.
5. *Income redistribution*
6. *Impacts on living standards*:
   - Housing.
   - Availability of safe water.
   - Sanitary facilities.
   - Electricity.
   - Availability of fuel (wood).

Water resource planners must be aware of the advantages associated with involving sociologists and anthropologists to obtain the basic social data needed to evaluate project impacts. Engineers, planners and economists, unless specially trained, rarely have the background and skills that are required to obtain adequate and valid social information. Such technical specialists are usually urban males (or expatriates) who have difficulty in communicating with rural people, especially women, and different ethnic groups. International expert teams generally do not include women, and few team members understand or appreciate local customs. Their local informants are usually government officials who, themselves, often do not understand rural people and their culture. Rural people frequently are very cautious in discussions with "outsiders", and the lack of common language and dialect often accentuates communication problems. Planners also must be aware that considerable time may be required to obtain the needed data (especially if the program is large and complex), and the data are often needed early in the planning process.

The economic, social and cultural characteristics of rural communities and rural people vary widely from country to country and even within a given country. Because of the hierarchical structure of rural people, those in power are usually those who are better off; they are often more open in expressing their views than the rural poor, but their views may differ significantly from those of the majority (especially those of the landless, women and other groups, who are reluctant to oppose the elites).

Social factors frequently affected by water resources programs in developing areas can be generally grouped in four categories:

1. socio-economic factors,
2. quality of life indicators,
3. agricultural factors, and
4. services.

Potential impacts on these indicators during construction and operation of water development programs must be carefully assessed and considered in planning, design, construction and operation. The extent to which these potential impacts are positive and the extent to which negative impacts can be mitigated may well determine whether or not outputs and benefits projects for a development program are achieved. Usually it is much cheaper to incorporate measures to provide a safe water supply, a healthy environment, etc., as a part of project design, than it is to add such measures to a project after construction has started.

The following tabulation lists indicators most likely to require evaluation in assessing impacts of a water project in a developing rural area. Conditions vary from country to country and from project to project; thus, the tabulation is not complete (see also Biswas, 1980).

1. Quality of life indicators
- Food supplies, food preferences, consumption.
- Nutritional status.
- Health, health services.
- Fertility.
- Infant and child mortality.
- Life expectancy.
- Housing.
- Distance to safe drinking water.
- Source and distance to water for laundry and bathing.
- Sanitation facilities.
- Type of fuel, distance to source of supply.
- Electricity.
- Education, literacy, school enrolment.
- Means of transport.

2. Socio-economic factors
- Household composition and demographic characteristics.
- Migration (nomadic, rural to urban).
- Ethnicity.
- Hierarchical village structure.
- Kinship patterns.
- Role of women.
- Farm size and type.
- Main economic activities.
- Sensitivity to change.
- Sensitivity to risk.
- Adult employment patterns (male and female).
- Child labour.
- Modification of cultural values and lifestyles.

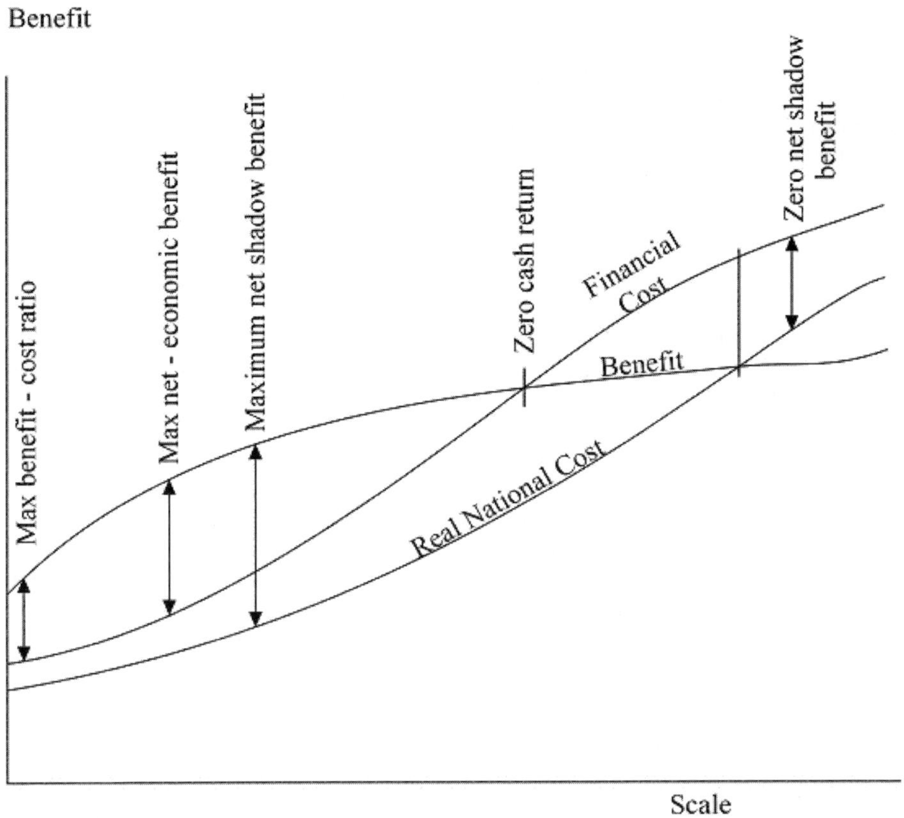

Figure 11.4 The effect of shadow prices on optimum project extent

3. *Agricultural factors*
   - Land fertility.
   - Subsistence acreage.
   - Cash crop acreage.
   - Age of tree crops.
   - Animals.
   - Fishing.
   - Farming tools and equipment.
   - Preservation and processing of crops and animal products.
   - Role of women and children.
   - Government extension services.
   - Land inheritance patterns.
4. *Services*
   - Transportation networks.
   - Health services.
   - Agricultural extension services.
   - Marketing facilities.

However the project is evaluated benefits and costs should be evaluated on a common basis. That is social benefits as well as adverse impacts should be weighed on a democratic basis to ensure a fair deal. The problem of redistribution of gains and losses should also be addressed. This is likely to be more complex than evaluation of benefits and costs. And sustainability needs to be weighed against development unless those preferring environmental factors subsidize the poor in wealth and potential.

REFERENCES

Bach, G.L. (1977). *Economics, an introduction to analysis and policy*, 9th Edn., Prentice Hall, Englewood Cliffs, Ch. 6, Supply, demand and market prices.

Baum, W.A. and Tolbert, S.M. (1985). *Investing in Developing, The World Bank.* Oxford University Press.

Biswas, K. (1980). *Water Resources Development*, Wiley.

Dantzig, G.B. (1963). *Linear Programming and Extensions*, Princeton Univ. Press.

Goodman, A.S. (1984). *Principles of Water Resources Planning*, Prentice-Hall Inc., New Jersey.

Hirschleifer, J.J., De Haveman, C. and Milliman, J.W. (1960). *Water Supply Economics, Technology and Policy.* Univ. Chicago Press.

Hollingworth, B.E. (1988). Course on Water Resources in Developing Areas, University of the Witwatersrand, Johannesburg.

ICE (Institution of Civil Engineers), (1980). *Engineering Economics*, London.

IWES (1983). *Water Supply and Sanitation in Developing Areas.*

James, L.D. and Lee, R.R. (1971). *Economics of Water Resources Planning.* McGraw-Hill Inc., New York.

King, J.A. (1967). *Economic Development Projects and Their Appraisal.* John Hopkins Press, Baltimore.

Stephenson, D. and Petersen, M. (1991). *Water Resources Development in Developing Countries*, Elsevier, Amsterdam.

Stephenson, D. (2001). Problems of Developing Countries, in *Frontiers in Urban Water Management – Deadlock or Hope*, Eds. C. Maksimovic and J.A. Tejada-Guibert, IWA Publishing for UNESCO.

United Nations (1968). *Planning Water Resources Development*, United Nations Office of Technical Co-operation.

World Bank Report (1980). *Poverty and Human Development*, Oxford University Press, New York.

World Bank, Operations Evaluation Department (1987). *The Twelfth Annual Review of Project Performance Results.*

CHAPTER 12

Administration of Water Projects

12.1 PLANNING PROCESS

The concepts in this book could not be implemented efficiently without an effective institutional infrastructure and good governance. The World Bank and other funding agencies have realized the importance of these and in fact the chain of steps necessary in successfully implementing a water resources project is now something like the following:

- Conceptualization
- Cooperation between national government and funding agency
- Feasibility studies
- Environmental impact study
- Socio-economic study
- Technical design
- Construction contract
- Institutional capacity building
- Good governance

The learning process has been expensive to aid organizations including the World Bank and its lenders. This may be because the learning curve has been discovered by trial rather than by pro-active forethought. It is possible that further new stumbling blocks to perfect projects will emerge, e.g.:

- Educational level of communities
- Social infrastructure establishment
- International agreements
- Changes in government
- Replacements by military regimes
- Guarantee of continuance of support by investing countries

Within the country or region developing water resources, a good administration is required to minimize wastage of time and money. This is also important to maintain good staff and in the end to ensure the success of a project.

Table 12.1 Duties of various authorities

Surface Water	Ground water	Water quality	Inhouse	By
Monitoring Assessment Data collection Planning Protection Legislation Negotiation Regulation Standards Funding Communication Allocation	Assessment Data collection Recharge	Monitoring Legislation Fine	Link to other government departments; Environmental, Weather Geology	National water resources agency
Treatment Distribution Storage Charge	Permits Monitoring	Water permits	Engineering	Water service provider
Design Construct Manage	Drilling Pumping	Remediation	Equipment	Contractor

The institutions involved will range from National Water Resource Agency through Provincial Administration to Local Government or establishment of a Project Water Authority. The latter may be necessary particularly in the case of international or cross watershed (interbasin) projects. The responsibility of each organization including Contractors needs detailed legal and technical specification. This may be even more important in the case of new or developing country institutions.

12.2 ASSET MANAGEMENT

In its broader sense, asset management is the assessment of valuables, investment and the management of them, such that their value is optimally enhanced while at the same time providing a benefit to the owner. In the water services sector, asset management is now applied to evaluate engineering installations, to ensure good maintenance and provide users (consumers) with the best possible service. The steps in asset management, namely compilation of asset registers, institution of best management practices and reporting, are of value in making the service more efficient and transparent. Aging infrastructure and the difficulties in replacement are likely to make asset management an increasingly important subject in the future.

12.2.1 Assets

To a water authority assets are not the same as to an investor. Water authority assets are grouped under the following headings:

Infrastructure: Often split into surface and subsurface hard assets, i.e.:

- Buried infrastructure such as pipelines
- Surface buildings
- Surface facilities, reservoirs and treatment works
- Stored or installed hardware such as valves, meters

Human resources: Maintaining an adequate level of trained, experienced and motivated staff, ensures efficient operation. The operational side is of particular concern in developing communities. They may rely on outside assistance (design and equipment) for installation, but operation and maintenance is an ongoing activity which the community has to manage.

Capacity building ensures sustainability and responsibility. Training in technology and accountability may require a sizeable budget, but is the only means of ensuring sustainability.

Financial: Profitability is measured in economic terms so budget control of financial assets (capital, investments) as well as cash flow (billing, expenditure) should be efficient to maximize dividends and minimize costs to consumers.

Past experience in the water industry has been typified by:

- Public or government ownership.
- Little consumer interaction.
- Financing by stocks or government loans.
- Reporting to government.
- Long-term capital interim planning.
- Ad hoc maintenance.
- Separate technical and financial departments.

In addition less attention was paid to asset management in the past because systems were newer and failures were fewer, it was easier to deal personally with problems on smaller scales and it was easy to raise capital for installations as well as maintenance and operation.

Staff employment in the water industry used to be long-term so experience and knowledge was available informally and consumers were less aware and less interactive. There was less use of and even access to computers and software for formal procedures. This made data processing poor.

Engineers tended water works from design to operation and they were often not trained in broader skills. Environmental responsibility was not enforced. Image was not important because capital was obtained from government. Reportability was minimal. And consumer reaction was rare so water works tended to be unapproachable institutions. With new management ideas this has certainly changed.

Low standards in the past, particularly for developing communities have included:

- Varying water quality (inadequate quality control).
- Poor quality materials - low class pipes of non-durable materials, mechanical breakdown, power failure.
- Lack of management structure, even collection of rates.
- Susceptibility to drought, lowering of water table, etc.
- Installation of communal instead of individual connections.
- Low flow.
- Low pressures.
- Lack of meters.

- Little storage for meeting peaks.
- Little surplus for firefighting.

It is often advisable that a regulator and a national system for monitoring, assessment and information be established. This will provide data on quantity and quality of water and the catchment. Of more relevance, the quantification of water uses and resources will require some formalisation and standard procedure. The data could be used to prioritise development and to establish budgets. Mass balances in catchments will be possible. Standard methods for monitoring include guidelines and procedures. The nature, type, time period and format of data may also be prescribed. Asset registers may become common for security as well as economic reasons. Asset registers are of benefit as follows:

- Better understanding of the core of the business
- Accountability
- Evaluation using an EAV (Equivalent Asset Value)
- Serviceability grading
- Maintenance planning
- Strategy planning
- Unit cost establishment
- Forecasting of investment requirements
- Rationalization
- Deferment of expenditure
- Investment planning
- Operating expenditure matching asset serviceability
- Work scheduling
- Operational efficiency gauging
- Statutory reporting
- Bargaining with regulators
- Raising capital
- Water balance
- Organizational review

The activities of water authorities which could benefit from such data include:

- Engineering planning for future installations
- Strategic planning for better operations
- Improvement of standards (quality, pressure, flow, reliability)
- Rehabilitation, disposal, replacement
- Business strategy for future funding
- Cash flow
- Accounting and auditing
- Operation
- Maintenance
- Monitoring
- Loss control
- Safety
- Reporting
- Establishing life cycle costs and viability

Some of the requirements of an asset register are the following:

- It should record the details necessary to identify each asset.
- It should record a basic set of information that is the same for every asset. This would include the identification, the location, the age, the assessments of the value, performance, condition and risk of the asset.
- It should record, for each type of asset, any information over and above the basic set of information that is necessary to effectively manage that asset. The information that should be recorded is any information for which the value of knowing the information is greater than the cost of obtaining the information.
- It must meet the organisation's management, planning, technical and financial needs, as well as any legislative requirements.
- It must be easy to operate and provide quick and accurate access to information, in the form required, to anyone who has a right to that information.
- It should facilitate accurate and confident decision-making.
- It must be secure so as to prevent unauthorised changing data.

Systematic methods are required to cope with large volumes of information and to ensure transparency in the face of new owners and operators.

12.3 PUBLIC VERSUS PRIVATE MANAGEMENT

Public organizations are notoriously inefficient and in some counties corrupt. Private industry is supposed to have a higher efficiency if it is to be competitive. But in the case of owning or managing public services there is a danger a monopoly could develop with lack of competition. Profits could be raised by increasing prices of water or by neglecting assets. Regulation and auditing are required to prevent this. Asset management was enforced in the UK primarily with a view to privatization. Generally private industry will continue to play a role.

This section describes various forms of PPPs (Public-Private Partnerships). It is taken from a study to develop an expert system to decide the best level of private involvement (Richardson, 2002). Specific characteristics from the descriptions have been identified and weightings were given to each characteristic according to the level of correspondence between the characteristic and the form of partnership.

Full Privatization:
Under this arrangement the municipality sells off all its assets to a private company. The private company legally owns the entire water and sanitation system and has legal responsibility to provide, operate and maintain services and collect revenue on a permanent basis. The municipality would assume a regulatory role in terms of monitoring water quality, adequacy of services and tariff setting. Few water and sanitation services have been fully privatized. There is likely to be a significant opposition from several stakeholders to full privatization.

Concession:
This form of partnership entails the municipality transferring the ownership of all its water and sanitation assets to a private company for a significant time period, usually between twenty and thirty years. Under the contractual agreement the private company would have to operate and maintain the entire system. It would also be required to invest significant amounts of money in expanding the services. The municipality would

regulate the performance of the concessionaire and also approve tariff structures. At the end of the contract the private company has to return the assets to the municipality in acceptable state.

Lease Contract:
Under this arrangement a private company leases the entire water and sanitation system from the municipality for a period typically of 20 years. The private company would operate and maintain all the systems. Depending on the arrangement, revenue may be collected by the municipality or the lessee, and the profit being shared between the two parties. Under a lease contract the lessee never owns the asset and never invests any money in new infrastructure. At the end of the contract the lessee should return the assets to the municipality in good order. The municipality still has a regulatory role.

Management Contract:
A management contract entails a private management team working with the existing municipal staff. Typically of five years duration, the management team would implement programs to improve cost recovery and control, asset management, staff training and systems efficiency. While a management contract should result in improved municipal revenue and efficiency, the responsibility to provide services ultimately still rests with the municipality.

Service Contracts:
A service contract would be entered into with a private company for a specific municipal task. Service contracts, normally two years long, are commonly used in meter reading and solid waste collection. They can also be utilized in more technical functions like the operation of purification and wastewater treatment works. Service contracts represent a good means of obtaining skills which the municipality does not posses themselves. The municipality needs good management skills and needs a financially sound footing in order to implement effective service contracts.

Corporatization:
Corporatization is the process of ring fencing and registering a certain municipal function (for example water supply and sanitation) as a private company. While the municipality would still wholly own the new company, corporatization serves to formalize the relationship between service provider (private company) and service authority (the municipality). This enables the private company to deal more easily with other private sector companies and reduces the political interference from the town council. Corporatization often is a step taken by a municipality before it enters into other forms of PPP.

Public-Private Partnerships:
A public-public partnership is a partnership between two public organizations, where one organization offers its services to the other under some sort of contractual arrangement. While it is not likely that this sort of partnership will occur at present in many municipalities owing to the complicated financial state that some municipalities are in, there is very real possibility of public-public partnerships between municipalities and water boards.

BOOT Projects:
BOOT refers to Build Operate Own and Transfer. They are more common in developing countries.

BOT projects
BOT refers to Build, Operate and Transfer. BOOT and BOT arrangements are usually project specific. They are utilized as a means of financing a new development. Under one of these arrangements a private company would supply the capital for a project, implement the project and operate the project for a certain time period, usually twenty to thirty years. Upon expiry of the contract the assets would be transferred to the municipality and the municipality would assume operation of the system.

Municipal Debt Issuance:
Municipal Debt Issuance is purely a mechanism of obtaining capital through selling public bonds. Currently in South Africa there is little prospect for municipalities obtaining capital through a municipal debt issuance due to bad credit ratings, and lack of institutional capacity.

Table 12.2 Comparison of privatization benefits

Type of PPP / **Activity** / **Characteristic** / **Indicate level of correspondence between type of PPP and Characteristic,(high 10, moderate 5, low 1)**	**Public-Public Partner-ship**	**Corporatisation**	**Service Contract**	**Management Contract**	**Lease Contract**	**Concession**	**Private Consultants**	**Build Own Operate Transfer (BOOT)**	**Built Operate Transfer (BOT)**	**Municipal Debt Issuance**	**Full Privatization**	**Municipality in Question** / **Immediate Requirements**	
Ability to Reduce Liabilities	1	2	1	2	8	8	2	6	4	8	10	0.00	Need to Reduce Liabilities
Ability to Improve Cost Recovery	5	3	1	7	7	7	5	5	5	1	8	10.00	Need to Improve Cost Recovery
Ability to improve Cost Control	6	6	4	8	8	9	4	5	5	1	8	10.00	Need to improve Cost Control
Ability to Raise Capital	4	2	1	2	8	9	4	9	9	10	9	#NAME?	Need for Capital
Ability to Improve Infrastructure Maintenance	4	2	8	7	8	8	2	2	2	2	8	10.00	Need to Improve Infrastructure Maintenance
Ability to Improve Customer Service	4	5	6	7	8	8	5	6	6	1	9	6.22	Need to Improve Customer Service
Ability to Improve Water Quality	6	3	8	7	7	7	9	2	2	1	8	5.00	Need to Improve Water Quality
Ability to Improve System Efficiency	4	2	2	8	9	9	7	5	5	1	8	5.39	Need to Improve System Efficiency
Ability to Improve Staff Ability	7	1	1	7	8	9	1	1	1	1	9	3.97	Need to Improve Staff Ability
Ability to adjust staff numbers	6	5	1	6	7	9	1	1	1	1	9	3.52	Need to Adjust Staff numbers
Level of Motivation for efficient service provision	2	5	9	6	9	10	7	7	7	1	10	7.25	Need to Improve Motivation
Totals:	49	36	42	67	87	93	47	49	47	28	96	#NAME?	

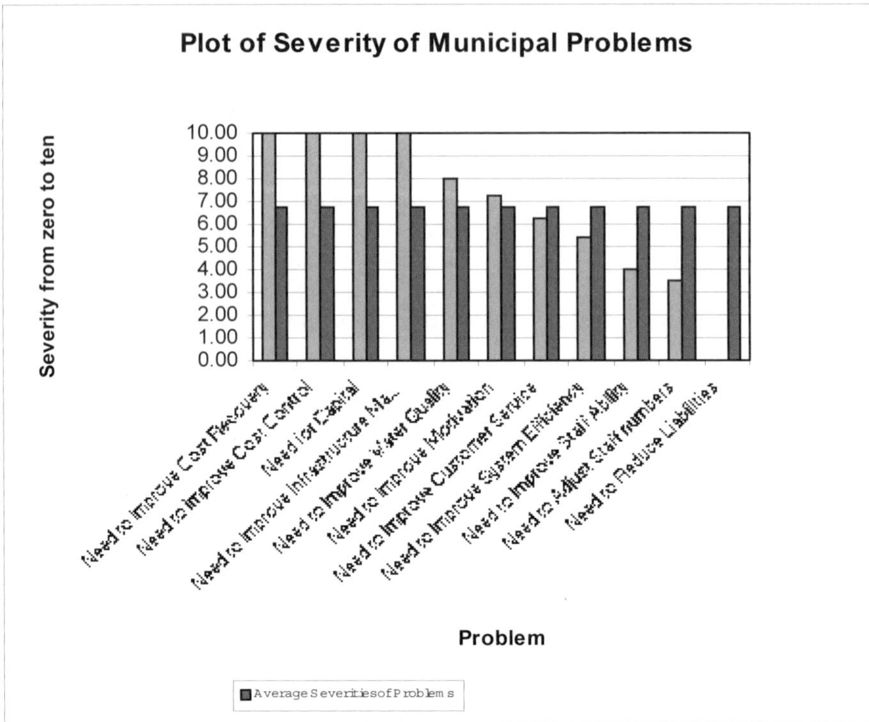

Figure 12.1 Comparison of factors affecting operations of a water company

Private Consultants:
Private consultants would be called in to work on specific specialized technical problems encountered at a municipality (also called outsourcing). The municipality would benefit immediately from the work done by the consultant but if the situation changed then they would have to seek the consultant's expertise again.

The business functions of water companies were evaluated by Richardson (2003), who plotted the relative efficiencies in Figure 12.1. Each parameter can be attached a weighting and the system optimized by allocating funds to different sections in the correct proportion. A cash flow series can be plotted to ensure a balance of revenue and expenditure, and a financial improvement as well as efficient asset management will result. Then the Expert system identified a Management contract as the most appropriate form of private partnership. This was decided from a study of the weighted deficiencies of the town.

12.4 LIFE CYCLE COSTING

All too frequently engineering construction contracts are awarded on the basis of tender price alone. However, the operating costs and life of the works have an important bearing on the best solution. It is often done in Mechanical Engineering practice but seldom in Civil Engineering that operation and maintenance costs are added for tender evaluation. And then after installation, deterioration can affect life and replacement costs. This explains a way of correctly evaluating different proposals. A spreadsheet for lifecycle costing and comparison of different pipe materials is presented to explain the concepts.

Whereas the procedure gives the most economic solution, there could be other constraints affecting the decision. Cash availability may dictate a low capital cost system. This should rarely arise as most capital projects are financed with loans. And the prevailing interest rate enters the formulae used to compute the optimum solution. Then there are taxes to be considered in the case of private installations. Capital may come from taxed profit, whereas operating costs are subtracted before paying tax. In that case an operating cost intensive system may be more economic. In the majority of cases pipelines are public services and tax is omitted in the calculations.

12.4.1 The life of a works

There are alternative definitions of 'life' when it comes to engineering works:

- The Physical life is the time the works exist whether in use or not.
- The Useful life is the time it can be used
- The Service life is the time it can be serviced, e.g. spares obtained or repaired.
- The Safe life is the time in which it can be safely used or is safe to leave in place.
- The Economic life is the time it can be used economically, i.e. before it is economically better to do something else.
- The Engineering life is the time it should be in use i.e. after which it should be closed down whether for economic or safety reasons.
- The Financial life is the time over which it is financed. It may be the period of the loan if a loan is made to pay for the construction. It is also possible to re-float a loan if the works is still serviceable.
- The Payback period is the time to pay off the loan and may not be the same as the original loan period.

The value of the works at any stage is also difficult to quantify. It could be any of:

- The original construction and design cost.
- Inflated cost, i.e. cost of a new works
- Replacement cost, which may be for a substitute.
- Insured value.
- Depreciated value allowing for wear and tear. This depends on level of maintenance kept up.
- Linearly decreased value over the life, whether it be financial or economic life.
- One of the above minus outstanding operating and maintenance costs discounted
- One of the previous plus value of product sold in the case of a non-public owner.

Insurance companies are well aware of the different values and use this knowledge in meeting claims. Unless the insured is able to distinguish between the different values and is aware of the terms in the documentation he may not receive his premium worth.

12.4.2 Economic evaluation

There are many ways of evaluating and optimizing assets. The method of discounting should be entered into with care. The standard practice is to convert annual payments to a present value by dividing them by $(1+i)^n$ where i is the interest rate as a fraction and n is the number of intervals from now. n is normally in years. The Present value is the sum of such numbers. Where the annual costs A are the same, the Present Value, PV, becomes $A((1+i)^n-1)/i(1+i)^n$ (see Chapter 11).

The discount rate to use is often taken as the prevailing interest rate paid on loans. It demonstrates a time rate of preference but purely financial. So it assumes no cash flow obstacles or taxation payments. In the long term interest rates vary so a time average rate would be desirable but the future is uncertain so prevailing rates are generally adopted. The source of funds often affects the interest rate. Development funds may be offered at cheap rates. Private sources generally want higher returns, to cover risk and to ensure profit.

Risk affects the outcome in a number of ways. There may be uncertainty regarding demands, reliability of source, finance or dangers ahead. A probability analysis is a good way of accounting for risk on large projects. Each scenario should be multiplied by its probability to obtain the probable optimum. Short cuts can be made as generally risk reduces the optimum capacity. So a higher interest rate or smaller capacity (shorter life) could be adopted.

When inflation enters into the picture the prices should be escalated at the inflation rate and the discounted. As an approximation the effective discount rate is the interest rate minus the inflation rate.

How to select the optimum solution is the next question. Optimization methods are available on spreadsheets, and generally the difference between benefits and costs is maximized using these methods. Where a public system is to be optimized the benefits are taken implicitly and the system with minimum cost is sought subject to an acceptable level of supply.

Other criteria for selecting the optimum have occasionally been used. These include the benefit/cost ratio, which is now only used for ranking results optimized prior using the maximum difference principle. Maximum internal rate of return is also a possibility used often in sensitivity studies of the results to interest rate. On national projects other features affect the answers and shadow values are often attached to commodities to allow for intangible or indirect benefits. The benefits of water supply range beyond the price of the water. Improved health, lifestyle and environment result. Less directly, new opportunities arise, e.g. better homelife and education, more employment including on the water scheme, and uses of water manifest, e.g. irrigation, hydro power, recreation or fishing.

Pipelines generally represent the major assets of a water supply organization. They have many of the components to illustrate the necessity of including running costs as well as capital cost. An example showing the costing of two types of pipe is made in Table 12.3.

Table 12.3 Spreadsheet for Life cycle costing of pipes

LIFECYCLE COSTING OF PIPES

PIPE A

Pipe Matl	PVC	**Hydraulics**		**Original costs**		**Energy etc**		**Financl**		**Risk%/akm**
Length m	1000	Fb Int l/s	50	PipeCos/m	300	EngyC/kW	20	Interest %	14	1
Dia Imm	250	Fb Finl l/s	80	Laying /m	200	Effcncy%	70	Inflation %	8	
Year instl	1992	MaintLev%	50	Coatg/m	0	Load fac%	80	Loan Yrs	15	
Year now	2002	EngLife Ys	50	WatHam /m	20	DarcyFricl	0.012	% inc F/a	0.1	

Year	**Yrs from**	**CapitalCo**	**LinFinDec**	**LinEngDe**	**Power/a**	**Maintnce**	**Inflated**	**Discountd**	**RemE-Op**	**TotalCos**
1992	-10	520000	520000	520000	2495	11700	14195	14195	236275	534195
1993	-9		485333	509600	2589	11700	15432	13537	240071	547732
1994	-8		450667	499200	2685	11700	16778	12910	243207	560643
1995	-7		416000	488800	2783	11700	18245	12315	245718	572957
1996	-6		381333	478400	2884	11700	19841	11748	247632	584705
1997	-5		346667	468000	2987	11700	21580	11208	248980	595913
1998	-4		312000	457600	3093	11700	23474	10694	249788	606607
1999	-3		277333	447200	3201	11700	25537	10206	250082	616813
2000	-2		242667	436800	3312	11700	27785	9740	249888	626553
2001	-1		208000	426400	3425	11700	30235	9297	249229	635851
2002	0		173333	416000	3541	11700	32904	8876	248126	644727
2003	1		138667	405600	3660	11700	35813	8474	246602	653201
2004	2		104000	395200	3781	11700	38984	8091	244676	661292
2005	3		69333	384800	3905	11700	42440	7727	242367	669019
2006	4		34667	374400	4032	11700	46208	7380	239694	676399
2007	5		0	364000	4162	11700	50316	7049	236674	683448
2008	6		0	353600	4294	11700	54795	6734	233323	690182
2009	7		0	343200	4429	11700	59679	6433	229657	696615
2010	8		0	332800	4568	11700	65006	6147	225690	702762
2011	9		0	322400	4709	11700	70816	5874	221437	708636
2012	10		0	312000	4853	11700	77153	5614	216911	714250

The table includes data input in shaded blocks and computational results. Thus under pipe materials the pipe length in metres, diameter in mm, year installed and present year of evaluation are inserted. In the next section under hydraulics, the start year flow rate in litres per second, then the final flow rate, the level of maintenance as a percentage of best possible, and engineering life are typed. Under original costs are the cost of pipe per metre, laying, coating and water hammer protection. In the next section under Energy the table requires energy cost in cents per kilowatt-hour (kWhr), pumping efficiency in %, operating load factor and Darcy friction factor at the start. Under Financial are interest rate as a %, inflation rate, life of loan in years and % increase in friction factor per year. On the right is the risk of damage by external sources per year per km of pipe.

12.4.3 Computation

To enable a true evaluation to be made a full lifecycle costing must be made. To this end the following costs are added to the installation costs of a pipeline:

- Capital cost, split into pipe material, laying, coating and water hammer protection.
- These costs obviously depend on pipeline length, diameter, pressure class etc. as well as the year of installation.

- The design flow rate influences the operating cost particularly in the case of pumping lines, but also indirectly for gravity lines since a smaller diameter is possible with larger heads. The average flow rate may initially be low, growing over the years so that discounting procedures are necessary.
- The engineering life relevant to the particular pipe material must be decided and this will be commensurate with the level of maintenance planned. A maximum maintenance cost of 2.5% of the capital cost is used here but it can be less if the level of maintenance is less.
- Energy cost, pumping efficiency and annual load factor affect pumping costs. And an allowance for pumpset and other operating related costs should be made in the energy cost.
- The friction factor affects pumping cost, and increase in friction over years can have a large effect as the higher flows occur towards the end of the life of the pipeline.
- Loan interest rate, inflation rate and period of loan are required for discounting or comparing annual costs with capital costs.
- Some pipe materials are at greater risk of damage and the factor is also required as it affects repair maintenance costs. A risk cost of 1% per annum is considered an upper limit.

The spreadsheet tabulates year by year over the engineering life, the costs, inflated and discounted to a present value at year 0, i.e. year of installation. Then it plots for each pipe material the decreasing value of the pipe as it gets older (Fig. 12.2). An interesting curve is the residual engineering cost minus outstanding operating costs, discounted to present value. This line indicates a much higher value for PVC than other, e.g. steel pipe, because of the low friction and low maintenance costs.

In some situations this curve could even be below zero indicating the pipe has a negative economic value. This situation could arise with corrodible steel for example. What this means is that the outstanding operating costs exceed the residual value of the pipe. If it were not for the fact that water must be delivered, the pipe should be scrapped at that stage. But we have not added the value of water supplied so unless a more economic system could be installed, i.e. with greater net value, the system will have to be retained. The same applies at the end of the engineering or economic life. It may be more economic to refurbish the old pipe rather than install a new pipeline.

Figure 12.3 shows the comparison of two pipes over their life. Here the present value of all costs expended are added consecutively, i.e. year by year pumping costs and maintenance costs add up, even though they are discounted to their present value. Whereas one pipe may have a lower capital cost, when all costs are added, friction and maintenance costs influence the final decision significantly. That is, lifecycle costing reveals the true picture of what is the most economic pipeline.

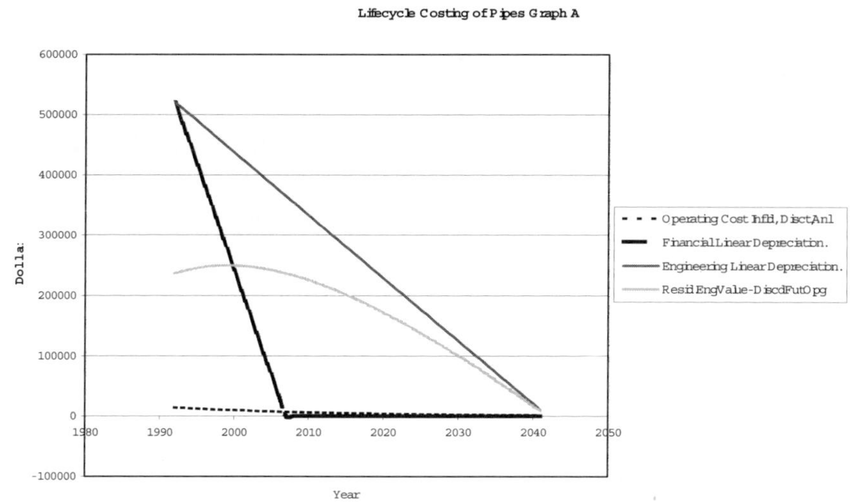

Figure 12.2 Depreciating value of pipe using different methods.

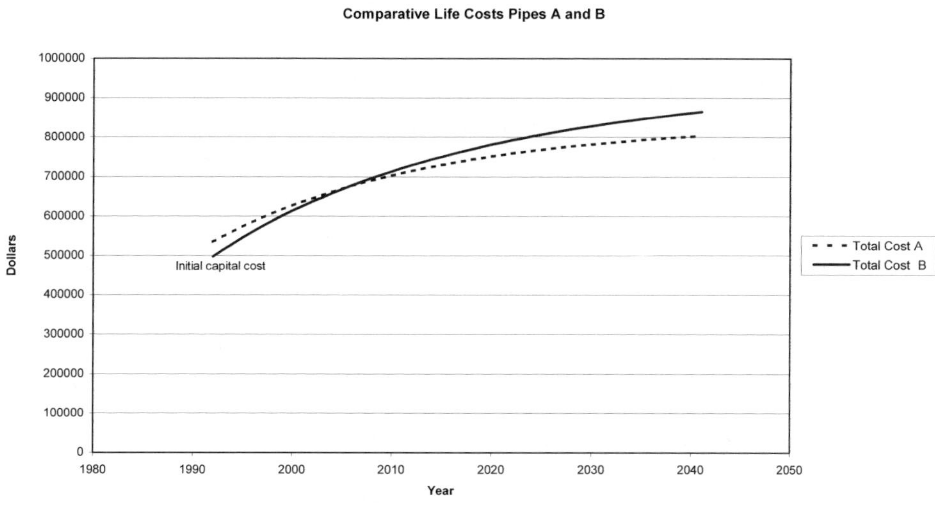

Figure 12.3 Comparative costs of different pipes adding operating costs.

The level of maintenance can have an important bearing on the life and value of an asset. Figure 12.4 illustrates the cost of different levels of maintenance over the years. Whereas no maintenance is foolish, too high an input can be expensive, and discouraging. The capital value continues to decrease and the optimum level of maintenance should be sought by comparison of the sum of the costs of maintenance and loss in value.

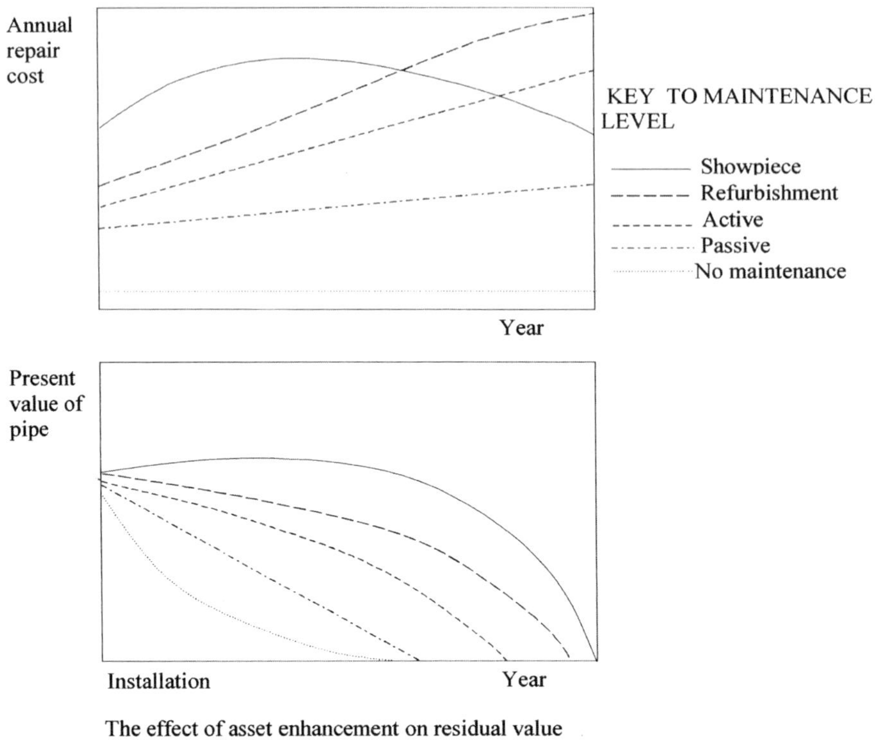

Figure 12.4 Costs and effects of different levels of maintenance

Thus by including operating costs in pipeline evaluation, the material selection is likely to change. Pipes with low friction coefficients, particularly in later years have a cost advantage. And the level of maintenance may be less, also reducing lifecycle costs. A high level of maintenance can enhance the life of a pipe as well as reduce friction losses, but the optimum level needs to be investigated.

The method of economic evaluation is also important as future costs can inflate, whereas those costs should be discounted. The effective discount rate, risk, and load factor influence the selection, so a sensitivity study is needed with all these factors considered.

12.5 VULNERABILITY

Since the attacks on the World Trade Centre increasing diligence has become the norm particularly in public services. Water services are of particular concern because poison dosing could reach millions of people in a short time. Attacks to structures and infrastructure are of less concern, but the vulnerability of computer systems could be a problem. There has long been a danger of attack to water infrastructure in some politically unstable countries and we could learn much from counties already in a state of preparedness, e.g. Israel. It is a fact that security costs money so the risk needs to be assessed, i.e. probability as well as the hazard.

The following programmes need considering or implementing:

- Water quality monitoring
- Water pressure monitoring
- Visual and audio surveillance
- Vulnerability assessment
- Likelihood assessment
- Hazard assessment
- Crisis management
- Security officer training
- Public awareness campaigning
- Source and reservoir protection
- Distribution network and fittings protection
- Review hydraulic, pneumatic, electrical, mechanical ,thermal, chemical, nuclear, systems
- Check sources of services, water and discharges of wastes
- Stresses, structures, links, materials, and time dependent strengths
- Natural hazards, flood, earthquake, geological, wind, fire
- Security hardware and access control installation
- Protective clothing, shields, nose, mouth, eye and ear protection
- Neutralizing and washing facilities
- Isolation rooms
- Staff and visitor identification and screening at different levels
- Communications protection
- Duplication of vital systems
- Data backup
- Computer and internet systems
- Vehicle protection
- Threat identification

The identification and prioritization of vulnerability in water services could be guided by studies done in the nuclear industry, for computer network firewalls, and by security planning done by water authorities in the past. The implications of damage, interference, accident or probability can be assessed from economic and danger points of view. Loss of facilities or data and danger to humans and the environment should be considered. Differentiation is needed between installations for water supply or sewerage, and operation. I.e. if vulnerability is considered at planning stage a different design may result, but there will always be the necessity for alertness during operation.

Risks due to intentional or malicious damage cannot easily be quantified but accidents and incorrect estimates can be attached probabilities and costs. Differentiation should be made between probability and damage, and hazard-risk indices are suggested (Stephenson, 2002). The same approach can be used for financial risks, IT risks, natural hazards (e.g. drought, earthquake) and economic risks (e.g. changes in water demand, AIDS, interest rates).

Vulnerability can be reduced by preparedness and another index will account for this. Margins of safety such as in structural design have cost implications and should be evaluated. Such studies have already been done. There has not been much published on vulnerability of water services, but a number of books exist on safety and security in general (e.g. Grimaldi and Simonds, 1984). Industrial safety has long been a concern and safety programmes and insurance schemes and legislation have developed for this.

Service authorities from large Water Boards to small local municipalities should be informed for introducing safety programmes. The cooperation of security officers and local disaster management organizations should be solicited. The experiences of security organizations may also prove useful in finalizing guidelines. The appropriate level of security may vary with the risks, community and affordability. However all avenues of attack or danger need to be sought out.

The consequences of attack are expensive, often far more than the cost of protection, but the probability needs evaluating before embarking on expensive protection systems. The following may result from attack:

- Structural damage
- Increased danger
- Human or animal death
- Injury or disability
- Illness
- Trauma (this may happen even due to stress of anticipation)
- Inconvenience and cost of disruptions
- Loss of information and know how
- Breakdown of computer, electrical or mechanical systems

12.6 RISK

By ignoring risk in water resources and water demand, engineers can make mistakes and thereby increase the cost of water. Water resources availability can be described with a probability distribution. The risk of insufficient water is a function of the capacity of the project. That is the larger the reservoir on a river the less the risk of it running dry. Even during the operational phase the probability of different inflows must be matched with an acceptable probability of being able to meet demands.

The probability of water demand reaching a target figure also influences the design capacity of the project. The effect of a lower demand than predicted could be to increase water costs and thereby reduce consumption even further. A time simulation is required to cost alternatives with different probabilities.

On the one hand uncertainty (in river flow) increases the cost of water storage (conversely to reduce risk in supply rate the cost must increase). On the other hand, higher uncertainty in demand increases costs and should dampen capital expenditure (see Fig. 12.5).

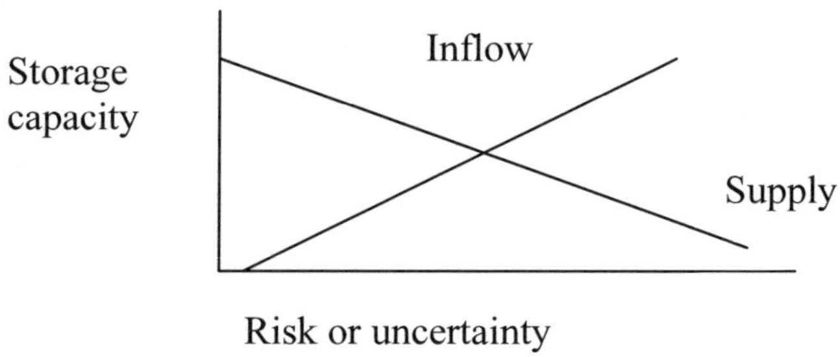

Figure 12.5 Uncertainty and storage

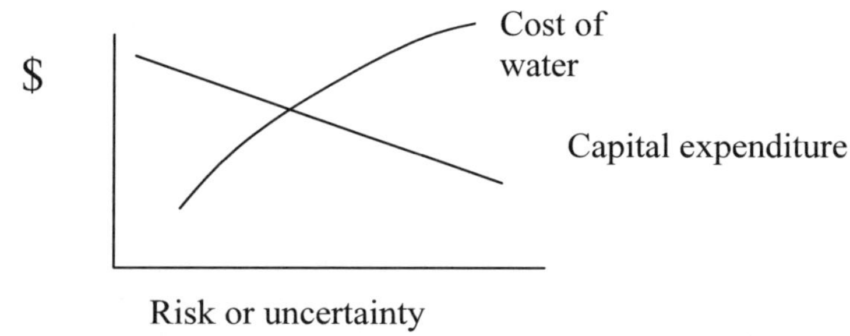

Figure 12.6 Uncertainty and cost

Although river flows are caused by natural meteorological and geophysical processes, we are not able to predict the flow easily and tend to regard the flow as a random variable superimposed on known trends. River flows have a random variation from year to year and month to month. However well we can predict climate they are still innumerable unknowns which makes the runoff and river flow even more of a random variable. We therefore often design storage reservoirs to meet a certain probability of drought associated with any level of assuredness as a cost. (Fig. 12. 7)

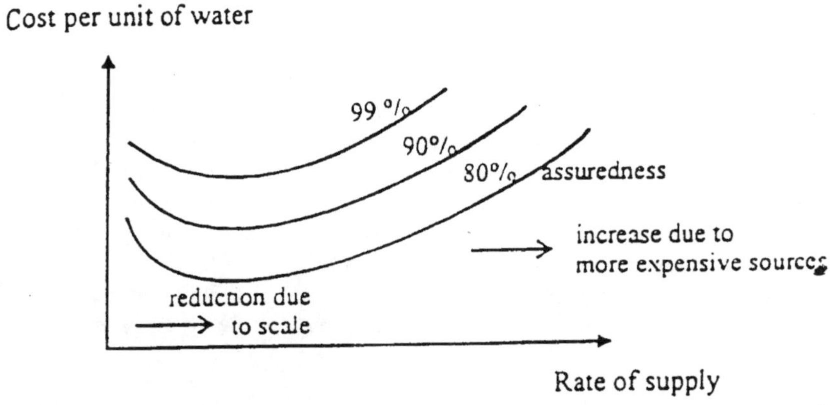

Figure 12.7 Effect of assuredness on cost of water

On a normal design basis when the drought of the selected risk does occur, then the reservoir stored volume would drop to zero. Similarly, if the reservoir were oversized spill would reduce and the evaporation increase so that there is also a reduced yield .

Other sources of water such as ground water have similar unknown probabilities of being recharged. Whether it be harvesting of rain water or cropping of rainfed farms, there is always a degree of uncertainty in the amount of water available.

One way of minimizing the risk of failing to supply is to reduce the release rate from the reservoir as the volume of water stored decreases. Such operating rules can be derived by trial and error or the releases at different storage levels can be optimized using economic principles (Stevens et al, 1997).

Variable draft operating rules are derived by considering both the probability of different inflows occurring and the economic value of water at difference rates of supply. Thus, if the demand has some elasticity, then higher rates of supply would be of lower unit value whereas the last amount available is of very high value to consumers. If the economic value at different supply rates and the hardship to consumers at different supply rates can be put in numbers the system can be optimized to minimize the probable economic loss or maximize the probable economic benefit.

Operating rules can be derived for conjunctive use systems where risks are diminished by looking at probability of independent events occurring simultaneously.

According to the Berthoux (1971) if the amount of uncertainty is high it pays to decrease the Investment in projects. However, his theory was intended primarily for the demand side and in the case of supply the effect of increase risk of availability of water could be to increase the capital size of the works.
However if a higher risk can be taken in having to drop the supply rate then the size of the reservoir can be reduced, i.e. cost savings can be achieved if the operating rule could be optimised allowing for the probability of having to decrease the supply rate.

12.6.1 Effect of uncertainty in demand estimates

The greatest unknown in planning of a water project is frequently the demands in the future. Where public works are being planned then the rate of increase in demand in the supply needs to be estimated and this may not have a finite limit as would be the case for irrigation areas. The time horizon needs to be decided as well as the rate of growth.

However, the old fashion method of projecting on a logarithmic scale the demand using historic growth rate and selecting an arbitrary time horizon can result in expensive incorrect estimates. Figure 12.8 illustrates what would happen if the demand were overestimated. This has occurred in South Africa after the planning of the Lesotho Highlands project, costing approximately US$1 billion to date. The expected demand has not materialized with the result that an expensive capital project has to be funded by dividing by a lower supply rate. The unit cost of the water increases with the result that the price increase reduces demand even more (see Dandy and Connerty, 1994).

One of the solutions of this dilemma is to plan in the face of high uncertainty with high operating, low capital cost projects. Whereas with a high certainty of demand, the higher capital cost project is more viable particularly if the capacity factor is high and the load factor is high.

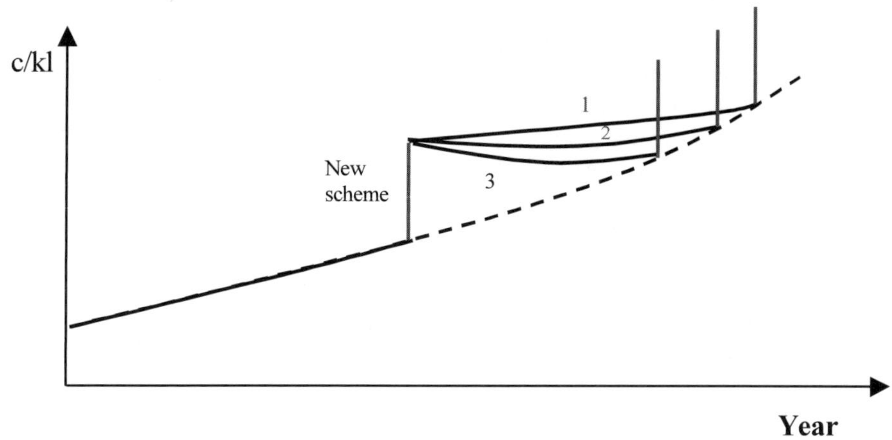

Figure 12.8 Effects of tariffs on demand (demand 1 is low, demand 3 is high)

The capacity factor is the range of average water consumption to the capacity of the project. The load factor is the ratio of the average water consumption at a particular time to the peak consumption at that time. The peak could be less than the capacity of the project for a number of years as the demand increases.

Thus, the effect of uncertainty is to reduce capital expenditure as indicated by Berthoux (1971). He also suggested another way of allowing for uncertainty. That is to artificially plan with a larger discount rate than would be used for a deterministic system. A third way is to do a probability study whereby a probable minimum cost project is selected (see Fig. 12.9).

The effect of under-designing, i.e. underestimating the demand, would also be to increase the cost. This is because a smaller project than necessary would be constructed. The demand would therefore reach the supply capacity of the system sooner than expected and another project would have to be planned. This may be a more expensive project increasing the marginal cost of water or it could be a parallel project. The cost of two pipelines to take a certain flow is considerably more expensive than one larger pipeline, i.e. the advantage of scale in hydraulic engineering is to decrease the marginal cost of water (Stephenson, 1998).

Even thus there are fluctuations in the water demand over time. Apart from the compound increase in water demand due to increasing population and increasing standards of living, there is also the variation in consumption due to seasonal factors.

Economic factors will also cause change. The world experienced a steady growth over the second half of the 20th century but resources may not sustain that growth into the 21st century. Greater interaction between developed and developing countries, due to changing human perspectives may speed investment in poorer countries.

There have been a number of other factors which have decreased the water consumption growth rate in South Africa. The AIDS epidemic is of particular concern. Estimates have indicated 10-25% of the population have HIV or AIDS and as a result the mortality rate is expected to increase in the future and the population growth rate will decrease. The economic effects of AIDS were compounded by the international economic slump being experienced. The high cost of AIDS is due to wasted cost on training, early retirement and deaths in the workplace, lack of skilled people, aging of the workforce and high medical care costs. Medicine has till recently been unavailable but low cost generics are still a drain on finances of poorer people.

The demographic movements due to economic and social pressures must also be added to the equation. Some businesses have become concerned by the extent of trained personnel becoming too il to be productive. And others consider emigrating. Their places are taken by less skilled people and the economic productivity of city dwellers is greatly decreasing. There is also a peri-urban migration from rural areas. These people are often unskilled and not adjusted for high pressure city life. They can frequently not afford the urban system, and informal housing and servicing develops around the cities in peri urban shanty towns. The inability to pay for services and the pressure of the richer people to minimize health risk drains economic resources. There is cross-subsidization resulting in discontent amongst some of the richer sectors and irresponsibility amongst the receivers.

The end result is not only a decrease in growth rate but also a possible decrease in water consumption in a decade or so. This is compounded by a great uncertainty in the future demand. Both factors therefore point to a lower planning horizon. Alternatives to large capital cost projects must therefore be sought. These can range from improved loss-control to more efficient water use through better plumbing and greater awareness amongst consumers. Demineralization and recycling are also relatively low capital but high operating cost sources. Alternative supply systems should be sought. These could include supply in containers to ensure water quality and payment. Alternatively communal standpipes have more than cost efficiency. They maintain community structure provided they have been planned correctly. Consultation with communities, especially women, is necessary to ensure acceptance and the best level of supply.

The tightening of the reins can only be done so much and then an expansion must occur again. In the medium term it solves some problems, but poor economic conditions will remain for many decades despite job creation and better education.

Drainage systems, which are more capital intensive than water supply systems, particularly in poorer areas, are even more susceptible than water supply to uncertainties. Therefore alternative drainage systems also need consideration, from the financing and health points of view.

To conclude, there are many uncertainties in planning for water supplies:

- Uncertainty in the river flow rate in the case of surface water resources.
- Unknowns in aquifer and catchment boundaries and runoff.
- Uncertainty in water demand in the future in demographics and economics.
- Economic uncertainty (international and local) affecting industrial demand.

The economic relationship between value of water and supply rate is non-linear, so hedging against risk at an early stage of drought could reduce the effects of deficits if the reservoir must dry, i.e. a hedging rule is advisable.

In the face of high uncertainty it is wise to avoid high capital cost expansions (e.g. dams, long transmission conduits) and instead revert to operating intensive schemes (e.g. demineralization, recycling, pumping). This can be affected by using high discount rates. Underdesign is advisable in the face of uncertainty. Underdesigning of reservoirs saves cost and unexpected demands can be met by rationing.

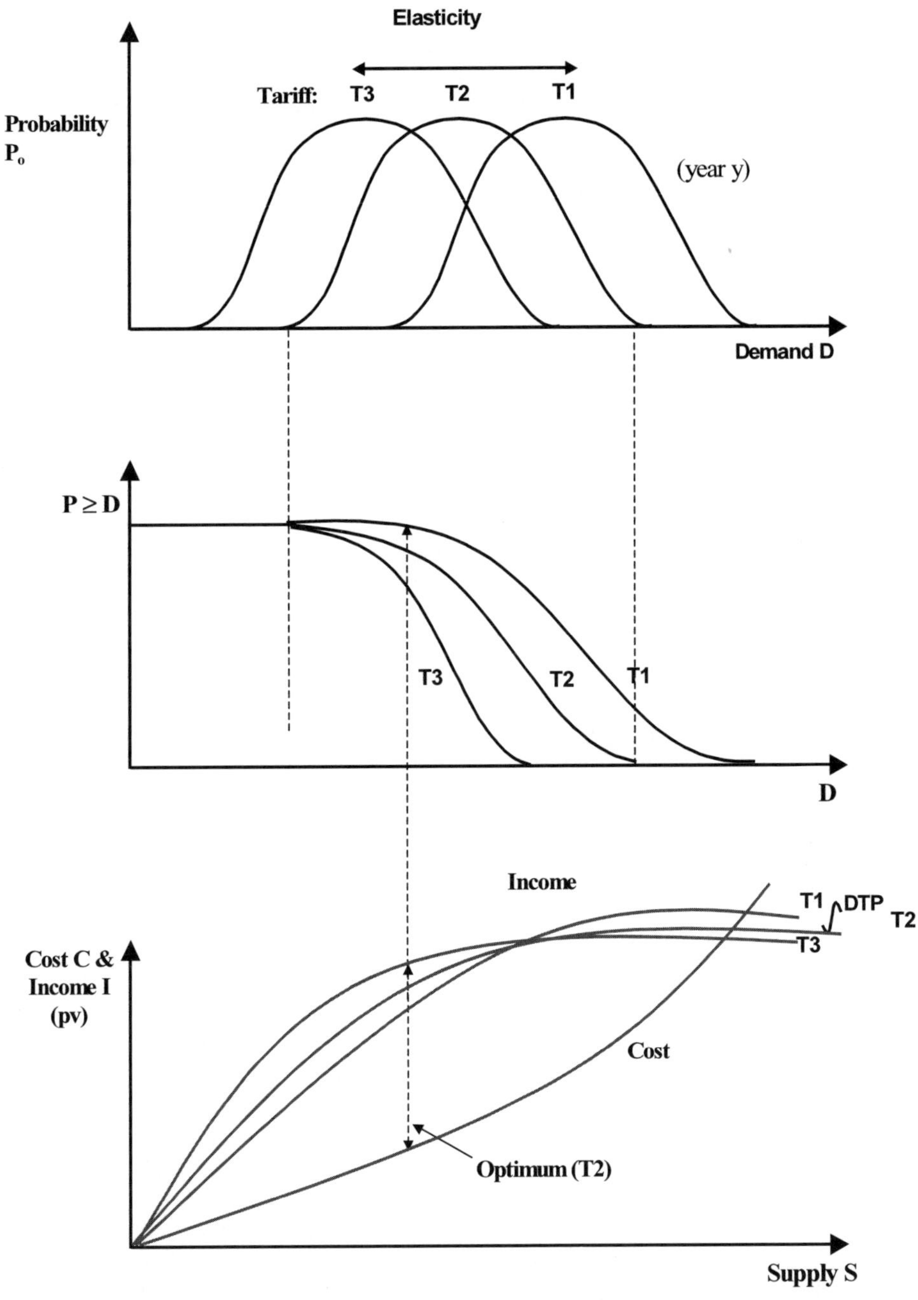

Figure 12.9 Probabilistic planning with different tariffs

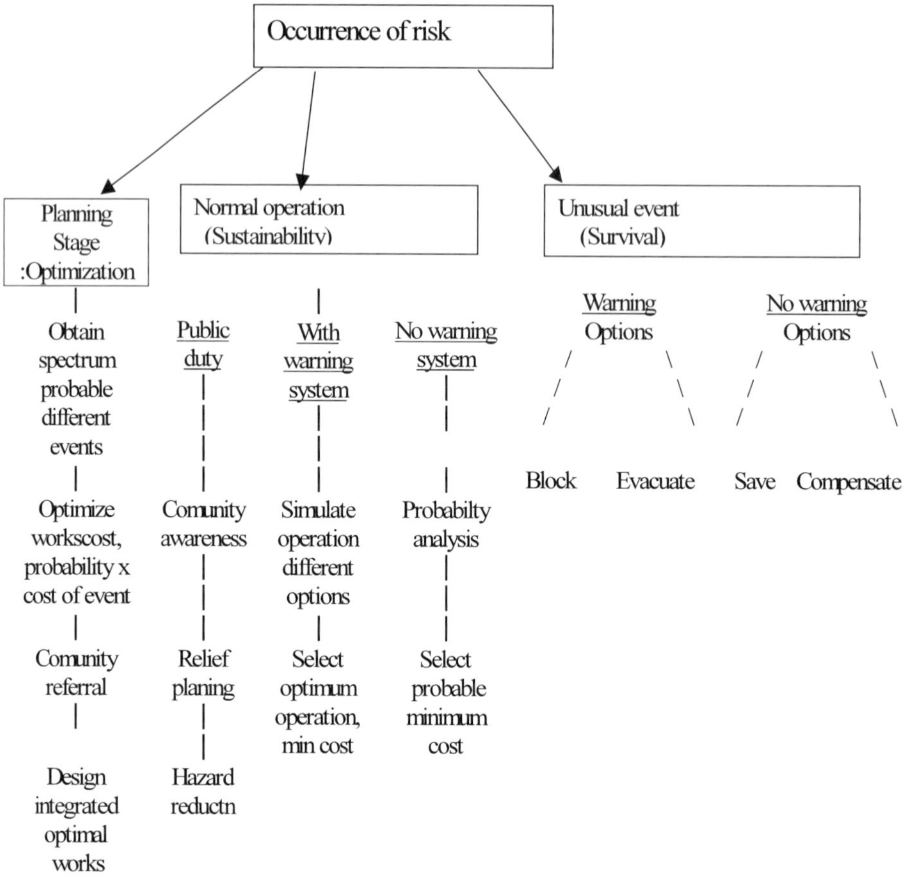

Figure 12.10 Risk decision tree

REFERENCES

Berthoux P.M. (1971). Accommodating uncertainty in forecasts. *J.AWWA*, 66(1), 14

Dandy, G.C. and Connerty, M.C. (1994). Interaction between water pricing, demand and sequencing water projects. *Proc. Water Down Under*, Adelaide, Australn. Inst. Eng. 219-224.

Grimaldi, J.V. and Simonds, R.H. (1984). *Safety Management*, 4th Ed., Richard Irwin Inc., Illinois.

Institution of Civil Engineers (1980). *Engineering Economics*, London.

Palmer Development Group (1990). Economic Assessment of Water Supply Projects in S.A., Water Research Commission, Pretoria.

Richardson A. (2002). Development of an expert system for evaluation of Public Private partnerships for the Water Industry. *Civil Engineering in SA.*.

Stephenson, D. (1998). *Water Supply Management*, Kluwer, 308pp.

Stephenson, D. (2002). Hazard risk indices for flooding. *Urban Water*, Sept.

Stevens, E., Stephenson, D., Chu, S. and Huang, W-L (1998). Management of a reservoir for drought. *Water S.A.*, 24 (4), 287-292

World Bank (2002). Water Resource Sector: Strategic Directions for World Bank, Washington.

CHAPTER 13

Computer Modelling and Optimization

13.1 TYPES OF MODELS

Computer modelling of hydrological processes and even other natural processes has become common and indeed essential for a proper understanding of system interactions and effects of changes in the systems. Provided the process can be understood and formulated it can be modelled. Computers can therefore predict results of change, in precipitation, runoff, control and use of water. They can predict water flows into the future if inputs such as precipitation can be forecast. They can model weather from earth and ocean temperature changes. They can model ocean temperature and current changes from solar radiation modelling. And they can model climate change from the knowledge of what influences the atmosphere.

The main restriction on accuracy is input data, i.e. the assumptions regarding the solar and global effects are often speculative and therefore the quantification of parameters is the most difficult. At the other extreme, ecosystems and biosystems are also difficult to model. There are so many unknown interactions and effects, that the subject is primarily a research topic rather than being able to provide answers regarding water quality and effects.

The more unknown the inputs are, the more modellers are prone to using stochastic models. For example the generation of a series of monthly river flow volumes using stochastic techniques is common. The statistics are derived from a limited sample of data and used to synthesize a longer series. This has its dangers but is still of use to depict the range of reactions of say a reservoir storage over a long period.

Computer models of hydrological processes are fairly reliable once coded correctly as there is a mass balance to be made. They can be used to calculate and depict storage changes and illustrate when shortages could occur. They are used to study floods and droughts, and by trial and error a management system can be designed to control flows. Complex simulation of conjunctive sources of water and multiple parameters, e.g. quantities and qualities can be modeled simultaneously. The following list illustrates some of the applications of computers in water resources modelling:

- Rainfall-runoff conversion for events or continuous series of flows
- Stochastic generation of flow series
- Streamflow prediction

- Operation of reservoirs and hydropower plant
- Flood routing and flood plain modelling
- River flow depth and velocity calculations
- Safe yield calculation
- Sustainability estimates
- Optimization of designs

One dimensional models are generally sufficient for hydrological computations, but flood plain models may be two dimensional and groundwater flow models should be two or three dimensional. Water quality can be modeled in one dimension for simple mass balances but two or three dimensions are required for dispersion or thermal effects. The equations obviously become more sophisticated and numerical methods for solving the equations require expert input. There are many commercially available packages for doing this.

Flow prediction models are of use for operation of reservoir systems. Depending on the source of input a long-term estimate can only be made with uncertainty but a short-term prediction can be quite reliable:

- Stochastic models can be used only for generating generalized operating rules. A probabilistic approach can be used.
- Weather prediction months ahead can be made within certain probability ranges or confidence bands. Then rainfall runoff models can calculate runoff within the same confidence bands.
- Remote sensing can estimate rainfall which actually happens (real time) and models can convert to runoff and river flow days or weeks ahead.
- Upstream river flows can be monitored and a model can calculate downstream flow and water levels days or hours ahead.

13.2 CONCEPTUAL RUNOFF MODELLING

Conceptual models may be subdivided into deterministic and parametric models. In the former case, the mathematical expressions used in the model purport to represent the physics of the actual processes modeled, i.e. the theoretical structure of the model is based on physical laws. Under these circumstances, input parameters to the model will be physically measurable quantities relating to the various processes such as roughness, slope, flow length and so on. Parametric modeling on the other hand is less rigorous than pure deterministic modeling, although some deterministic components may be included, and the model parameters are not necessarily defined as measurable physical quantities (Foster, 1982).

Hydrological models range from statistical to conceptual, embracing probabilistic, curve fitting, black boxes, analogue, e.g. cell type (Diskin et al., 1984), through to the more hydrodynamically correct theory. Even the latter range from simplistic, e.g. time area through first approximation kinematic type, diffusion equations and hydrodynamic equations as in SWMM (Huber et al., 1982). For overland runoff, accelerations and backwater effects are generally not significant.

Viesmann et al. (1977) list some twenty programs in use for event, continuous or urban runoff modeling. They describe some of the widely used ones in considerable detail, in particular the Stanford Watershed Model (Crawford and Linsley, 1966), which may be considered to be the first comprehensive, continuous digital model available. Green and Stephenson (1985) describe and compare the performance of a number of event models for application in an urban environment, notably the stormwater management model (Huber et al, 1982), the Illinois urban drainage area simulator (Terstiep and Stall, 1974) and a two-dimensional kinematic model (Brakensiek, 1967).

On the other hand, the time-area approach which derived from the rational method, does not accommodate the effect of water depth on concentration time

It is into the hydraulically based models that the majority of research has now been directed. By suitable selection of module arrangement, one-dimensional flow can be assumed. For the kinematic approximation, flow is assumed uniform down the reach. There may be local backup which does effect system storage, however. Since inflow is spread over the full length of a reach each time step, the routing effect can be unrealistic unless the time step is sufficiently large. Methods of minimizing numerical routing (or using it to approximate hydraulic routing) have been investigated (Holden and Stephenson, 1988; Ponce, 1986).

Continuous models can generate long time series and therefore usually summarize results of flows in monthly totals. They are primarily for use in optimizing reservoir operation for drought storage. Event models work with much shorter time intervals. They may be for generating hydrographs and computations and result tables are in hourly or shorter time intervals. The computational interval for continuous modeling may be shorter than the printout interval for accuracy purposes.

13.2.1 Sub-catchment arrangement

The interconnection of one sub-catchment or element with another can be done in various ways (Fig. 13.1):

1. *Finite different grids*. These assume a continuous system can be modeled with numerical approximations to the differential equations of motion. In the case of a homogeneous catchment a rectangular grid can be superimposed. Flows and depths are computed at grid points. Either one or two directional flow can be assumed.
2. *Finite elements*. Size and shape of elements can be varied to suit the topography if a finite element approach is used. In general, a two-direction flow pattern is assumed. Equations are established for continuity at element boundaries.
3. A simple and versatile model is one made up of *modules* which can be linked. Generally the flow is assumed one directional along the module but two dimensional catchments can be made up of modules in parallel and series, i.e. the orientation of the module is ignored because the directional momentum of the water is not considered

13.2.2 Continuous simulation

For long term simulations the correct interaction between surface and subsurface conditions is important. Groundwater flow modeling capability with aquifer modules makes possible long-term simulation of catchment yield. Groundwater contributions can lag surface runoff by hours or even months.

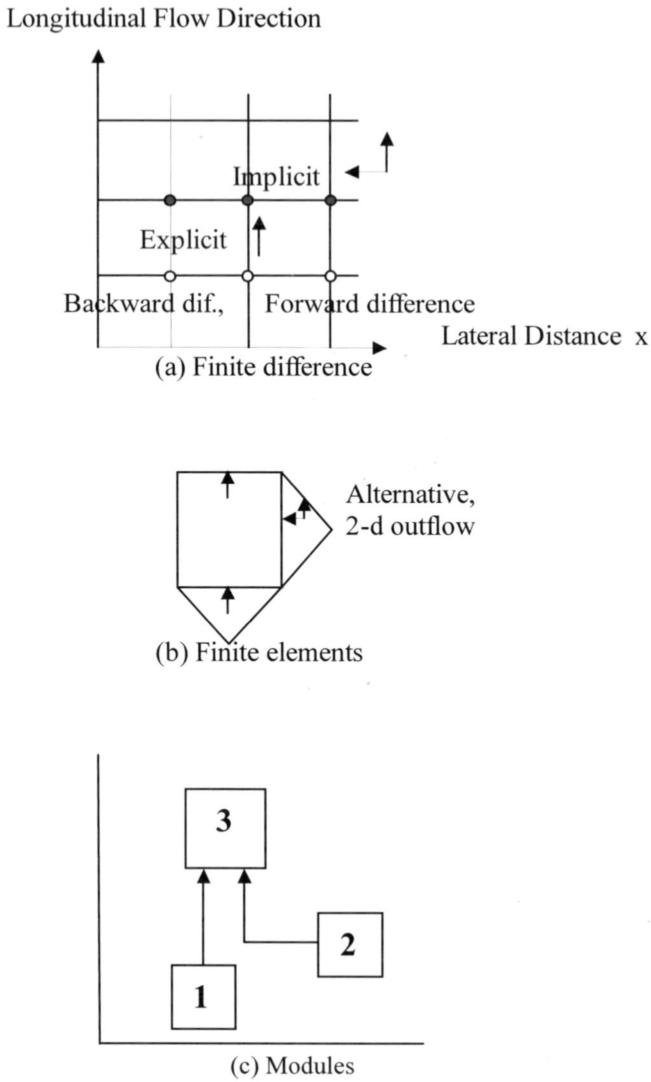

Figure 13.1 Alternative grids for catchment models

Recharge of surface layers is however important from the point of view of antecedent moisture and permeability for forthcoming storms. The continuous simulation capability therefore improves estimation of surface storm runoff. Surface layer moisture is also important for estimating evaporation and losses.

The time scale of flow from deeper aquifers may be much longer than from the higher water tables, and a longer time step could be used once surface runoff is reduced.

13.2.3 Routing Process

Although simplistic flood routing methods such as the Muskingum method are often used for manual calculation of changes in hydrograph shape as flow occurs along channels, there is nowadays no excuse for not using hydraulic equations to model flows in waterways. Various approximations to the hydraulic equations of motion can be used depending on the circumstances. Kinematic waves are theoretically not subject to diffusion, i.e. spreading and attenuation, as no dynamic effects are included in the equation. There may be changes in wave shape since dx/dt is a function of depth, but there can be no change in peak flow unless there is an inflow. The advantage of taking large distance increases with the kinematic method therefore results in a sacrifice in accuracy. Explicit solution of the kinematic equations is often employed in preference to implicit solution as the friction equation is non-linear, and explicit schemes such as the backward centered, or semi explicit such as the 4-point scheme of Brakensiek (1967) are reasonably accurate and fast.

Explicit schemes can be subject to numerical instability unless the time increment is small enough, i.e. $\Delta t < \Delta x/(dx/dt)$, (the Courant criterion) where $dx/dt = \alpha\ my^{m-1}$. On the other hand, the smaller Δt the greater the numerical diffusion as the numerical effect travels at speed $\Delta x/\Delta t$. The optimum compromise is for $\Delta x/\Delta t = dx/dt$. This is not always possible in an equispaced grid as dx/dt varies. Ponce (1986) attempted to reproduce actual diffusion in kinematic equations by writing the finite difference equations for flow in a way similar to the Muskingum Cunge routing.

Adopting a more practical approach, the kinematic diffusion process can be explained as follows. The routing process which occurs with kinematic modelling is similar to reservoir routing where discharge depends only on the stage at the outlet. A unique stage discharge relation is assumed, i.e. no allowance is made for accelerations or water surface gradient.

The resulting effect is similar to that employed in the Muskingum method and in addition allows for non-linearity in the stage-discharge relationship. It also has the advantage that the parameters in the equations are physically measurable and not empirical.

13.2.4 A Continuous Runoff Model

A model (RAFLER) described here is a conceptual deterministic model which converts rainfall data to runoff over a length of time eg years. RAFLER is an acronym for Rainfall Flow Erosion. It is based on the simplified hydrodynamic equations for overland flow and the simplified Green Ampt (1911) infiltration model for vertical flow into the soil and underlying aquifers. The parameters required by the model are physical factors which can be measured or are available in literature. In the case of some factors which are difficult to observe, e.g. aquifer characteristics, they can be obtained by calibration. For overland flow, factors such as the roughness, the slope and width and overland flow length of the catchment, are required. Since real runoff is not in the form of a perfect sheet flow, a factor to account for flow in rills is required. The permeability of the soil is required, but since this is highly influenced by the fact that most soils are semi-saturated, further guides are needed for assessing this.

With hydraulic models, many parameters are not the same as would be measured in a laboratory. This is particularly so in the case of surface roughness and infiltration rate

(Stephenson, 1989). Lumped models (i.e. where model sub-catchments are average conditions over a large area) generally require higher roughness factors than engineers will be accustomed to under laboratory or channel flow conditions. This could be because the depth of overland flow is less, resulting in lower Reynolds numbers or because the relative roughness for shallow flow is greater or because overland flow is not direct but tortuous.

Infiltration rates on a macro scale also appear lower than under laboratory conditions. A prime factor is that the soil is unsaturated in nature. The slightest obstacle tends to have a determining effect on aquifer properties. It could also be because the runoff concentrates in rills or rivulets. The soil surface area available for infiltration is therefore less than the total catchment area. The latter flow concentration or canalization effect results in a deeper and faster flow than for overland flow and more rapid concentration times.

The program attempts to reproduce runoff and silt yield on a monthly basis, using basic catchment parameters and monthly rainfall records. In order to estimate surface runoff rates and soil erosion rates, it is, however, necessary to operate the model using a very much shorter time interval, e.g. hourly instead of monthly. An estimate of the distribution of monthly rain is therefore made, using the average number of rainy days in a year.

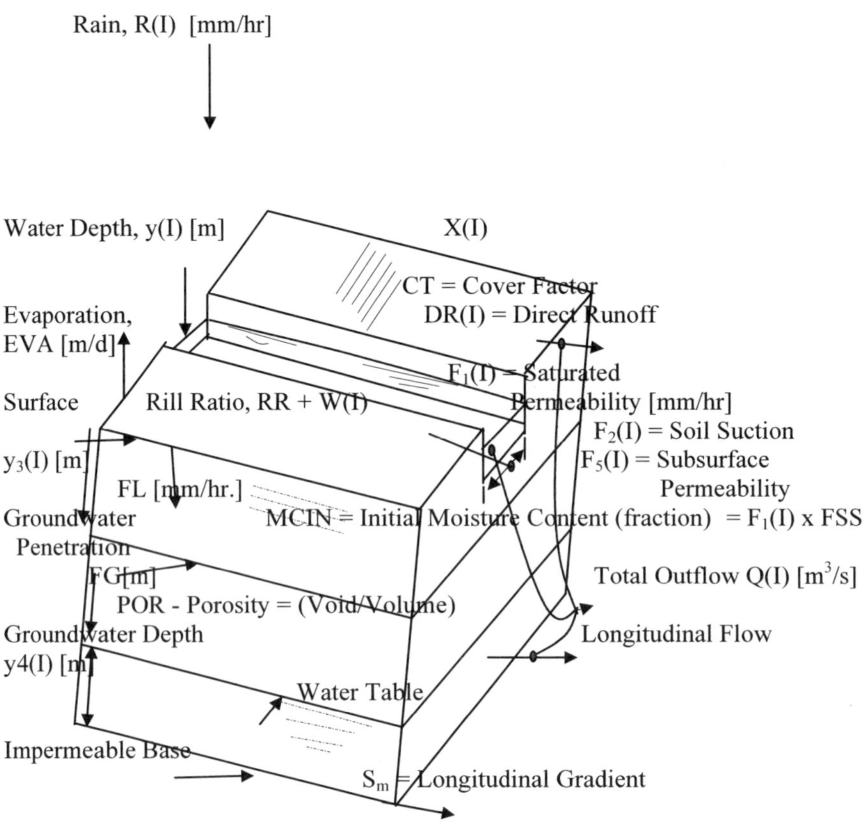

Figure 13.2 Elements in the runoff model

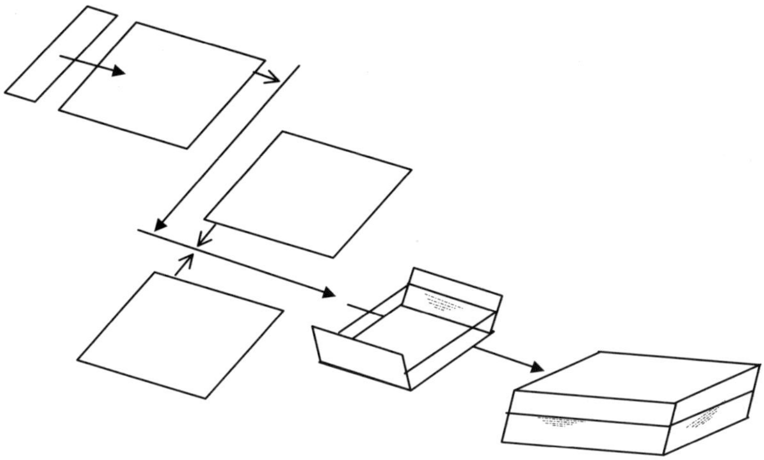

Figure 13.3 Diagrammatic assembly of modules

The number of hours of rain a month is taken to be proportional to the number of rain days per year multiplied by the ratio of months rainfall to average rainfall all to the power of 0.75. The rainfall intensity is therefore increased to the power of 0.25 for precipitation greater than mean. The factor 0.75 was found by experiment on rainfall data from South Africa.

The program is able to accommodate various combinations of three elements or modules (Fig. 13.3):

- Plain rectangular catchments: with rill factor, cover factor and permeability
- Uniform rectangular channels with erodible or stable beds
- Reservoirs

In conceptual models the method of routing of overland flow, and the routing method for water conduits and reservoirs determine largely the degree of realism that can be attained. The continuity principle applied to one-dimensional flow can be written as:

$$\frac{\partial Q}{\partial x}+\frac{\partial A}{\partial t}=q_i \qquad (13.1)$$

Where:

Q = volumetric flow rate
x = distance in direction of flow
A = cross-sectional area
t = time
q_i = lateral inflow rate per unit length along the x-axis.

The conservation of momentum for one-dimensional, unsteady non-uniform flow is given by the following equation:

$$S_f = S_o - \frac{\partial y}{\partial x} - \frac{v}{g}\frac{\partial v}{\partial x} - \frac{1}{g}\frac{\partial v}{\partial t} \tag{13.2}$$

Where:

S_f = friction gradient
S_o = bed slope
y = water depth
v = mean water velocity
g = gravitational acceleration
t = time

The equations for the conservation of mass and momentum are known as the St Venant equations and describe a hydrodynamic wave. Often the change in water velocity over time during the passing of surge is gradual. The flow can thus be considered near-steady, under which conditions the acceleration terms will play only a minor role.

If, in addition, the cross-section of the conduit is fairly constant, a uniform flow may be assumed. This will reduce the conservation of momentum equation to:

$$S_f = S_o \tag{13.3}$$

which means the equation of motion can be approximated by a uniform flow formula of the general form $Q = ay^b$, where a and b are constants. The combination of the continuity equation and a uniform flow formula describes kinematic flow.

If the flow over land areas as a result of a storm is assumed to be evenly distributed, this flow is effectively steady and uniform. Consequently, kinematic theory can be applied.

13.2.5 Rainfall-runoff simulation

The rainfall input requirements for a continuous monthly simulation model can be less stringent than for event simulation. If rainfall records are available, daily data are generally readily obtainable, but they may pose some economic and practical drawbacks. For a long simulation period and the use of records from several rainfall stations, the computational effort can be overburdened. Furthermore, the data input effort could become prohibitively expensive, while daily records are also prone to more errors than monthly ones. Therefore, in this model historical monthly rainfall records form the basis for the simulated rainfall intensity and duration.

13.2.6 Sediment yield calculation

Silt yield from catchments is calculated using Yalin's theory (1963) which is based on Shield's critical shear slope criterion for erodibility. Total erosion is summated over each month and deposited into the downstream element each time step.

In the case of channels, only what comes into the channel from plains is assumed removable. Accretion of silt will occur if inflow is greater than potential erosion in the channel and the channel will flow clean if erosion potential is greater than silt inflow.

13.2.7 Infiltration into the unsaturated zone

The infiltration is based on a conceptual model utilizing Darcy's law as proposed by Green and Ampt (1911). Darcy's law can be written as:

$$v = \frac{f}{n} = k(Lf + h + Sw)/Lf(m/s) \tag{13.4}$$

Where:

v = infiltration velocity (m/s)
f = infiltration rate (m/s)
n = porosity
k = permeability = hydraulic conductivity (m/s)
Lf = depth to the wetting front (m)
h = surface ponding depth (m)
Sw = suction at the wetting front (m)

Several assumptions were necessary to write Darcy's law in the form above, namely (Stephenson and Meadows, 1986):

- There exists a distinct and precisely definable wetting front.
- Suction at the wetting front remains essentially constant, regardless of time and depth.
- Above (behind) the wetting front, the soil is uniformly wet and of constant hydraulic conductivity.
- Below (in front of) the wetting front, the soil moisture content is relatively unchanged from its initial moisture content.

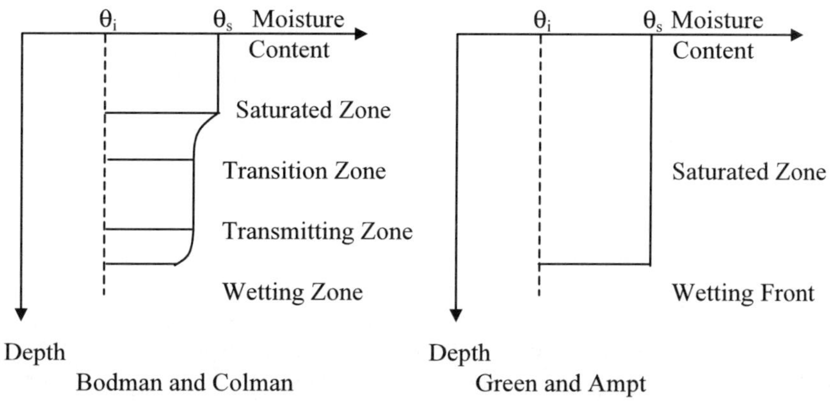

Figure 13.4 Comparison of Green and Ampt soil moisture profile with Bodman and Coleman profile

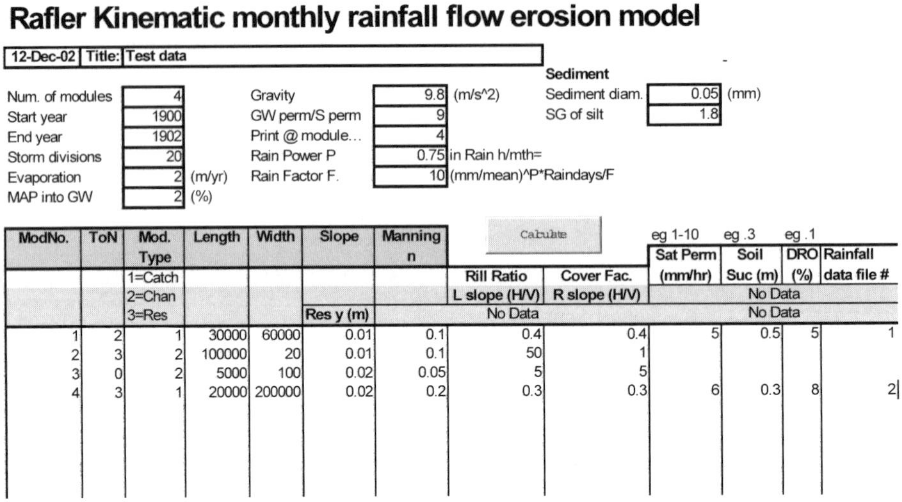

Rafler Kinematic monthly rainfall flow erosion model

12-Dec-02 | Title: Test data

Num. of modules 4
Start year 1900
End year 1902
Storm divisions 20
Evaporation 2 (m/yr)
MAP into GW 2 (%)

Gravity 9.8 (m/s^2)
GW perm/S perm 9
Print @ module... 4
Rain Power P 0.75 in Rain h/mth=
Rain Factor F. 10 (mm/mean)^P*Raindays/F

Sediment
Sediment diam. 0.05 (mm)
SG of silt 1.8

Calculate

ModNo.	ToN	Mod. Type 1=Catch 2=Chan 3=Res	Length	Width	Slope Res y (m)	Manning n	Rill Ratio L slope (H/V)	Cover Fac. R slope (H/V)	Sat Perm (mm/hr) eg 1-10	Soil Suc (m) eg .3	DRO (%) eg .1	Rainfall data file #
									No Data			
						No Data			No Data			
1	2	1	30000	60000	0.01	0.1	0.4	0.4	5	0.5	5	1
2	3	2	100000	20	0.01	0.1	50	1				
3	0	2	5000	100	0.02	0.05	5	5				
4	3	1	20000	200000	0.02	0.2	0.3	0.3	6	0.3	8	2

Figure 13.5 RAFLER input form

The approximate nature of the Green and Ampt (1911) model is illustrated by a comparison with the actual soil moisture profile as given by Bodman and Coleman (1943) (Fig. 13.4). After the surface water film has disappeared, the pending saturated zone starts moving downwards.

The capillary suction at the wetting front S_w is the difference of the capillary potential at the soil surface and that at the wetting front. Values for this parameter can vary between 50mm for sand and 500mm for clay.

13.2.8 Application

The model has been used for river flow series and sedimentation modeling (Stephenson and Paling, 1992) and monthly flow forecasting for hydro power dam operation (Stephenson et al, 2000).

13.3 STORAGE ANALYSIS

The modeling of storage fluctuations in a reservoir is a typical simple programming exercise. Inflows can be routed through the reservoir with selected spillway characteristics to study the effect on flood hydrographs. Low flow operational rules can be optimized by computer modeling by trial and error. An initial layout and design may be possible using direct optimization followed by simulation of the operation to study variants in inflow and operation. Computers are not only the most suitable tool to do the modeling, they can depict the results in graphical and tabular forms for ease in interpreting results.

13.3.1 Multiple reservoirs

Complex water resources systems involve many dams along the same river system. These may exist for different users or to achieve the optimum control for flood management or to catch as much water as possible in arid or highly variable hydrology. There are limits to the effectiveness of greater and greater volumes of storage and some guides will indicate the maximum useful storage corresponding to different upstream configurations.

Reservoirs in parallel would be on adjacent rivers and each reservoir and catchment could be modeled individually. Generally the optimum storage volume for each river will be approximately equal to the variance in annual flow for drought storage, and the sum of average monthly deficits for seasonal storage.

If reservoirs exist in series down a river with inflow upstream and between the reservoirs the storage yield relationship is more complicated. Assuming the hydrology of the entire basin is consistent then a general storage-yield analysis could be generated for a single reservoir as indicated in Chapter 3. Now generalize this relationship by expressing reservoir volume S in terms of mean annual runoff MAR, i.e. S/MAR, and yield Y for the selected drought probability, also in terms of MAR i.e. Y/MAR. This relationship would apply to the upstream reservoir on its own, i.e. if reservoir 1 has a capacity S1 and mean upstream inflow F1, obtain Y1/F1 from the point on the curve corresponding to S1/F1, and for successive downstream reservoirs the following holds (see Fig. 13.6):

Let S2 be the capacity of reservoir 2 and F2 the inflow between reservoir 1 and 2, etc. Then

1. If S2/F2<=S1/F1, then Y2/F2 is indicated by the point on the storage-draft curve with abscissa equal to S2/F2. Then the total draft available at the downstream dam is Y2 plus whatever yield from the upstream reservoir 1 is not consumed.
2. If S2/F2>S1/F1 then (Y1+Y2)/(F1+F2) is indicated on the storage-draft curve with abscissa (S1+S2)/(F1+F2).

It is assumed the river flows are proportional to each other at all times and the storage - evaporation loss relationships of the reservoirs are similar. Alternatively new storage-draft curves could be generated for each site. This type of relationship gives a starting point for computer simulations and saves a lot of trial and error computations.

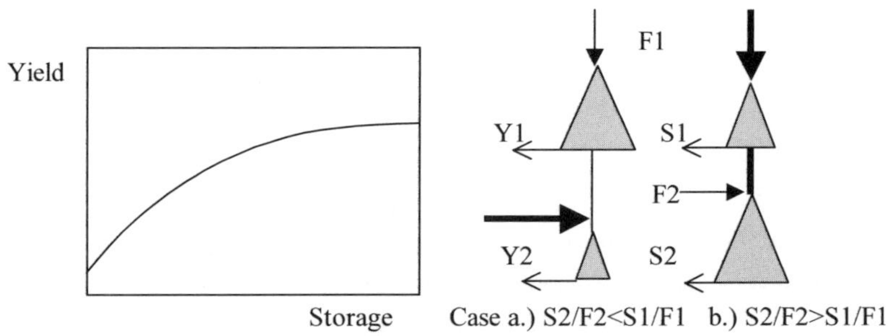

Figure 13.6 Yield of multiple reservoirs

13.4 WATER DISTRIBUTION SYSTEMS

Water distribution is not the subject of this book but there are numerous computational approaches in the field of pipe network analysis and design which can be applied to the transport of water across catchments and which justify their knowledge. The typical urban water distribution network comprises a supply pipeline, a balancing reservoir and distribution pipes. The relationship between flow rate, pipe diameter and head loss is non linear and solution for flow rate in each pipe and heads at each node has to be performed by iterative or simultaneous non linear equations. This type of analysis is required for ensuring adequate flows and pressures throughout the system. Pressure zoning and reduction is also used to manage leakage. A balance between computational efficiency and effort in data assembly is required and a number of programs exist for this (see Lansey and Mays, 1989). A time varying simulation can also be done using the network analysis sub program. This can be used for reservoir sizing and pump selection.

The problem of network design or optimization is also complicated by the non-linearity of the equations. Gradient methods are generally used but local optima have to be avoided. The selection of optimum network design can be done in two steps (Stephenson, 1984). The first step is to select the best layout and flow along each pipe. The next step is to select the optimum pipe diameters, and ones which are commercially available. An alternative method is to choose an optimum hydraulic gradient throughout the network (see Fig. 13.7). Then if flows are inserted by trial, the pipe diameters can be calculated. An iterative solution may be necessary, varying flows, grade lines and choosing commercially available diameters.

In Figure 13.8, a typical pipe reticulation network is depicted for analysis or design. Models of the system operation can produce flows or pressure fluctuations over time for sizing reservoirs, planning pressure zones or optimizing the relationship between pipe diameters and reservoir capacities. Such simulation models are essential for redesigning the system after an initial simplistic optimization. In Figure 13.9, the pressure heads over a week, with flow variations, as indicated by such a model are depicted.

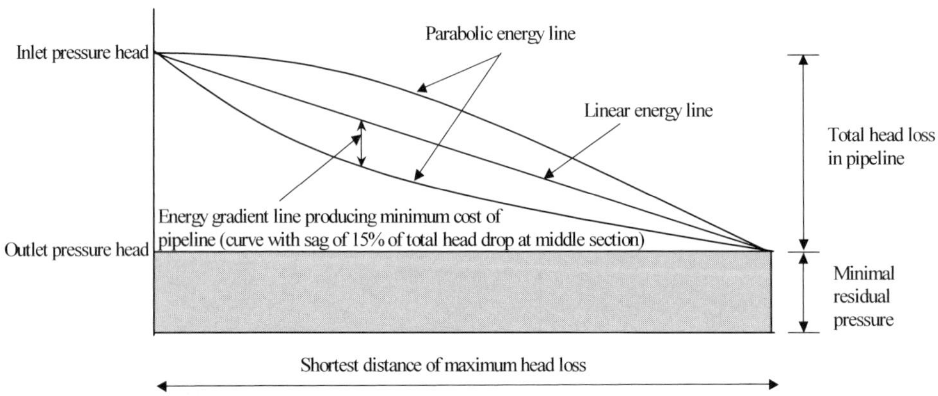

Figure 13.7 Optimum hydraulic gradient throughout network

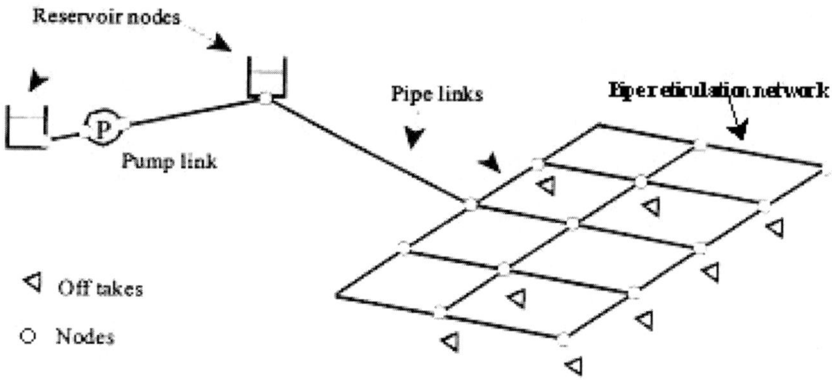

Figure 13.8 A pipe network layout

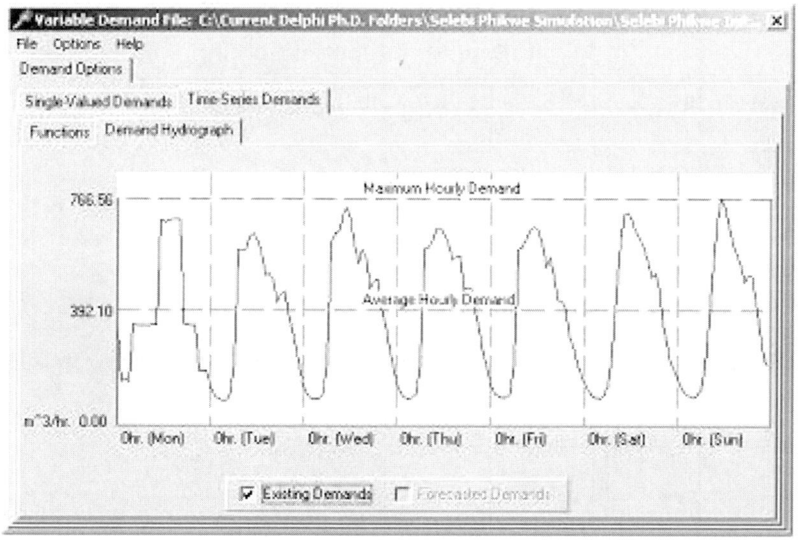

Figure 13.9 Pressure fluctuations in a reticulation system (Ilemobade, 2003)

13.4.1 Transportation programming

A simple technique for optimizing distribution systems is presented below. It can be used to optimize distribution of any resource from a number of suppliers to users. The only limitation is that costs should be linearly proportional to flow along each route. The method used, Transportation Programming, is amenable to manual solution i.e.:

1. Set out tableaux with sources and users.
2. Put in max possible trial flows at bottom left of each cell starting with top left
3. Type costs in top right of cells.
4. Using 0 for first coefficient, allocate coefficients to each row and column.
5. Calculate next coefficient knowing total for occupied cells is 0.

TRANSPORTATION PROGRAMMING EXAMPLE

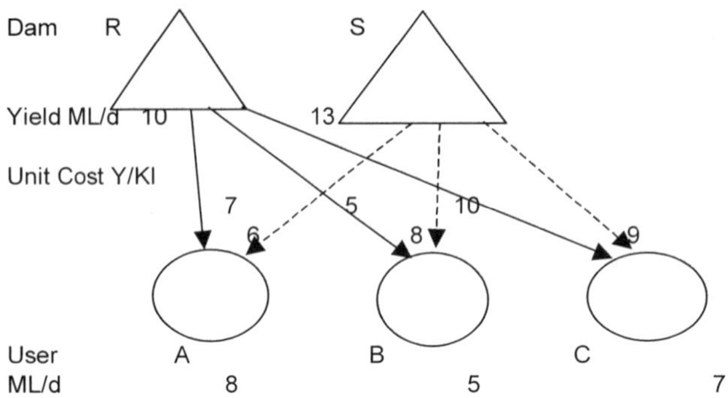

Tableaux 1:

User:	A	B	C	Surplus	Yield	Coefficient
Source R	7 - Try 8	5 + 2	10 6	0 -3	10	Use 0
S	6 10 -4 +	8 - 3	9 7	0 3	13	3
Demand	8	5	7			
Coefficient	7	5	6	-3		

Tableaux 2:

User:	A	B	C	Surplus	Yield	Coeff.
Source R	7 - 5	5 5	10 10	0 1 -1 +	10	Use 0
S	6 + 3	8 4	9 7	0 - 3	13	-1
Demand	8	5	7			
Coefficient	7	5	10	1		

Tableaux 3:Optimal solution

User:	A	B	C	Surplus	Yield	Coef
Source R	7 2	5 5	10 6	0 3	10	Use 0
S	6 6	8 4	9 7	0 0	13	-1
Demand	8	5	7			
Coeff	7	5	6	1		

Figure 13.10 Transportation example

6. Find max cost reduction by comparing coefficient and cost, i.e. cell SA is best.
7. Allocate maximum possible to the cell, i.e. 3.
8. Repeat from the top until there is no more cost reduction.

13.5 SYSTEMS ANALYSIS TECHNIQUES

The following section outlines some techniques of systems analysis, or operations research, which are useful in analysing water resources systems. The methods are useful for preliminary analysis although a simplification is often necessary. There is not a widespread use of the techniques in water resources practice (IAHS, 1989) although simulation has gained acceptance, perhaps because it is simpler and more comprehensive generally for complex problems.

Queuing theory and dynamic programming have become routine techniques in inventory control, and mathematical optimization is used extensively in business management. Linear programming has been used to optimize industrial systems and military manoeuvres, while transportation programming, as its name implies, can be used for seeking least-cost systems for transporting goods between sources and demand centres. Examples of the application of some of the techniques are given by Stephenson and Petersen (1991).

Whereas simulation is useful for analysis of pre-defined systems, the methods described here are direct optimization techniques, i.e. they lead automatically to an optimum design. The theory behind the techniques is not verified rigorously; instead, the methods are demonstrated with the aid of simple examples, and the reasons for steps proved informally. Books such as Dorfman et al. (1958) illustrate applications of the methods and Maass et al. (1962) were the first to write up applications of the techniques in water resources.

13.5.1 Economic policy

The usual criterion for selection of an optimum system is that the difference between economic benefit and cost, discounted to a common time base, is a maximum (Eckstein, 1961). This is the logical basis for planning for private enterprise, which usually wants to maximize net income, or for national planning for developed countries where market prices adequately reflect true value. However, for countries which have not yet reached full maturity, such comparison does not necessarily yield the plan which would be of most benefit to the economy in the long run.

13.6 LINEAR PROGRAMMING BY THE SIMPLEX METHOD

The Simplex method of linear programming is one of the most powerful techniques for optimizing linear systems. If a system can be defined by a set of linear equations or inequalities and an objective can be expressed as a linear function of the variables, then there exists a direct method of reaching an optimum combination of the variables (Loomba, 1964).

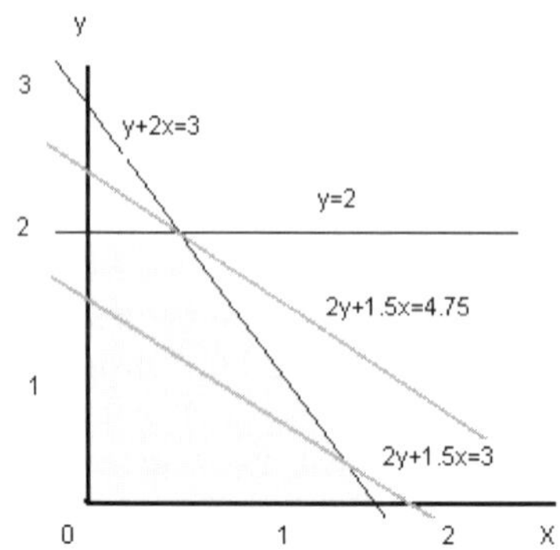

Figure 13.11 Two-dimensional example

A two dimensional example (i.e. two variables) will illustrate why a linear system as opposed to a non-linear system, can be solved by a mechanical process. Suppose a combination of two variables, X and Y, is to satisfy the following constraints:

$$\begin{aligned} Y &\leq 2 \\ Y + 2X &\leq 3 \\ X, Y, &\geq 0 \end{aligned}$$

X and Y could represent two possible forms of development, e.g. irrigation and industry, receiving water from a reservoir. It is desired to maximize the function 2Y + 1.5Xm, i.e. two units of Y are worth 1.5 units of X. The permissible domain for the variables is indicated by the shaded area in Figure 13.11.

Any point in the shaded domain conforms with the constraints, and the boundaries are the constraints with the inequality signs replaced by equal signs. Two values of the objective function are plotted on the same figure. The maximum value of 2Y + 1.5X occurs at the intersection of the lines Y = 2 and Y + 2X = 3. The value of the objective function at any other point in the domain is less than this.

By comparing the values of the objective function at neighbouring intersections, it will be found that its value always increases around the boundary in the direction of the optimum. By proceeding between successive intersections, the optimum must eventually be reached, and the objective function will always increase each step until the optimum is reduced.

The domain of a linear system must always be convex, i.e. bulging outwards at each junction. (It is impossible to construction a junction which points inwards, since the bounding line would cross the domain). Therefore, there are no local maximum at which the objective function ceases to increase when proceeding around the boundary, and once a point is found at which the objective function ceases to increase, that must be the true optimum. This would not necessarily be so if the boundaries were non-linear since a local maximum could exist between two intersections. The principles could be extended to more than two dimensions but would be difficult to illustrate graphically. Note that this theory assumes linear constraints and objective function as it is designed to be solved simply and by hand. Nowadays non-linearity is not much of a problem as non linear programming and search methods are available in spreadsheets and other software packages.

13.7 DECOMPOSITION OF COMPLEX SYSTEMS

In principle linear programming could be applied to the optimization of any linear system. However in some cases, the problem could become unwieldy if treated using normal linear programming techniques. It may thus be desirable to sub-divide the problem into reasonably sized components. A number of sub-programs could be set up. These would be linked by a master program. The penalty for such simplification is that the master program and sub-programs will have to be solved successively a number of times (Dantzig, 1963).

The method is illustrated by an example involving one transportation sub-program and a set of master constraints. The example involves optimizing an irrigation plan. To follow the reasoning requires a knowledge of linear programming and dual functions.

Assume there are two irrigable areas, referred to as K and L, and two possible sources of water, A and B. Reservoir A can supply 300 Mℓ/d (megalitres a day) and reservoir B can supply 700 Mℓ/d. There is a tract of land 14600 hectares (ha) at locality K which needs 1m of water per annum per ha and another 12166 ha at locality L which needs 1.5m of water per annum per ha. Thus the maximum requirement of area K is 14600 x 10000 x 1/365 x 1000 = 400 Mℓ/d, and of area L, 12166 x 10000 x 1.5/365 x 1000 = 500 Mℓ/d. The cost of water conveyance to either area from either source is indicated in million dollars per annum per 100 Mℓ/d, at the top of the relevant grid positions in Table 13.1. The objective is to minimize the total conveyance cost.

In Table 13.1, each column represents a demand, and each row source of water. A slack column takes up the surplus water from the dams. An artificial slack row is used to allow for area which may not be irrigated.

The problem as so far outlined is a simple transportation programming problem. However, additional constraints which cannot be incorporated into the transportation sub-program, are imposed for various reasons. In order to ensure balanced agricultural development in relation to the remainder of the economy, we require the area irrigated to be 23000 ha. As the available irrigable area totals 26700 ha, not all the area need be irrigated, so slack, S, is introduced into each of the columns in Table 13.1. Stated algebraically in terms of variables S_K and S_L, this is $(400 - S_K)\ 36.5/1 + (500 - S_L)\ 36.5/1 = 23000$, or $L.5S_K + S_L = 155$

Prior to expansion of the project, there existed 10950 ha in area K under irrigation. It is desirable that this area shall not be abandoned. Hence a second constraint becomes $(400 - S_K)\ 36.5/1 \geq 10950$, or $S_K \leq 100$. In simplex form, this becomes $S_K = x = 100$, where x is a slack variable with zero value.

The problem is split into two: It is simpler to solve these separately than to attempt to solve one giant linear programming problem. The transportation program is referred to as a sub-program, and the other constraints as the master program. The method of solution is to assign artificial costs to the slack in columns K and L of Table 13.1 to control the quantity of water made available for irrigation. If zero cost were attached to the column slacks, as much as possible of the water would be allocated to those cells during the transportation programming process. Whereas if too large a cost were assigned, the quantities in those cells would decrease and as a result the master constraints might be violated. The optimum cost to be assigned to those cells is determined from the dual to the master program. This yields the shadow price of the master constraints. The process of optimization involves integrating successive sub-programs and master programs until there is no further improvement in the objective function. Each sub-program solution yields a new variable to be introduced into the master program, and the premium costs are then revised.

Suppose that a number of feasible but not necessarily optimal solutions to the sub-programs are available. Corresponding to solutions 1 and 2, exist transportation sub-programs 1 and 2. Provided the allocations in each of the transportation programs satisfy the demand and availability constraints, then any weighted combination of them will also satisfy the constraints. Thus if each entry in transportation sub-program 1 is multiplied by 1/3 and added to 2/3 the corresponding entry in sub-program 2, the result will still satisfy the constraints. This suggests that each sub-program could be represented by a weight in the master program and the sub-program constraints could be replaced by the single constraint that the sum of the weights should total unit. The master program would then become:

$$
\begin{aligned}
&(C_1)V_1 + (C_2)V_2 + (C_3)V_3 && = Z \text{ (objective function)}\\
&(1.5S_K + S_L)_1V_1 + (1.5S_K + S_L)_2V_2 + (1.5S_K + S_L)_3V_3 && = 155\\
&(S_K)_1V_1 + (S_K)_2V_2 + (S_K)_3V_3 + x && = 100\\
&V_1 + V_2 + V_3 && = 1
\end{aligned}
$$

where the C's are total conveyance costs and the V's are weights. Subscripts of V refer to the sub-programs 1, 2 and 3.

Computations are initialized by generating an arbitrary, not necessarily optimum, basic solution to the sub-program. The initial assignments in Table 13.1 were calculated by proceeding from the top left to bottom right as outlined in the section on transportation programming.

A linear programming problem is formulated with the initial sub-program solution and the master constraints. The variables to be considered are the weights, V, by which the solution to the sub-program is to be multiplied before it is included in the master problem.

Table 13.1 Sub-program 1

Area	K	L	Slack	Yield Mℓ/d	Cost		1.55_K	$+S_L$	S_K
Source	1.2	0.6	0				0	0	0
A	**300**			300	300x1.2 =	360			
	1.5	1.0	0						
B	**100**	**500**	**100**	700	100x1.5 =	150			
Slack S			0						
			1000	1000	500x1.0 =	<u>500</u>			
						1000			
Limit Mℓ/d	400	500	1100						

The objective function is to minimize $1010V_1 = Z$

subject to the master constraints $(1.5S_K + S_L)_1V_1 = 155$

$(S_K)_1V_1 \leq 100$

In addition, the sum of the weights must total unity: $\Sigma V_1 = 1$

Expressed in Simplex form,

$$1010V_1 = Z \text{ (min)}$$
$$0V_1 + a = 155$$
$$0V_1 + x = 100$$
$$V_1 + b = 1$$

where a and b are artificial slack variables with very large cost coefficients called M. The maser program is optimized by linear programming. This is a minimization case, so the largest negative number in the net evaluation row indicates the variable to be brought into the program.

The variables a, x and V_1 appear in the optimum program. To determine whether it is worthwhile to introduce other sub-programs into the master program, the corresponding opportunity costs must be determined. Note that opportunity costs are calculated by comparing the actual price on any particular variable with the sum of the column coefficients multiplied by numbers p_1, p_2 and p_3 and which correspond to rows 1, 2 and 3 in the program.

The values of the multipliers are not immediately apparent from the final tableau of the above linear program, since the original constraint coefficients have been transformed. Values of the multipliers may be calculated by using the fact that opportunity costs of the optimum variables in the program are zero. Opportunity costs corresponding to the variables V_1, a and x in Table 13.2a are calculated as follows:

$$1010 + Op_1 + Op_2 + 1p_3 = 0$$
$$M + 1p_1 + Op_2 + Op_3 = 0$$
$$0 + 1p_1 + Op_2 + Op_3 = 0$$

Table 13.2a Master program 1

				1010	M	0	M	
Variable	Profit	Quantity		V_1	a	x	b	
a	M	155		0	1	0	0	155/0
x	0	100		0	0	1	0	100/0
b	M	1		1	0	0	1	1/1=1*
				1010 -M*	0	0	0	

Table 13.2b Sub-program 2

			1010	M	0	M
Variable	Profit	Quantity	V_1	a	x	b
a	M	155	**0**	1	0	0
x	0	100	**0**	0	1	0
V_1	1010	1	**1**	0	0	1
			0	0	0	M – 1010

Coefficients in the above equations are the numbers in Table 13.2a under the variables V_1, a and x respectively. Solution of the above equations, which comprise the dual program, yields the multipliers p_1, p_2 and p_3, as follows:

$p_1 = -M, p_2 = 0, p_3 = -1010.$

If another sub-program represented by V_2 can be found which, when these multipliers are used, indicates an opportunity cost less than zero, then according to the Simplex criterion that sub-program should be introduced into the master program. A second sub-program will be represented by weight V_2, and the coefficients in the new column in the master program will be:

$$(1.55_K + S_L)_2$$
$$(S_K)_2$$
$$1$$

The opportunity cost of the new sub-program will be the actual cost together with the products of the coefficients and the corresponding multipliers p_1, p_2 and p_3. To discern the sub-program associated with minimum opportunity cost, the sub-program is optimized with new prices. In addition to actual prices, premium prices are generated as follows:

Corresponding to row 1: $(1.5\ S_K + S_L)_2 \times p_1$
Corresponding to row 2: $(S_K)_2 \times p_2$
Corresponding to row 3: $1 \times p_3$

Thus the prices placed on S_K and S_L for the new sub-program become:

for S_K : $0 + 1.5p_1 + 1p_2 = -1.5M$
for S_L : $0 + 1p_1 = -M$

These prices are used in optimizing sub-program 2. Table 13.1 is taken as a starting array for the transportation programming procedure which is solved and the results are given in Table 13.3.

Table 13.3 Sub-program 3

	K	L		Cost	$1.5S_K + S_L$	S_K
A	1.2	0.6	0			
			300	0	1100	400
B	1.5	1	0			
			700			
S	-1.5M	-M	0			
	400	**500**	**100**			

13.8 A PLANNING MODEL

Water needs associated with rapid development can be restricted the water resources of a country. More and bigger water supply schemes may continuously have to be planned and implemented as the economy expands. If planning is to cope with economic development, bold steps should be taken. Developments in systems analysis and availability of computers provide the necessary tools. Not only will computers speed up analysis, but automatic data assembly and analysis by computer is becoming essential to cope with the vast volumes of data requiring analysis.

This section describes a method whereby plans for all levels of development could be compiled. A mathematical model representing the various sectors of the economy at different levels would be set up as illustrated in Figure 13.12.

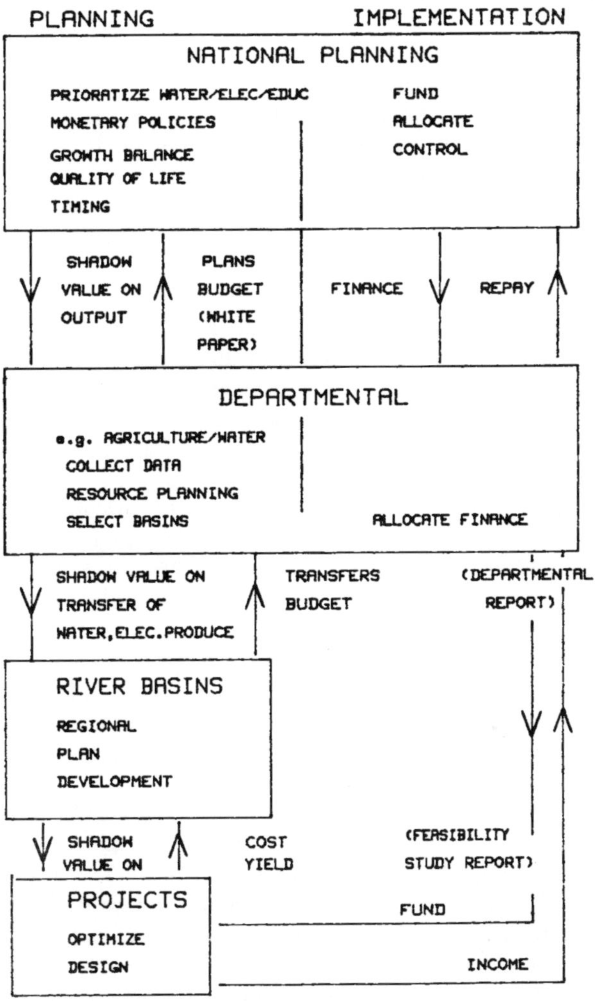

Figure 13.12 Master plan flow diagram

A national master plan would control more detailed departmental planning, such as mineral and water resources, power, communication, industry and services. These plans in turn could be sub-divided into regions or river basins, and so on.

The national master plan would incorporate factors such as capital availability, population, foreign exchange, and economics growth rate. It would control or encourage outputs of the various sectors by applying shadow values. Social and environmental factors can be included although the present example refers to economics objectives only.

The objective of the individual departments would be to optimize the net economic benefit, applying the shadow values suggested by the national master plan to output. For instance, the Department of Water Affairs would plan to maximize the difference between value of agricultural produce, hydro-electricity and other benefits of water, and the cost of dams and conduits. The department would in turn control the transfer of water or produce between river basins by means of shadow values imposed on the river basin plans.

The whole process is ideally suited to computer solution using the principle of decomposition of linear programs (Dantzig, 1963). Each plan would be represented by the national master program of which the departmental plans would be sub-programs. These departmental programs would actually also be master programs which would have river basin or regional sub-programs, and so on.

In fact, it would not be necessary to use linear programming to optimize each program. Any other planning technique, such as dynamic programming, transportation programming, computer simulation or incremental analysis could be employed.

To compile the plans, much data on resources (water, minerals, manpower), economics policy and geography will have to be stored in computer memory. A comprehensive filing index will be required for access by all departments.

The approach could be adapted to the 5-year plan concept. Capital availability would be determined after each 5-year plan, and after adjusting for private consumption and balance of payments (Kindelberger, 1965), a plan making optimum use of the available resources could be derived.

The output of water resources development projects, namely agricultural produce, industrial and domestic water supply, hydroelectric power generation and recreational facilities, would be controlled by a national master plan. Outputs would be evaluated by the master program in order to decide shadow values for this purpose.

The objective of the national master plan would be to optimize the economic growth rate of the country within the limits of available capital while maintaining balanced development of sectors. The national master program would comprise a set of equations relating the production of various sectors. For instance, the water requirements for irrigation would have to be limited to that available from various sources after accounting for urban and other consumption.

The master program would comprise a weighted combination of individual sector sub-programs. The optimum combination of sector sub-programs would be selected by linear programming. (Linear programming is a mechanical method of selecting an optimum combination of variables from a system which could be described in terms of linear equations or constraints.) The effect on the overall cost of the system is studied while successively introducing elements of each variable. That variable which would result in maximum benefit is brought into the program and the process repeated until there is no further improvement.

By comparing alternative sector plans, the national master program computes marginal values, or shadow values, for the outputs from the sector programs. These shadow values ensure that balance is maintained between sectors. Thus, a high shadow value would encourage production and a low one would limit production. In effect, shadow values represent the value of output to the country. They emerge from the linear programming computation as opportunity costs, but they can also be specified by the development authority and added to values.

The departments would then proceed to optimize their plans using the shadow values, and re-submit the plans (output and capital requirements) to the national master program. The process of selecting sector plans and generating shadow values would be repeated until no further improvement in the national economic growth rate could be achieved.

The method of optimizing the master program and solving the shadow values is based on the principle of decomposition of linear programs and is identical to that for the solution of the water resource master program. However there are various ways of optimizing the individual problems ranging from manual methods, through various optimization methods, to computer models. This is also a formalization of interactive planning which is becoming more common with integrated management and transparent planning process.

13.9 SOLVER

Nowadays the manual optimization procedure is not needed except for research and software development. Spreadsheets such as EXCEL have algorithms for solving optimization problems. The following example was solved by EXCEL. On the Menu Bar the Tools button has Solver, or if it does not appear, it can be added in. Then the problem can be set up for direct solution on the spreadsheet. Cells are allocated for the variables and the constraints. The formulae are typed into cells and then the Solve algorithm sheet is pulled down and the required data entered. Remember to specify non-negative values in Options.

Consider a possible reservoir R supplying two potential users with water. The dam would cost \$1 per cubic metre capacity. The storage capacity-yield curve is given in the figure below. User A is an irrigation farmer wanting up to $2m^3/s$ and to whom the value of water is \$1.20 per cubic metre, and user B can use 1 m^3/s at a value of \$2.30 per cubic metre, but he has an alternative source of water S costing \$1.50 per cubic metre. What size dam should be constructed? Use a present value factor of 10 x annual cost and assume 5000 irrigation hours per year. Hence the Present Value of \$1 per second is 10 x 5000 x 3600= \$18 million.

The problem is set out in the spreadsheet as shown in Figure 13.13 The spreadsheet originally shows zero in the shaded cells and when Solve button on the Solve pop-down is pushed the answers appear. There are various ways of presenting the results and plotting relationships between variables using the spreadsheet.

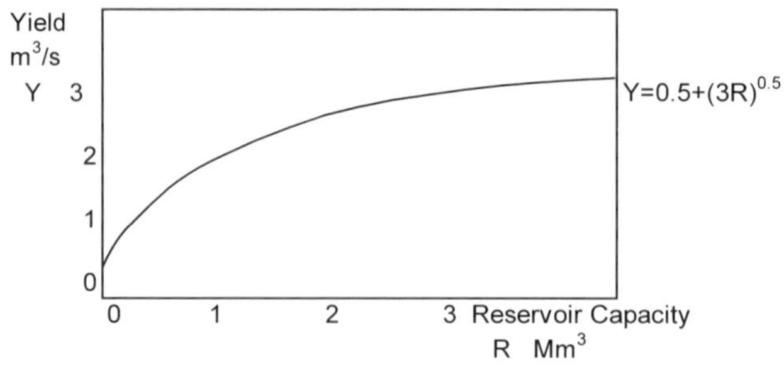

Variable:	Y	YA	YB	R	S	Type Formula		Constant
Answer:	2.30	1.30	1.00	1.08	0.00	below	<=>	
Constraints:					Y-(3*R)^.5	0.50	=	0.5
					YA+YB-Y	0.00	=	0
					YA+YB-Y	1.30	<=	2
					YB+S	1.00	<=	1
Objective fn:			1.2YA+2.3YB+(2.3-1.5)xS -1R			2.78	=	max
						Values appear above		

Figure 13.13 Solving a dam sizing problem

13.10 COMPUTER PACKAGES

Nowadays it is seldom necessary to write a computer program for water resources analysis. There are a number of software companies devoted to hydraulic and water resources systems analysis. Some programs can be found on the Internet and others can be purchased from reputable merchants. The packages cost from $50 to $5 000 and more depending on the scope of analysis and size of system to be modeled. Some programs were developed at Universities and these are often available as freeware. Others were developed by Government agencies, such as the modeling of river flow by HEC (Hydrological Engineering Centre) and SWMM (Stormwater Management Model).

Various adaptations of these models with attractive interfaces are available on the market. Some packages can be split eg runoff and /or river hydraulics and / or water quality. Special modifications may also be made to software for special purposes. Other river basin type models were developed by privatized research institutes, e.g. Danish Hydraulics and Hydraulic Research, Wallingford. These companies offer backup and training in the use of the models. This is essential in some cases, for an understanding and efficient use of the models. Model calibration, verification and subsequent sensitivity studies can be time consuming, but result in impressive reporting. Some model outputs from a range of commercially available programmes are illustrated in the figures below.

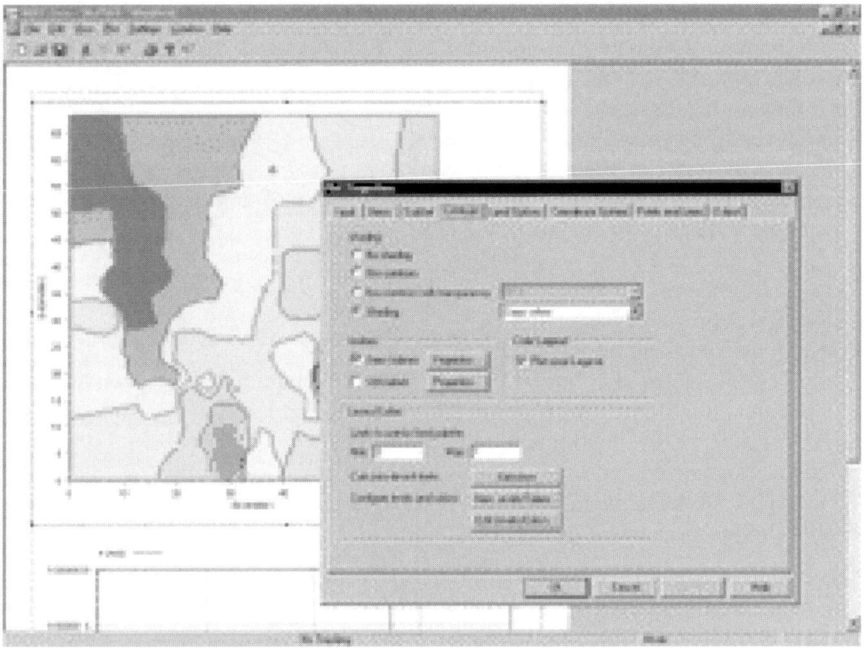

Figure 13.14 Printout from the model MIKE-SHE (DHI Software)

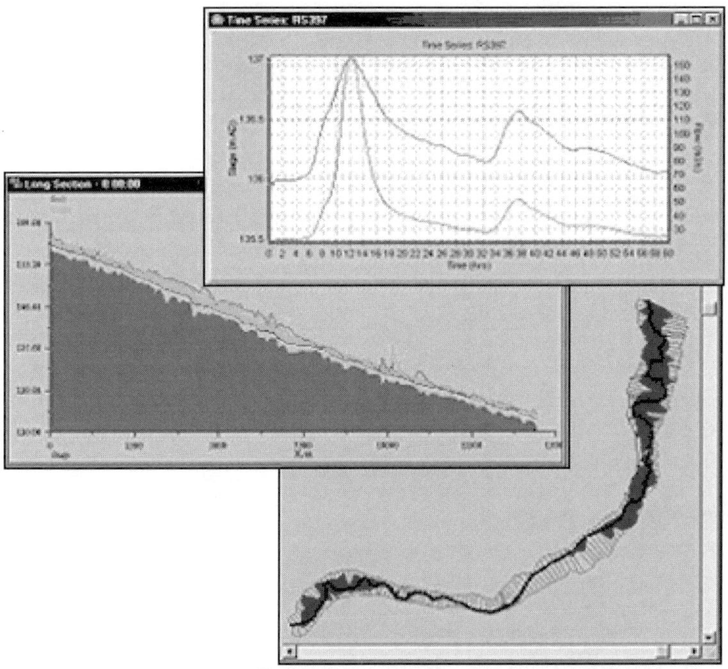

Display of model results over time, in profile and in plan view

Figure 13.15 Printouts from the model ISIS (HR, Wallingford)

The capability of computer models in water systems analysis is now comprehensive. Although hydrological and hydraulic systems can be fairly well modeled, atmospheric water movement and climate are more difficult. Some aspects of water quality and ecohydraulics may also require further research.

The following list illustrates some types of models available:

- Rainfall-runoff conversion for storm events.
- Long-term rainfall-runoff series generation.
- Statistical analysis of rainfall, runoff and synthesization of time series for models.
- Catchment and reservoir operational models.
- River hydraulics and flood plain models.
- Water body and sea two-dimensional models for currents and waves.
- Groundwater flow analysis.
- Pollutant transport and dispersion.
- Analysis of pipes and distribution systems for steady and time varying conditions.
- GIS (Geographic Information Systems) for mapping and location of resources.
- SCADA (Supervisory, control and data acquisition)
- Sediment erosion and deposition models including coastal protection.
- Topographic modeling including DTM (Digital terrain mapping).
- Global climate modeling (GCM)

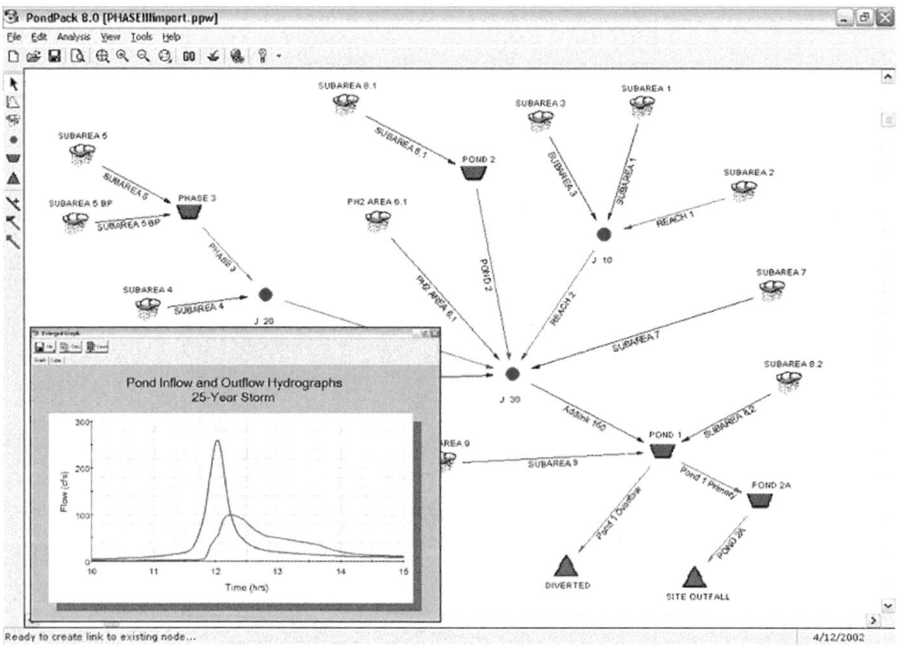

Figure 13.16 Software for modeling drainage (Haestad Methods)

Many of the packages produce output in a form suitable for processing by other models. For example DTM (digital terrain models) data could be used for modeling catchments, and GCM (global climate models) output could be used for flow modeling and prediction. Spreadsheet graphics capabilities can be used to present results. VBA (Visual Basic Application) software is based on existing spreadsheet capability. While the modeling is done in a hidden macro, the results can be manipulated by the spreadsheet. Input is also facilitated and checked on a spreadsheet.

It is not always necessary to select the most advanced or sophisticated model for a study. There is a long learning period involved in getting the most out of a model, and the models themselves sometimes are costly. The cost and time may not be justified and the effort in data acquisition and calibrating may either be expensive or could result in corner cutting and incorrect assumptions. The larger the model the more difficult it is to see the sensitivity to input assumptions and to interpret the results. So often a simpler mass balance on a spreadsheet on using a simple program can produce adequate results which can be studied by the designer, client and public more readily.

REFERENCES

Bodman, G.B. and Coleman, E.A. (1943). Moisture and energy conditions during downward entry of water into soils. *Proc. Soil Sci. Soc. of America*, 7, 116-122.

Brakensiek, D.L. (1967). Kinematic flood routing. *Trans. Am. Soc. Agr. Engrs.,* 10 (3), 340-343.

Constantinides, C.A. (1982). Two-dimensional kinematic modelling of the rainfall-runoff process. Report No. 1/82, Water Systems Research Programme, Univ. of the Witwatersrand, Johannesburg.

Dantzig, G.B. (1963). *Linear Programming and Extensions*. Princeton Univ. Press, Princeton, Ch. 23.

Crawford, N.H. and Linsley, R.K. (1966). Digital simulation in hydrology: Stanford watershed model IV. Tech. report 39, Dept. Civil Eng., Stanford Univ., California.

Diskin, M.H., Wyeseure, G. and Fayen, J. (1984). Application of a cell model to the Bellebeek watershed. *Nordic Hydrol.*, 15.

Dorfman, R., Samuelson, P.A. and Solow, R.M. (1958). *Linear Programming and Economic Analysis*. McGraw Hill, N.Y.

Eckstein, O. (1961). *Water Resources Development*. Harvard Univ. Press, 300p.

Foster, G.R. (1982). Hydrological modelling of small watersheds. In Haan, C.T., Johnson, H.P. and Brakensiek, D.L. (eds). ASAE Monograph 5, 297-370.

Green, W.H. and Ampt, G.A. (1911). Studies of soil physics. 1. The flow of air and water through soils. *J. Agric. Science*, 4 (1), 1-24.

Holden, A.P. and Stephenson, D. (1988). Improved 4-point solution of the kinematic equations. *J. IAHR*, 26 (4), 1-11.

Huber, W.C., Heaney, J.P., Nix, S.J., Dickenson, P.E. and Pilman, D.J. (1982). Stormwater Management Models, User Manual, Dept. Environ. Eng. Sciences, Univ. of Florida, Gainesville, Florida.

Illemobade, A.A. (2003). *Decision Support System for Rural Water Distribution*, PhD dissertation, Univ. Witwatersrand, Johannesburg.

International Association of Hydrological Sciences (1989). From Theory to Practice. *Proc. Workshop on Systems Analysis in Water Resources Planning*, Baltimore.

Kindelberger, C.P. (1965). *Economic Development*. McGraw Hill, N.Y., 395p.

Lansey, E.K. and Mays, L.W. (1989). Optimisation Models For Design of Water Distribution Systems. *Reliability Analysis of Water Distribution Systems*, Edited by Mays, L. W., ASCE, 37-84.

Loomba, N.P. (1964). *Linear Programming*. McGraw Hill, N.Y.

Maass, A., Hufschmidt, M.M., Dorfman, R., Thomas, H.A., Marglin, S.A. and Fair, G.M. (1962). *Design of Water Resource Systems*. Macmillan, London.

Ponce, V.M. (1986). Diffusion wave modelling of catchment dynamics. *J. of Hydraulic Engineering, ASCE*, 112 (8), 716-727.

Stephenson, D (1984). *Pipeflow Analysis*. Elsevier, Amsterdam.

Stephenson, D. and Meadows, M.E. (1986). *Kinematic Hydrology and Modelling*. Elsevier Science Publ., Amsterdam, 250 pp.

Stephenson, D. (1989). Selection of stormwater model parameters. *ASCE J. Environ. Eng.*, 115 (1), 210-220.

Stephenson, D. (1989). Planning model for water resources development in developing countries. *Proc. Symp. IAHS.* Baltimore, Publ. 180, 63-72.

Stephenson D. and Petersen, M.S. (1991). *Water Resources Development in Developing Countries*, Elsevier, Amsterdam, 289p.

Stephenson, D. and Paling, W.A.G. (1992). An hydraulic based model for simulating monthly runoff and erosion. *Water S.A.*, 18 (1), 43-52.

Stephenson, D., Randell, B. and Held, G. (2000). Prediction model for the Caledon River. *Proc. 4th Biennial Congress of IAHR*, Africa Division, Windhoek, Namibia.

Stephenson, D.(2002). A kinematic modular runoff model, in Singh, V.P.(Ed.) *Modelling of small watersheds*, Water Resource Publications, Boulder, Col.

Terstiep, M.L. and Stall, J.B. (1974). The Illinois urban drainage area simulator, ILLUDAS, Illinois State Water Survey, Urbana, Bulletin 58.

Viessman, W., Knapp, J.W., Lewis, G.L. and Harbaugh, T.E. (1977). *Introduction to Hydrology*. Harper and Row Publ., New York.

Yalin, Y.S. (1963). An expression for bed-load transportation. *ASCE J. Hydraul. Civ.*, 89 (HY3), 221-250.

Yeh, W.W.G. (1985). Reservoir Management and Operations Models: A State-of-the-Art Review. *Water Resources Research*. Vol. 21 (12). 1797-1818

Subject Index

A

B

C

D

E

F

G

H

M

N

O

P

T

U

V

W

Y

Z